Soil Degradation, Soil Pollution and Ecological Restoration

Soil Degradation, Soil Pollution and Ecological Restoration

Editors

Bo Sun
Ming Liu
Yan Chen

Basel • Beijing • Wuhan • Barcelona • Belgrade • Novi Sad • Cluj • Manchester

Editors

Bo Sun
Ecological Experimental
Station of Red Soil Academia
Sinica
Institute of Soil Science,
Chinese Academy of Sciences
Nanjing
China

Ming Liu
Ecological Experimental
Station of Red Soil Academia
Sinica
Institute of Soil Science,
Chinese Academy of Sciences
Nanjing
China

Yan Chen
Ecological Experimental
Station of Red Soil Academia
Sinica
Institute of Soil Science,
Chinese Academy of Sciences
Nanjing
China

Editorial Office
MDPI
St. Alban-Anlage 66
4052 Basel, Switzerland

This is a reprint of articles from the Special Issue published online in the open access journal *International Journal of Environmental Research and Public Health* (ISSN 1660-4601) (available at: https: //www.mdpi.com/journal/ijerph/special_issues/soil_degradation_ecological).

For citation purposes, cite each article independently as indicated on the article page online and as indicated below:

Lastname, A.A.; Lastname, B.B. Article Title. *Journal Name* **Year**, *Volume Number*, Page Range.

ISBN 978-3-0365-8959-6 (Hbk)
ISBN 978-3-0365-8958-9 (PDF)
doi.org/10.3390/books978-3-0365-8958-9

Contents

About the Editors

Bo Sun

Bo Sun was a professor at the Institute of Soil Science, Chinese Academy of Sciences, and held several leadership positions in soil science research and practice. He was the Director of the National Agro-Ecosystem Observation and Research Station in Yingtan, the Director of the Key Laboratory of Farmland Conservation of the Ministry of Agriculture and Rural Affairs, the Vice President of the China Plant Nutrition and Fertilizer Society, and the President of the Jiangsu Soil Science Society. His research focused on soil nutrient cycling and soil quality evolution in agro-ecosystems from field to regional scales. He led the Soil Biological Special Survey of the Third National Soil Condition Census, two National Key R&D Programs, a key project of NSFC, and a National Key Technology R&D Program of the Ministry of Science and Technology of China. He also contributed to over 20 national and provincial-level projects. He published over 110 international academic papers (indexed by SCI) in journals such as *Nature Communications* and *PNAS*, over 170 papers in Chinese (indexed by CSCD), 5 monographs, and 3 national standards. He also obtained three invention patents. He received two second-class prizes of the National Science and Technology Progress Award in 2004 (ranked sixth) and 2011 (ranked fourth), a second prize for scientific and technological progress in Jiangxi Province in 2021 (ranked first), and five other provincial- and ministerial-level awards.

Ming Liu

Ming Liu is currently an associate professor at the Institute of Soil Science, Chinese Academy of Sciences. He has long been engaged in research on the utilization of organic resources in red soil, soil improvement and fertilization, soil ecological restoration, and ecological planting. He has presided over or participated in nearly 20 national and provincial-level projects. He won one second prize for scientific and technological progress in Jiangxi Province in 2021 (ranked fourth), published more than 80 papers, served as an editorial board member of the "Terrestrial Microbiology" column of the journal *Frontiers in Microbiology*, authorized three invention patents and three software copyrights, formulated one local standard in Jiangxi Province, and participated in the compilation of one monograph.

Yan Chen

Yan Chen is an associate professor at the Institute of Soil Science, Chinese Academy of Sciences. Her research focuses on plant rhizosphere microbial ecology, with a particular emphasis on interspecific plant root-microbial interactions. She has presided over or participated in three national and provincial-level projects and has published high-quality papers in international journals such as *Microbiome, Soil Biology and Biochemistry*, and *mBio*.

Preface

Soil degradation is defined as a change in the soil's health status resulting in a diminished capacity of the ecosystem to provide goods and services for its beneficiaries. Ecological rehabilitation is required when the soil is degraded to such an extent that the land becomes unproductive. However, the types, causes, and characteristics of soil degradation are complicated and require effective and up-to-date technology for soil degradation and soil pollution restoration. The Special Issue aimed to provide a platform for researchers to communicate their research results or latest insights in related fields. It can also serve as a scientific reference for readers engaged in soil degradation remediation.

In this Special Issue, Sun et al. provide an overview of China's experiences and challenges in achieving sustainable agricultural development. Zhao et al. proposed a new standard describing the surface cracking conditions of soda-saline-alkali soil. Cai et al. found that the application of biochar and manure to soil can improve micropore volume and aggregate stability and increase crop yield. According to Qian et al., soil erodibility (K factor) and saturation conductivity (Ks) are the key indicators to evaluate the quality of land reclamation. The Dorocki and Korzeniowska survey showed that heavy metals in the soil of Jaworzyna Krynicka, in the Beskidy Mountains of Poland, were not highly contaminated. Chen et al. combined geostatistics and chemometric methods to analyze the sources and spatial patterns of heavy metal pollution in soil. Wyszkowska et al. suggested that grass composting and corn planting could mitigate the toxic effects of tetracycline on soil bacterial communities. Lin et al. found that farmland consolidation had a significant effect on soil bacterial community structure. Gong et al. showed that different land use types and soil depths had significant effects on soil mineral element content, soil enzyme activity, and the fungal community in karst areas. Li et al. conducted a seven-year fertilization reclamation experiment in a coal mine subsidence area and found that fertilization had a significant impact on soil bacterial community composition and diversity.

In the future, more attention will be paid to the following aspects of soil degradation and restoration: (1) Evaluation criteria and key indicators of soil degradation. (2) Clarification of soil degradation processes from micro-scale to field-scale to regional-scale. (3) Soil–plant microbial responses and feedback in soil degradation and restoration processes. (4) The critical role of soil degradation and restoration under global climate change.

The Special Issue was successfully completed under the organization of Professor Bo Sun, who had a high academic level in soil quality, soil health, and soil biology. Unfortunately, due to an accident, Professor Sun had to leave his love of soil science research forever. This article is dedicated to his memory.

Ming Liu and Yan Chen
Editors

International Journal of
Environmental Research and Public Health

Review

Coordinative Management of Soil Resources and Agricultural Farmland Environment for Food Security and Sustainable Development in China

Bo Sun [1,2,*], Yongming Luo [1], Dianlin Yang [3], Jingsong Yang [1,2], Yuguo Zhao [1,2] and Jiabao Zhang [1,*]

1   State Key Laboratory of Soil Science and Sustainable Agriculture, Institute of Soil Science, Chinese Academy of Sciences, Nanjing 210008, China
2   Key Laboratory of Arable Land Conservation, Ministry of Agriculture and Rural Affairs of the People's Republic of China, Nanjing 210008, China
3   Agro-Environmental Protection Institute, Ministry of Agriculture and Rural Affairs of the People's Republic of China, Tianjin 300191, China
*   Correspondence: bsun@issas.ac.cn (B.S.); jbzhang@issas.ac.cn (J.Z.)

Abstract: Major problems in China's pursuit of sustainable agricultural development include inadequate, low-quality soil and water resources, imbalanced regional allocation and unreasonable utilization of resources. In some regions, overexploitation of soil resources and excessive use of chemicals triggered a web of unforeseen consequences, including insufficient use of agricultural resources, agricultural non-point source pollution and land degradation. In the past decade, China has changed its path of agricultural development from an output-oriented one to a modern, sustainable one with agricultural ecological civilization as its goal. First, the government has formulated and improved its laws and regulations on soil resources and the environment. Second, the government has conducted serious actions to ensure food safety and coordinated management of agricultural resources. Third, the government has planned to establish national agricultural high-tech industry demonstration zones based on regional features to strengthen the connection among the government, agri-businesses, scientific community and the farming community. As the next step, the government should improve the system for ecological and environmental regulation and set up a feasible eco-incentive mechanism. At the same time, the scientific community should strengthen the innovation of bottleneck technologies and the development of whole solutions for sustainable management in ecologically fragile regions. This will enhance the alignment between policy mechanisms and technology modes and effectively promote the sustainable development of agriculture in China.

Keywords: cultivated land quality; agricultural environment; soil and water resources; agricultural engineering; synergetic development strategy

Citation: Sun, B.; Luo, Y.; Yang, D.; Yang, J.; Zhao, Y.; Zhang, J. Coordinative Management of Soil Resources and Agricultural Farmland Environment for Food Security and Sustainable Development in China. *Int. J. Environ. Res. Public Health* **2023**, *20*, 3233. https://doi.org/10.3390/ijerph20043233

Academic Editors: Paul B. Tchounwou and Xiao-San Luo

Received: 14 June 2022
Revised: 6 February 2023
Accepted: 6 February 2023
Published: 12 February 2023

1. Introduction

Agriculture is a social and economic activity of mankind that has the longest, widest and closest connection with nature. Sustainable agriculture belongs to the Sustainable Development Goals (SDGs) launched by the United Nations [1]. The essence of agricultural production is the development and utilization of agricultural resources, a process with great ecological and environmental impacts. Meanwhile, agricultural production is affected and restricted by resources and the environment [1].

China is faced with huge population, resource and environmental pressures. In its long history of agricultural development, China attaches great importance to relations between agricultural production, resources and the environment. Since the 1960s, China has established ecological agriculture models, such as the pig raising–biogas production–fruit-tree plantation model and multistory cropping-raising patterns in paddy fields and terraced fields (Figure 1) [2,3]. Since the 1980s, China has developed highly intensive agriculture;

some inefficient management measures have led to increasingly serious environmental and ecological problems and weaker sustainability [4].

Figure 1. The ecological agriculture models in Southern China: pig raising–biogas production–navel orange (**a**) or citrus (**b**) plantation models in hilly regions, and duck-raising model (**c**) and crayfish culture model (**d**) in paddy fields.

Therefore, sustainable agricultural development has become one of the highest priorities for agricultural policy makers in China. To achieve the Sustainable Development Goals (SDGs), China abides by the principles and patterns of the ecological system when engaging in agricultural production in order to enhance the mutual–complementary relations between agricultural production and environmental protection, optimize the structure and functions of agricultural ecosystems, and constantly improve its overall agricultural production capacity and sustainable development capacity. China has developed an efficient path for agricultural resources and environmental protection by formulating annual and long-term plans for modern agricultural development; constantly enhancing laws and regulations on resources and environmental management; setting up an implementation mechanism for the coordinated management of resources and the environment; and implementing various types of projects for agricultural development in different regions based on coordination between policy systems and technology modes. In this study, we review the keys to success from China's experience, which could offer models for others engaged in developing sustainable agriculture. First, we address the main problems facing China in sustainably utilizing soil resources and protecting the agricultural environment. Second, the pathway to develop sustainable agriculture and the impact of major national programs since the 1980s are discussed. Finally, we suggest future priorities for both the government and scientific community.

2. Methods

First, we undertook a review of the recent literature to evaluate the main problems related to soil resources and the environment for sustainable agricultural development in China. We then undertook a comprehensive review of sustainable agricultural development using four aspects: action plan, regulation system, implementation system and achievement. We used a multi-source review approach that involved assembling information from published scientific and gray literature, including scientific journal papers and reports, official government reports, yearbooks, statistics and webpages. Many of these sources were published in Chinese.

To evaluate the achievements of cultivated land quality improvement projects in China since the 1980s, soil organic carbon density data in the topsoil (0–20 cm) were collected from the National Earth System Science Data Center (http://soil.geodata.cn, accessed on 10 May 2022). The soil pH data in the topsoil (0–20 cm) were collected from the National Cultivated Land Quality Big Data Platform (http://www.farmland.cn/, accessed on 10 May 2022). The soil organic carbon contents and pH values in the topsoil were investigated in the second national soil survey between 1979 and 1984. The Chinese Academy of Sciences launched the "Climate Change: Carbon Budget and Relevant Issues" project and conducted a soil carbon stock survey at a national scale in 2011. The Ministry of Agriculture and Rural Affairs conducted the cultivated land quality survey including soil pH evaluation at a national scale during 2015–2017. SOC stock (tonnes ha^{-1}) was calculated by: SOC $\times$ BD $\times$ Depth $\times$ $(1 - RF)/10$, where SOC is the organic C content (fine soil fraction that passes through a 2-mm sieve, $g\ kg^{-1}$), BD is the soil bulk density ($g\ cm^{-3}$), Depth represents topsoil thickness (20 cm in this study), and RF represents the volume fraction of rock fragments (>2 mm). Based on the 1:4,000,000 soil database, the spatial analysis of soil organic carbon stock and soil pH was conducted using the GIS software ArcView 3.2 (ESRI, Redlands, CA, USA).

3. Problems with Soil and Water Resources for Sustainable Development in China

In general, the overall agricultural production capacity of China has been constantly improving, with gross grain output ranging from 616 to 669 million/from 2016 to 2020. However, demands for forage grain and industrial grain are increasing rapidly as a result of China's growing and more urbanized population, optimized dietary mix, and wider use and structural change of agricultural products for industrial purposes. The total grain demand will reach 725 million tonnes by 2030 when considering a population of 1.45 billion and per capita grain demand of 500 kg. Structure-wise, China falls short of the supply of certain agricultural products; the already large gaps may continue to expand in the future. In 2020, China imported 142.62 million tonnes of grain, including 100.33 million tonnes of soybean, 11.3 million tonnes of maize and 8.38 million tonnes of wheat. In addition, the increases in consumption per capita of vegetables, fruits, poultry and milk are estimated to be 5.7%, 13.7%, 14.8% and 44.0%, respectively, from 2020 to 2030. Therefore, a shortage of supply will be normal for China's agricultural development in the long run [5]. The pressure to ensure food security will lead to the intensive use of soil and water resources.

3.1. Shortage of Cultivated Land and Water Resources

China's cultivated land was 134.87 million hectares in 2019. There is only 0.087 hectares of cultivated land, which represents 36% of the global average (0.24 hectares). In addition, the large majority of arable land has been utilized, and only approximately 5 million hectares of reserved arable land can be developed and utilized on scale in China [6].

China had a total water resource amount of 2904.1 billion m^3 in 2019, with an annual per capita amount of 2124 m^3, which was 25% of the world average [7]. On average, agriculture in China faces a water shortage of more than 30 billion m^3 every year, and some regions are troubled by polluted water resources and shortage of water supply induced by poor infrastructure. In the North China Plain, groundwater overexploitation has led to a decline in the groundwater level in irrigated areas and formed a 4.1-million-hectare funnel-shaped area. In the region with the deepest groundwater level, i.e., in Hengshui city, the deep groundwater level experienced an annual decrease of 1.16 m from 2000 to 2014; however, it experienced an increase of 2.76 m from 2015 to 2016 and an increase of 0.26 m from 2019 to 2020 as a result of the exploitation–supplement balance strategy.

3.2. Low Quality of Soil Resources

In 2019, low-quality cultivated land—i.e., levels 7–10, as cultivated land is classified into 10 levels in China, with level 1 being the best and level 10 being the worst—and mid-quality cultivated land (levels 4–6) accounted for 31.2% and 46.8%, respectively, of China's national total [8]. Overall, the contribution rate of basic land productivity to the total grain yield (wheat, corn and single-season rice) is 45.7% to 60.2% [9], and there is a yield gap ranging from 20% to 40% [10,11]. If measures are taken and a piece of mid-quality cultivated land is upgraded by 1 level, there will be an increase of at least 80 billion kilograms of total grain production. The overall water quality of surface water was good in 2020. Among the 1940 state-controlled surface water monitoring sections, 12.6% were Grade IV and above (surface water is classified into Grade I, II, III, IV, V and above V in China, with Grade I being the best) [12].

3.3. Uncoordinated Distribution of Soil and Water Resources

First, there is an uncoordinated spatial distribution of natural factors such as light, temperature, soil and water. Regionally, with a humid climate and abundant water resources, eastern China (47.6% of the national total area) has 90% of the country's cultivated land, while arid, semi-arid or Alpine western China (52.4% of the national total area) has a mere 10% of the country's cultivated land. In addition, there is a mismatch between water resources and soil resources. Southern China has more than 80% of the country's water resources and less than 40% of the cultivated land, while northern China has 20% of the water resources and 60% of the cultivated land [13,14].

Second, there is a mismatch between agricultural production layout and natural resource distribution. In recent decades, the national grain production center has gradually shifted from the southern region with abundant water and heat resources to the northern region facing severe water shortages [15]. At present, with only 15% of China's water resources, Northeast China and the Huang-Huai-Hai region contribute 53% of the national total grain output. The export of commodity grain to the southern region caused a deficit in water resources. In 2012, the total grain transportation amounted to 79.45 million tonnes from northern to southern China, which equals a virtual water flow of 82.6 billiom m^3 [16]. However, the "South-to-North Water Diversion Project" transported approximately 6.43 billion m^3 of water annually from 2014 to 2020. Thus, the resource gap cannot be closed.

3.4. Inefficient Utilization of Soil and Water Resources

According to FAO statistics, with only 8% of cultivated land in the world, China feeds 19% of the global population. China's cultivated land utilization intensity is high: 2.2 times that of the U.S. and 3.3 times that of India. While the U.S. has a well-established fallow system covering a large amount of its cultivated land, China is taking intensive measures such as increasing the multiple-cropping index to ensure that we produce more with fewer land resources. As a result, China's high agricultural output comes at the price of a low resource utilization rate and low labor productivity. At the same time, agricultural infrastructure in China remains weak and is not fully capable of withstanding natural disasters. The effective irrigation area accounts for only 51.8% of the total, and the average water productivity is approximately 1.0 kg/m^3, which is much lower than the developed country figure (1.2~1.5 kg/m^3) [17].

4. Problems of Habitat Soil and Environment Safety for Sustainable Development in China

At present, China's agricultural resources and environment are affected by both exogenous and endogenous pollution, which aggravates the pollution of soil and water resources and increases risks to agricultural product quality and security. On the one hand, industrial, mining and domestic pollutants are released into the agricultural system, leading to poorer environmental quality and worse pollution problems. On the other hand, the overuse of chemicals (such as fertilizers and pesticides) in agricultural production as well as the inappropriate disposal of agricultural wastes (such as livestock/poultry manure, crop straw and plastic film residues) result in severe non-point source pollution.

4.1. High Fertilizer Consumption and Low Utilization Rate

In 2019, China used 54.04 million tonnes of chemical fertilizers, which was 303.2 kg ha^{-1} if taking into account the agricultural planting area (including orchards) [18]. In recent years, China has taken measures in regard to high fertilizer consumption, including an optimized fertilizer mix based on soil tests for main grain crops. Furthermore, the average nitrogen consumption of wheat, rice and corn is 210, 210, and 220 kg ha^{-1}, respectively, which is already below 225 kg ha^{-1}, the upper limit set by developed countries to avoid water pollution. However, the average chemical fertilizer consumptions in orchards and protected vegetable fields (including greenhouses, tunnels and padding) were 555 kg ha^{-1} and 365 kg ha^{-1}, far more than the safety ceiling [19].

The high fertilizer consumption is accompanied by a low utilization rate. The average seasonal utilization rates of nitrogen, phosphorus and potash in wheat, corn and rice in China were 33% (30~35%), 24% (15~25%), 42% (35~60%), and 10~20%, respectively [20,21]. With the implementation of Zero Growth Action in the use of chemical fertilizers and pesticides since 2015, the average utilization rates of chemical fertilizers increased from 35.2% in 2015 to 39.2% in 2019—and there is still much room for improvement [22].

4.2. High Pesticide Consumption and Low Utilization Rate

Green and comprehensive measures for pest prevention/control are not widely used in China. The total consumption of pesticides grew from 0.765 million tonnes in 1991 to 1.808 million tonnes in 2014 (physical volume, including active ingredients and auxiliaries; active ingredient amount is 1/7 of the global figure) and the consumption per unit area increased from 5.12 kg ha^{-1} to 11.0 kg ha^{-1}, representing 2.5 times the global average [23]. Additionally, the total consumption of pesticides dropped to 1.456 million tonnes in 2019 [22]. China's average pesticide utilization rate for maize, wheat and rice crop systems was 39.8% in 2019, which is 20~30% lower than developed countries. Problems including soil and water degradation and loss of biodiversity still exist as a result of the residual pesticides that contaminate water via precipitation, surface runoff and infiltration [24,25].

4.3. Low Utilization Rate and Recovery Rate of Agricultural Waste

Livestock/poultry raising in China produces approximately 3.8 billion tonnes of animal waste every year, emitting an amount of nitrogen and phosphorus larger than fertilizer consumption does. Livestock/poultry COD emissions account for more than 90% of China's agricultural non-point source pollution COD. According to the first national pollution source census in 2007, livestock and poultry farms produced 243 million tonnes of organic waste and 163 million tonnes of urine. Furthermore, the total nitrogen and phosphorus discharge from animal excretion reached 1,024,800 and 160,400 tonnes, respectively [26] (The census includes 1,963,624 discharge sources from medium- to large-intensive units and do not include discharges from small producers). Overall, with the increased consumption of fertilizers and the growing amount of animal waste, nitrogen and phosphorus enter the water through surface runoff, leaching and volatilization, which leads to agricultural non-point source pollution [27].

There were 718.8 million tonnes of straw produced from main crops in 2015 [28], of which 43.2% was used for fertilizer. China consumed 2.41 million tonnes of agricultural plastic films in 2019. The mulch films amounted to 1.38 million tonnes and covered a total area of 17.6 million ha [29]. The widely used ultra-thin plastic film (<0.008 mm) is easily aged, fragile, hard to recover and has a recovery rate of less than 60%. As a result, approximately 580,000 tonnes of film residues are left in the soil every year, causing "white pollution," impacting soil structure and permeability and eventually hindering crops from utilizing water and nutrients in the soil.

4.4. Regional Land Degradation and Soil Pollution

In the past 30 years, China's cultivated land soil fertility has been enhanced in general. According to long-term monitoring from 1987 to 2015, with conventional fertilization management, the amounts of organic matter and nutrient contents in farmland soil showed an upward trend. The average contents of soil organic matter, total nitrogen, available phosphorus and available potassium were approximately 24.7 g/kg, 1.45 g/kg, 27.7 mg/kg and 133 mg/kg—approximately 13.8%, 2.8%, 123.4% and 46.2%, respectively [30,31].

However, more than 40% of the cultivated land was degraded with intensive agricultural land use [8,32]. Approximately 24 million hectares of cultivated land suffered from soil erosion, 7.6 million hectares suffered from salinization, 17.4 million hectares suffered from acidification, and 2.56 million hectares suffered from desertification. In addition, there was approximately 2 million hectares of cultivated land and 3.3 million hectares of protected vegetable land suffering from continuous cropping obstacles.

Soil pollution in China is mainly caused by heavy metal and pesticide pollution and the unreasonable use of untreated organic manure and mulch films. According to a national survey conducted in 2014, 16.1% of investigated sites were polluted. The main pollutants were cadmium, nickel, copper, arsenic, mercury, lead, DDT and polycyclic aromatic hydrocarbons [33,34].

5. Achievements of China's Sustainable Agricultural Development

5.1. Formulating Sustainable Agricultural Development Plans and Goals

In 2015, the Ministry of Agriculture and Rural Affairs (MARA) issued the National Sustainable Agricultural Development Plan (2015–2030), which proposed five key tasks to promote sustainable agricultural development: (a) to optimize the development layout and enhance agricultural production capacity; (b) to protect cultivated land and facilitate sustainable utilization of farmland; (c) to save water resources, enhance the water utilization rate and guarantee agricultural water safety; (d) to control environmental pollution and improve agricultural and rural environments; and (e) to restore agricultural ecology and improve ecological functions [35]. The nation is then divided into optimized, moderate and protected development areas, where different management measures will be taken to optimize the agricultural production layout, focus on main varieties and dominant producing areas, and implement precision management based on permanent basic farmland.

In terms of cultivated land protection, 120 million hectares of cultivated land were preserved to guarantee grain production, in which 53.3 million hectares of high-standard cultivated land were constructed by 2020.

In terms of coordinated management of agricultural resources and the environment, water-saving agriculture was developed to ensure an irrigation water consumption of 372 billion m^3 and an effective irrigation water utilization coefficient of 0.55 by 2020. The "optimized fertilizer mix based on soil test" approach and green prevention/control technology over pests were used to reach coverage rates of 90% and 30%, respectively. The utilization rate of chemical fertilizers and pesticides reached 40%. In total, 75% of the large-scale livestock/poultry farms were supported by waste treatment facilities, the crop straw comprehensive utilization rate reached more than 85%, and the recycling rate of the agricultural plastic films reached more than 80%.

In 2016, the State Council issued the Soil Pollution Prevention and Control Action Plan, which proposed the following actions: (a) identify the soil pollution status of agricultural land through a detailed survey in 2018 and prioritize the protection of unpolluted and slightly polluted cultivated land; (b) take measures including agronomic regulation and alternative planting to guarantee appropriate parameters of agricultural products before 2020 and safely utilize 2.6 million hectares of mildly and moderately polluted cultivated land; (c) identify no-production zones to strictly manage the utilization of severely polluted cultivated land; and (d) restructure the planting mix so that 1.3 million hectares of land will be recovered or rebuilt into forest/grassland [36].

5.2. Improving Regulations and Establishing Research and Administration Systems for Agricultural Resource Management and Environmental Protection

China has formulated a number of environmental laws to clarify the requirements for the protection of agricultural resources and the environment. The Environmental Protection Law, the Environmental Pollution Law of Solid Waste and the Prevention and Treatment of Air Pollution Law and Water Pollution Prevention Law have stipulated the requirements for the entry of solid waste, atmosphere and irrigation water into the agricultural environment. In particular, the Agricultural Product Quality Safety Law issued in 2006 sets up specific chapters on agricultural habitat environmental protection. The State Council promulgated the Basic Farmland Protection Regulations and the Livestock and Poultry Scale Aquaculture Pollution Prevention Regulations to facilitate cultivated land protection, comprehensive utilization and non-polluted treatment of animal waste.

In 2015, the Ministry of Agriculture (now called the Ministry of Agriculture and Rural Affairs) set up 37 ministry-level key laboratory systems for discipline clusters, which comprised 42 key comprehensive laboratories, 297 professional (regional) key laboratories and 269 scientific observation stations. At the same time, the MOA improved modern agricultural technology systems for 50 products; each system consists of a national industrial technology research and development center and a number of integrated experimental stations in major agricultural areas. Linked with the research teams in other national research systems, these systems conduct joint research, set up pilot programs, and offer technical training sessions, policy consultation and contingency services on agricultural green development [37].

In terms of resources and environmental protection, the MOA has been establishing an agricultural resources and environmental protection system since the 1980s, which consists of 2 national stations, 33 provincial stations, more than 300 local-level stations and more than 1700 county-level stations, with a total of more than 12,000 practitioners. In 2017, the MOA set up the Cultivated Land Quality Monitoring and Protection Center, which is responsible for setting up the national monitoring network, carrying out cultivated land quality investigation and evaluation, and developing technologies and products for improving cultivated land productivity.

5.3. Carrying Out Fertile Soil Projects and Comprehensive Remediation Projects for Heavy-Metal-Contaminated Soil to Improve Soil Quality

China has perfected the quality monitoring/protection network for cultivated land and improved cultivated land quality. China has established a four-level (national, provincial, city and county), long-term quality monitoring network for cultivated land, which includes 357 national level monitor points and covers 35 major types of cultivated land [31]. Major focuses are middle- and low-yield fields with soil obstacles and soil degradation, such as black soil (Phaeozem) in Northeast China with soil erosion problems, Chao soil (Cambisol) in northern China with groundwater overexploitation problems, and red soil (Acrisol) and paddy soil (Anthrosol) in southern China with soil acidification and heavy metal pollution problems [38–41]. The construction pathway of high-standard farmland to reduce soil obstacles and improve soil structure, soil nutrient pools and soil biological functions was proposed in different regions (Figure 2) [42,43].

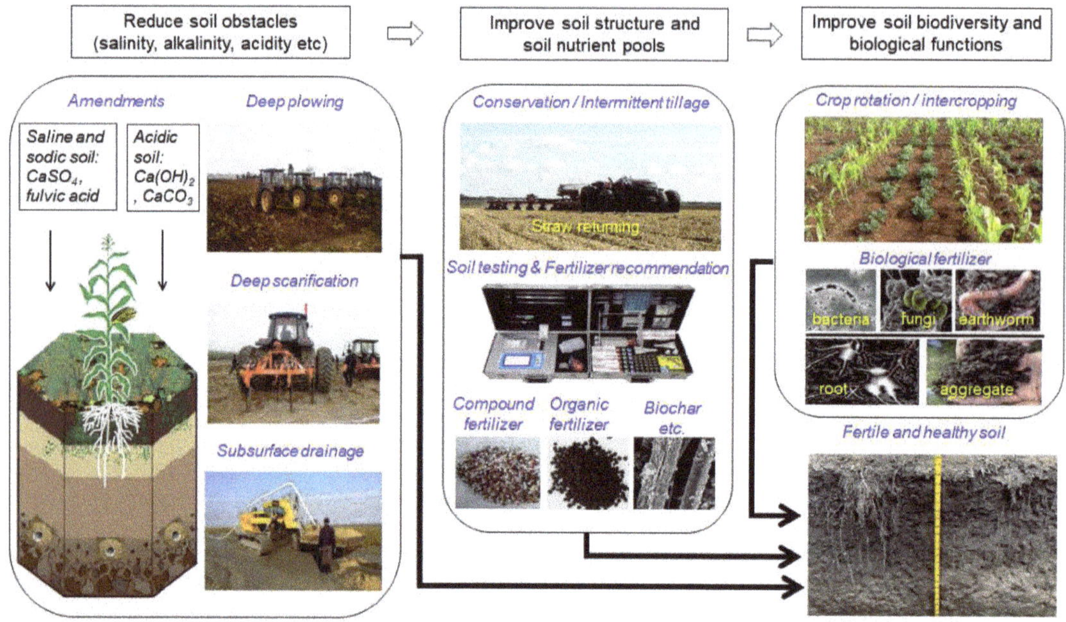

Figure 2. The construction pathway of fertile and healthy soil.

China has implemented a strict cultivated land protection system and comprehensively improved the quality of cultivated land and the ecological environment [44]. Since the 1980s, China has successively implemented the actions of the Fertile Soil Project [45], High-Standard Farmland Construction Project [46], and Northeast Black Soil Conservation Tillage Action [47]. In 2020, organic fertilizers were used in 36.7 million hectares in total, and the planting area of green manure crops was approximately 3.87 million hectares. The average soil organic carbon stock (SOCS) in 0–20 cm of cultivated land was 28.3 tonnes ha^{-1} in China in 1980 (Figure 3), which increased to 33.1 tonnes ha^{-1} in 2011 (Table 1). Only in Northeast China did the average SOCS decrease by 1.05 tonnes ha^{-1}. The main reasons for this are the decomposition of soil organic matter with cultivation and soil erosion in rolling hill regions. In other regions, the higher residue inputs following the large-scale implementation of the crop straw return policy increase soil organic carbon density [48]. In general, the average cultivated land quality increased from 4.73 in 2014 to 4.75 in 2019 (a total of 10 levels) [8].

China has carried out investigations on the heavy metal pollution of cultivated land and set up a classified management system. The Ministry of Agriculture released opinions on implementing the *Soil Pollution Prevention and Control Action Plan.* A total of 1.3 million sampling sites have been set up in 108.2 million hectares of cultivated land for dynamic monitoring of heavy-metal-caused soil pollution in agricultural habitats [36]. Integrated physicochemical–biological technology was developed for the remediation of heavy-metal-contaminated soils [49]. Since 2014, the Ministry of Agriculture and Rural Affairs has set up pilot zones for remediating slightly and moderately polluted farmland in 10 provinces. For example, the VIP + n model was adopted in Hunan province, which is represented using rice varieties with low cadmium accumulation (V), adopting reasonable irrigation (I), adjusting the soil pH value (P), and matching assistant measures such as adsorbents and foliar spray inhibitors for reducing the crop uptake of heavy metals (n) [50].

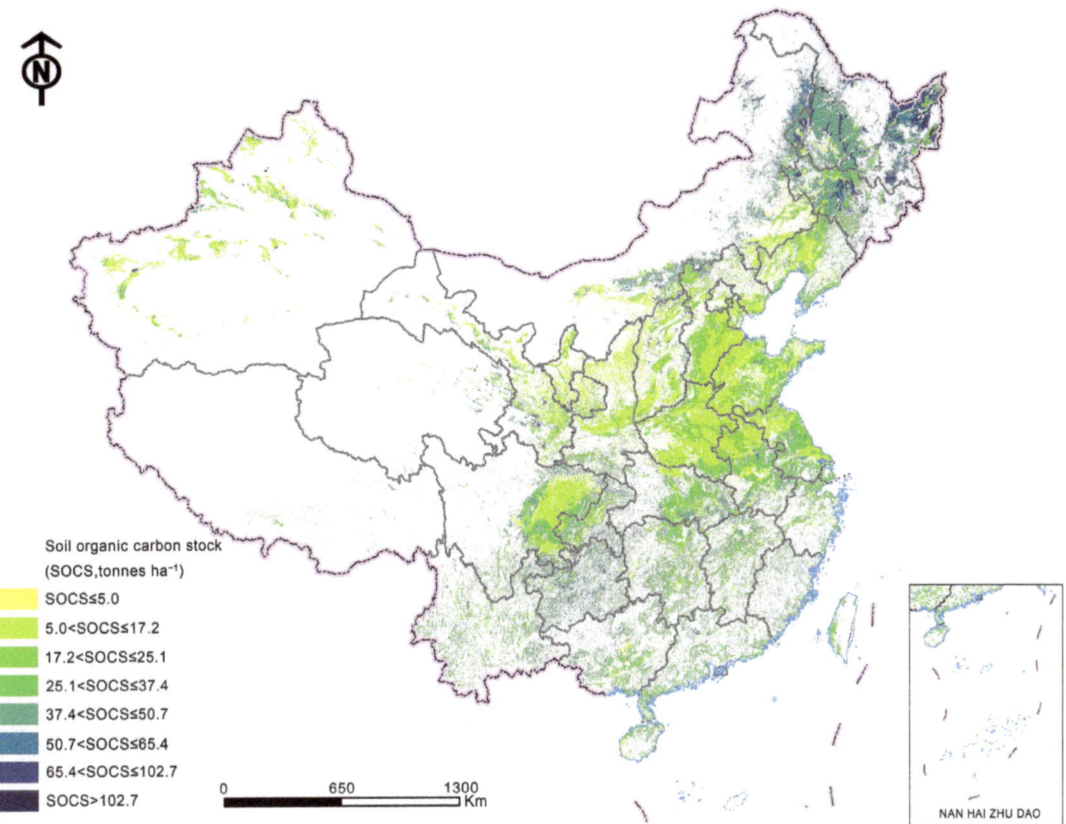

Figure 3. Soil organic carbon stock in the 0–20 cm layer of cultivated land in 1980 in China.

Table 1. Changes in the soil organic carbon stock in the 0–20 cm layer of cultivated land from 1980 to 2011 in China.

Vegetation Type	Soil Organic Carbon Stock in 1980 (Tonnes ha^{-1})	Change in SOCS from 1980 to 2011 (Tonnes ha^{-1}) **
One crop annually in cold temperate zone	33.2 ± 23.0 *	−1.05 ± 5.75
Two crops annually or three crops for two years and deciduous orchards in warm temperate zone	22.0 ± 16.6	9.75 ± 4.04
Two crops containing upland and rice annually and deciduous and evergreen orchards in transitional subtropics	31.8 ± 15.9	4.64 ± 7.99
One- or double-cropping rice followed by a cool-loving crop or three upland crops annually and evergreen economic crops and orchards in subtropics	28.6 ± 11.9	7.03 ± 4.13
Double-cropping rice annually followed by warm-loving crops and evergreen economic crops and orchards in tropics	29.8 ± 17.1	3.57 ± 5.88

* Data represents the mean value ± standard error. ** Soil organic carbon stock in 2011 was calculated from Table S1 [48].

5.4. Conducting Water-Saving Agriculture Projects and Zero Growth in Fertilizer/Pesticide Consumption Actions to Improve the Use Efficiency of Resources

China established water-saving agriculture demonstration areas to promote water-saving varieties and technologies, including sprinkler irrigation and water-fertilizer integration [51]. The demonstration program covers more than 26.7 million hectares of farmland. The average effective utilization coefficient of farmland irrigation water in China reached 0.559 in 2020, indicating an increase of 7.7% over 2012 [52].

China has expanded the coverage of the "Soil Testing and Formulated Fertilization Project" from 73.3 million ha in 2010 to 106.7 million hectares in 2016 [53]. Integrated soil-crop system management increased the average yields from 7.2 tonnes ha^{-1} to 8.5 tonnes ha^{-1} for rice; 7.2 tonnes ha^{-1} to 8.9 tonnes ha^{-1} for wheat; and 10.5 tonnes ha^{-1} to 14.2 tonnes ha^{-1} for maize, without any increase in nitrogen fertilizer [54]. By 2020, the coverage rate of green prevention/control for major pest-led diseases reached 41.5% and the use of professional services reached 35.5%. The fertilization rate decreased from 338.3 kg ha^{-1} in 2015 to 303.2 kg ha^{-1} in 2019 (based on agricultural planting area). The total consumption of fertilizers decreased by 10.3% from 2015 to 2019. The consumption of chemical pesticide technical decreased by 19.5% from 2015 to 2019, from 596,500 tonnes to 480,000 tonnes. The utilization rates of chemical fertilizer and pesticide reached 40.2% and 40.6%, respectively, for the three major grain crops of rice, wheat and maize [52].

Since 2017, China has conducted the action for replacing chemical fertilizer with organic fertilizer for fruit, vegetable and tea replacement in 238 counties [55]. The large-scale biogas natural gas pilot project could support a biogas fertilizer production capacity of more than 400 million tonnes every year. China is providing higher subsidies for straw returning, organic manure and thick mulch films (>0.01 mm) and implementing the straw comprehensive utilization project and mulch film recycling project in 143 and 229 counties, respectively. In 2020, the national comprehensive utilization rate of straw and livestock and poultry manure reached 86.72% and 75%, respectively. The residual film processing capacity increased by 186,300 tonnes.

However, soil acidification has increased since the 1980s. The average topsoil pH declined significantly by 0.13–0.80 for most soil types from the 1980s to the 2000s, while it remained unchanged for the Aeolian soils with high pH, which accounted for only 9.8% of the total cultivated land in China [56]. For the paddy soils, the average topsoil pH declined by 0.29–0.58 in most major rice-production areas from the 1980s to the 2010s, while it increased by 0.14 in the southeast region (Table 2). The areas of strong (pH < 5.5) and weak (pH from 5.5 to 6.5) acidic paddy soil accounted for 12.9% and 32.5% of the total cultivated land in the 2010s, respectively (Figure 4). The significant soil acidification was mainly caused by the high N surplus, which increased from 142.8 kg ha^{-1} in 2004 to 168.6 kg ha^{-1} in 2015 [57]. Reducing the N surplus while meeting the food demand in 2050 requires an increase in nitrogen use efficiency from approximately 40% to 60% in China [58].

Table 2. Changes in paddy soil pH in the 0–20 cm layer from 1980 to 2010 in China.

Region	Soil pH 1980s	Soil pH 2010s	Soil Acidification Rate from 1980s to 2010s (pH Unit/a)
Northeast China	6.41 ± 0.0074 *	6.07 ± 0.1856	0.0163 ± 0.0006
Middle and lower Yangtze River	5.99 ± 0.0044	5.70 ± 0.0243	0.0006 ± 0.0002
Southwest China	6.15 ± 0.0046	6.29 ± 0.0331	−0.0028 ± 0.0004
Southern China	5.93 ± 0.0040	5.35 ± 0.0243	0.0078 ± 0.0003

* Data represent the mean value ± standard error.

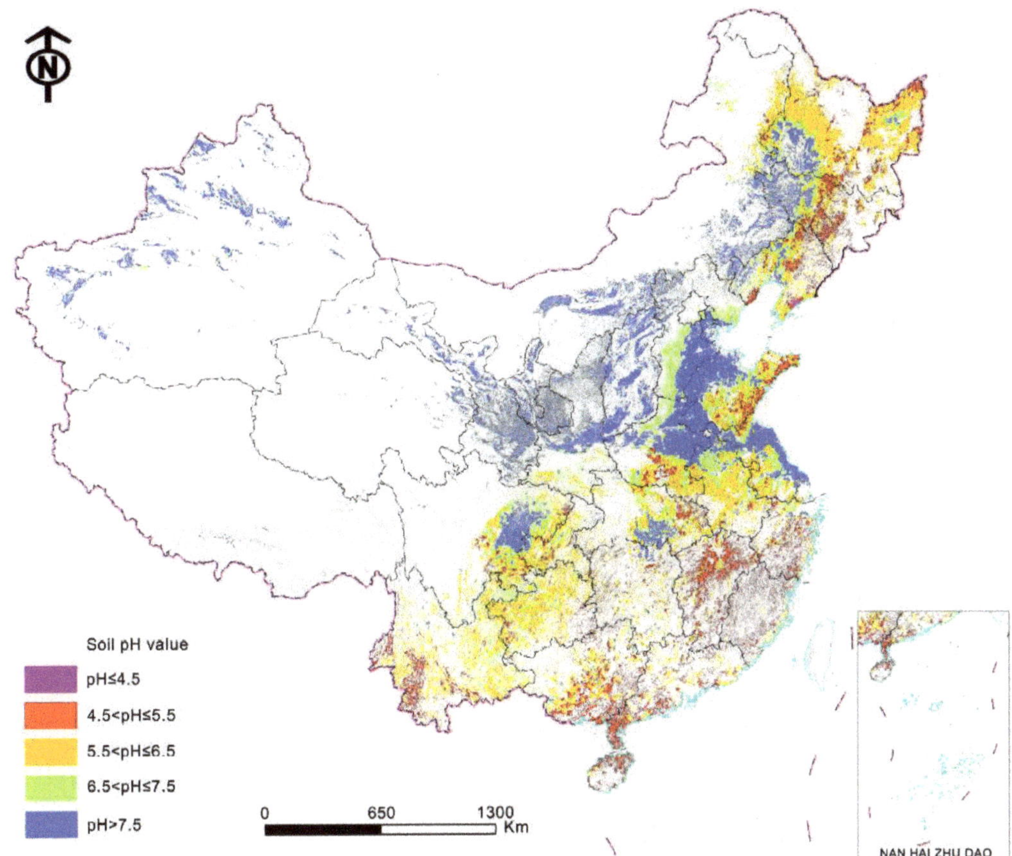

Figure 4. Soil pH in the 0–20 cm layer of cultivated land in the 2010s in China.

6. Future Priorities for the Coordinated Management of Soil Resources and the Agricultural Farmland Environment in China

In recent years, realizing soil health has become a global priority to sustain humans, animals and the environment [59,60]. A large across-the-board increase in soil and environmental quality is required for long-term and sustainable agricultural production in China. While guaranteeing food safety and the supply of major produce, China is paying more attention to the protection of the quantity, quality and ecology of cultivated land through high-standard farmland construction programs. By 2030, 80 million hectares of high-facilitated farmland (62.6% of total national farmland) will be established, with concentrated contiguous land, guaranteed harvests in drought or flood, stable and high yields, and a sound ecology [61]. During this process, a satellite–aviation–ground integrated soil health monitoring network for farmland should be established based on remote sensing, proximal sensing, and prediction models. Moreover, an intelligent decision-making system should be built to propose management strategies for healthy farmland construction.

China has built the Science and Technology Backyard (STB) platform to connect the government, agri-businesses, and scientific community with the farming community, enabling smallholders to sustainably achieve yield and economic gains [10]. China will establish approximately 30 national agricultural high-tech industry demonstration zones in 2025. Each demonstration zone will have a theme that aims to resolve a prominent problem that restricts agricultural green development in China. The main tasks of the demonstration zone include: (a) supporting the entrepreneurship and innovation of new agricultural business entities such as family farms and farmers' cooperatives; (b) improving innovative service platforms, such as various R&D institutions, testing and testing centers, new rural development research institutes and modern agricultural industry science and technology innovation centers; (c) guiding the scientific and technological resources and talents of colleges and universities and scientific research institutes to gather in the demonstration area; (d) building farmers' training bases with regional characteristics; (e) improving the ability of information management and service in the whole process of agricultural production; and (f) developing circular ecological agriculture and promoting the efficient utilization of agricultural resources and ecological environment protection.

The government should improve the system for ecological and environmental regulation and set up a feasible eco-incentive mechanism. First, the government should develop a monitoring and evaluation system for sustainable management that covers grain output, produce quality, and eco-resources. Second, the government should compensate and incentivize clean production technology and green production factors that are eco-friendly and resource-saving. Finally, the government should carry out pilot subsidy programs for the use of cycle agriculture technology in ecologically sensitive regions. This will help agricultural producers, agricultural cooperative organizations and eco-professional service systems in performing their role in sustainable management.

For the scientific community, the priority is breaking bottleneck technologies and developing whole solutions for sustainable management in ecologically fragile regions. Ecologically fragile regions such as Northeast China suffer from soil erosion and a decrease in soil organic matter; Southeast China suffers from soil acidification and heavy metal pollution, while Northwest China suffers from soil salinization and climatic aridity. On the one hand, scientists need to develop modern biological technology to solve the bottleneck of nutrient transformation technology in soil-root-microbe interfaces to improve soil quality and nutrient-use efficiency. On the other hand, scientists need to build ecological barriers for controlling and remediating soil degradation and implement a synergetic development strategy for mountain–river–forest–farmland–lake–grassland ecosystems (Figure 5).

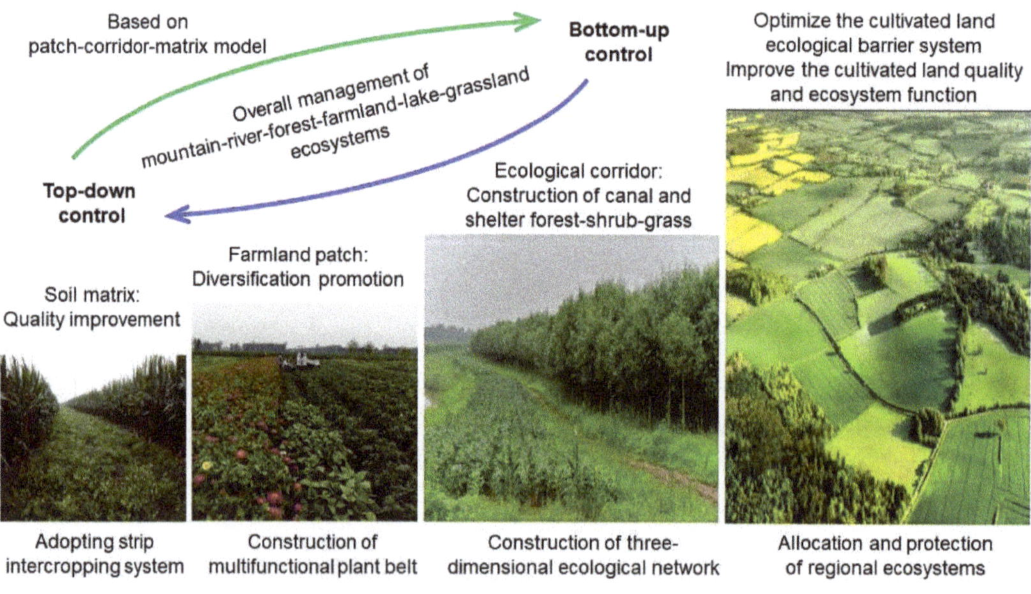

Figure 5. Implement a synergetic development strategy of mountain–river–forest–farmland–lake–grassland–sandy land ecosystems.

In conclusion, the insufficient use of agricultural resources, agricultural non-point source pollution and land degradation are the main problems facing China's sustainable agriculture. China has improved its laws and regulations and implemented serious actions to ensure food safety and coordinated management of agricultural resources. As a next step, the connections between the government, agri-businesses, scientific community and farming community should be strengthened by building demonstration zones. The government should improve regulation and set up a feasible eco-incentive mechanism, while scientists should break bottleneck technologies and develop solutions for the sustainable management of resources and the environment. This will facilitate sustainable agricultural development in China.

Supplementary Materials: The following are available online at https://www.mdpi.com/article/10.3390/ijerph20043233/s1, Table S1: Changes in soil organic carbon stocks (SOCS) from 1980 to 2011 for the 58 investigated counties across China.

Author Contributions: All authors contributed intellectual input and assistance to this study and manuscript; B.S. and J.Z. developed the research framework; B.S. and Y.Z. contributed to the analysis of soil and environmental problems, soil organic carbon storage and soil pH value in China; Y.L., J.Y. and J.Z. contributed to the review of the achievements related to the soil pollution control program and cultivated land quality protection program; B.S. and D.Y. contributed to the synergetic development strategy; B.S. and J.Z. wrote the manuscript. All authors have read and agreed to the published version of the manuscript.

Funding: B.S. was supported by the National Key R&D Program of China (2022YFD1900600) and the National Key Agricultural Science and Technology Project of China. J.Z. was supported by the Strategic Research and Consulting Project of the Chinese Academy of Engineering (2022-XY-96) and the China Agriculture Research System of MOF and MARA (CARS-52). D.Y. was supported by a project from the Chinese Academy of Agricultural Sciences (CAAS-XTCX2016015).

Institutional Review Board Statement: Not applicable.

Informed Consent Statement: Not applicable.

Data Availability Statement: The soil organic carbon data in the topsoil (0–20 cm) in China are available in the National Earth System Science Data Center (http://soil.geodata.cn, accessed on 10 May 2022) which was built by Institute of Soil Soil Science, Chinese Academy of Science. The soil pH data in the topsoil (0–20 cm) in China are available in the National Cultivated Land Quality Big Data Platform (http://www.farmland.cn/, accessed on 10 May 2022) which was built by Yangzhou Cultivated Land Quality Protection Station, Department of Planting Management, Department of Farmland Construction Management, Ministry of Agriculture and Rural Affairs.

Acknowledgments: We would like to thank three anonymous reviewers and the managing editor for constructive comments and suggestions that helped to improve the manuscript. We also would like to thank Xianli Xie and Wenxi Li for data assistance.

Conflicts of Interest: The authors declare no conflict of interest.

References

1. Lichtfouse, E.; Navarrete, M.; Debaeke, P.; Véronique, S.; Alberola, C. *Sustainable Agriculture*; Springer: Dordrecht, The Netherlands; Heidelberg, Germany; London, UK; New York, NY, USA, 2009.
2. Xue, D.; Dai, R.; Guo, L.; Sun, F. *The Modes and Case Studies of Eco-Farming in China*; China Environmental Science Press: Beijing, China, 2012.
3. Xie, J.; Hu, L.; Tang, J.; Wu, X.; Li, N.; Yuan, Y.; Yang, H.; Zhang, J.; Luo, S.; Chen, X. Ecological mechanisms underlying the sustainability of the agricultural heritage rice-fish coculture system. *Proc. Natl. Acad. Sci. USA* **2011**, *108*, 19851–19852. [CrossRef] [PubMed]
4. Zhang, X.-H.; Zhang, R.; Wu, J.; Zhang, Y.-Z.; Lin, L.-L.; Deng, S.-H.; Li, L.; Yang, G.; Yu, X.-Y.; Qi, H.; et al. An emergy evaluation of the sustainability of Chinese crop production system during 2000–2010. *Ecol. Indic.* **2016**, *60*, 622–633. [CrossRef]
5. Zhao, Q.G.; Huang, J.K. *Agricultural Science & Technology in China: A Roadmap to 2050*; Science Press: Beijing, China, 2011.
6. Ministry of Natural Resources of the People's Republic of China (MNR). The 2016 Annual Bulletin on Land Resources in China. 2017. Available online: http://www.mnr.gov.cn/sj/tjgb/201807/P020180704391918680508.pdf (accessed on 30 August 2021).

7. Ministry of Water Resources of the People's Republic of China (MWR). The 2019 Annual Bulletin on Water Resources in China. 2020. Available online: http://www.mwr.gov.cn/sj/tjgb/szygb/202008/P020200803328847349818.pdf (accessed on 30 April 2022).

8. Ministry of Agriculture and Rural Affairs of the People's Republic of China (MARA). The 2019 Annual Bulletin of National Cultivated Land Quality Grade in China. 2020. Available online: http://www.ntjss.moa.gov.cn/zcfb/202006/020200622573390595236.pdf (accessed on 30 April 2022).

9. Tang, Y.-H.; Huang, Y. Spatial distribution characteristics of the percentage of soil fertility contribution and its associated basic crop yield in mainland China. *J. Agro-Environ. Sci.* **2009**, *28*, 1070–1078.

10. Zhang, W.; Cao, G.; Li, X.; Zhang, H.; Wang, C.; Liu, Q.; Chen, X.; Cui, Z.; Shen, J.; Jiang, R.; et al. Closing yield gaps in China by empowering smallholder farmers. *Nature* **2016**, *537*, 671–674. [CrossRef] [PubMed]

11. Deng, N.; Grassini, P.; Yang, H.; Huang, J.; Cassman, K.G.; Peng, S. Closing yield gaps for rice self-sufficiency in China. *Nat. Commu.* **2019**, *10*, 1725. [CrossRef]

12. Ministry of Ecology and Environment of the People's Republic of China (MEE). The 2020 Annual Bulletin on Ecological Environment in China. 2021. Available online: http://www.mee.gov.cn/hjzl/sthjzk/zghjzkgb/202105/P020210526572756184785.pdf (accessed on 30 April 2022).

13. Liu, Y.S.; Wang, J.Y.; Guo, L.Y. The spatial-temporal changes of grain production and arable land in China. *Sci. Agric. Sin.* **2009**, *42*, 4269–4274.

14. Yang, Z.H.; Cai, H.Y.; Qin, C.; Liu, H.G. Analysis on the spatial and temporal pattern of china's grain production and its influencing factors. *J. Agric. Sci. Technol.* **2018**, *20*, 1–11.

15. Yang, X.; Mu, Y.-Y. Spatial-temporal matching patterns of grain production and water resources. *J. South China Agric. Univ. (Soc. Sci. Ed.)* **2019**, *18*, 91–100.

16. Wu, P.T. Warning and strategy of North-to-South Water Transfer Project. *Adv. Sci. Technol. Water Resour.* **2015**, *35*, 121–123, 180.

17. Zhang, T.L. Strengthening soil and environment management of agricultural producing area, promoting sustainable development of agriculture in China. *Bull. Chin. Acad. Sci.* **2015**, *30*, 435–444. (In Chinese)

18. National Bureau of Statistics (NBS). *China Statistical Yearbook 2020*; China Statistics Press: Beijing, China, 2020. Available online: http://www.stats.gov.cn/tjsj/ndsj/2020/indexeh.htm (accessed on 30 April 2022).

19. Yang, L.Z.; Sun, B. *Cycling, Balance and Management of Nutrients in Agro-Ecosystems in China*; Science Press of China: Beijing, China, 2008.

20. Zhang, F.; Wang, J.; Zhang, W.; Cui, Z.; Ma, W.; Chen, X.; Jiang, R. Nutrient use efficiencies of major cereal crops in China and measures for improvement. *Acta Pedol. Sin.* **2008**, *45*, 915–9249.

21. Yan, X.; Jin, J.-Y.; Liang, M.-Z. Fertilizer use efficiencies and yield-increasing rates of grain crops in China. *Soils* **2017**, *49*, 1067–1077.

22. Ministry of Agriculture and Rural Affairs of the People's Republic of China (MARA). The Utilization Rate of Chemical Fertilizers and Pesticides of the Three Major Grain Crops in China Exceeded 40%. 2021. Available online: http://www.kjs.moa.gov.cn/gzdt/202101/t20210119_6360102.htm (accessed on 30 April 2022).

23. Rural Social and Economic Research Division, National Bureau of Statistics of the People's Republic of China (RSERD, NBSPRC). *China Rural Statistical Yearbook*; China Statistics Press: Beijing, China, 2015.

24. Cai, Q.Y.; Mo, C.H.; Wu, Q.R.; Katsoyiannisc, A.; Zeng, Q.Y. The status of soil contamination by semi-volatile organic chemicals (SVOCs) in China: A review. *Sci. Total Environ.* **2008**, *389*, 209–224. [CrossRef]

25. Gao, J.; Liu, L.; Liu, X.; Lu, J.; Zhou, H.; Huang, S.; Wang, Z.; Spear, P.A. Occurrence and distribution of organochlorine pesticides-lindane, p,p'-DDT, and heptachlor epoxide in surface water of China. *Environ. Int.* **2008**, *34*, 1097–1103. [CrossRef] [PubMed]

26. Ministry of Environmental Protection of the People's Republic of China (MEP); National Bureau of Statistics of the People's Republic of China (NBS); Ministry of Agriculture of the People's Republic of China (MOA). Bulletin on the First National Survey of Pollution Sources. 2010. Available online: http://www.gov.cn/jrzg/2010-02/10/content_1532174.htm (accessed on 30 April 2022).

27. Sun, B.; Yang, L.Z.; Zhang, L.L.; Zhang, F.S.; Norse, D.; Zhu, Z.L. Agricultural non-point source pollution in China: Causes and mitigation measures. *AMBIO* **2012**, *41*, 370–379. [CrossRef] [PubMed]

28. Song, D.-L.; Hou, S.-P.; Wang, X.-B.; Liang, G.-Q.; Zhou, W. Nutrient resource quantity of crop straw and its potential of substituting. *J. Plant Nutr. Fertil.* **2018**, *24*, 1–21.

29. Rural Social and Economic Investigation Department, National Bureau of Statistics (RSRID, NBS). *China Rural Statistical Yearbook-2019*; China Statistics Press: Beijing, China, 2020.

30. Xu, M.G.; Zhang, W.J.; Huang, S.M. *Changes of Soil Fertility in China (Second Edition)*; China Agricultural Science and Technology Press: Beijing, China, 2015.

31. Cultivated Land Quality Monitoring and Protection Center, Ministry of Agriculture and Rural Affairs of the People's Republic of China (CLQMPC, MARA). *Change Law of Cultivated Land Quality in 30 Years*; China Agricultural Press: Beijing, China, 2019.

32. Ministry of Water Resources, Chinese Academy of Sciences, Chinese Academy of Engineering (MWR, CAS, CAE). *Prevention and Control of Soil Erosion and Ecological Security in China (General Volume)*; Science Press: Beijing, China, 2010.

33. Ministry of Environmental Protection of the People's Republic of China (MEP); Ministry of Land and Resources of the People's Republic of China (MLR). Bulletin on Soil Pollution Survey in China. 2014. Available online: http://www.mee.gov.cn/gkml/sthjbgw/qt/201404/W020140417558995804588.pdf (accessed on 30 August 2021).

34. Chen, H.; Teng, Y.; Lu, S.; Wang, Y.; Wang, J. Contamination features and health risk of soil heavy metals in China. *Sci. Total Environ.* **2015**, *512–513*, 143–153. [CrossRef]

35. Ministry of Agriculture (MOA); National Development and Reform Commission (NDRC); Ministry of Science and Technology (MST); Ministry of Finance (MOF); Ministry of Land and Resources (MLR); Ministry of Environmental Protection (MEP); Ministry of Water Resources (MWR); State Forestry Administration (SFA). China Agricultural Sustainable Development Planning (2015 to 2030). 2015. Available online: http://www.moa.gov.cn/ztzl/mywrfz/gzgh/201509/t20150914_4827900.htm (accessed on 30 April 2022).

36. The State Council of the People's Republic of China (SCPRC). Soil Pollution Prevention and Control Action Plan. 2016. Available online: http://www.gov.cn/zhengce/content/2016-05/31/content_5078377.htm (accessed on 30 August 2021).

37. Cui, Z.; Zhang, H.; Chen, X.; Zhang, C.; Ma, W.; Huang, C.; Zhang, W.; Mi, G.; Miao, Y.; Li, X.; et al. Pursuing sustainable productivity with millions of smallholder farmers. *Nature* **2018**, *555*, 363–366. [CrossRef]

38. Cao, Z.H.; Zhou, J.M. *Soil Quality of China*; Science Press of China: Beijing, China, 2008.

39. Sun, B. *Control and Ecological Rehabilitation of Red Soil Degradation*; Science Press of China: Beijing, China, 2011.

40. Zhang, X.Y.; Hui, L.J. *Achievements in Comprehensive Control of Soil Erosion and Water Loss*; China Water & Power Press: Beijing, China, 2015.

41. Zhou, W. *Theories and Approaches of Amelioration and Management of Low Yield Paddy Soils*; Science Press: Beijing, China, 2015.

42. Sun, B.; Zhang, X.D.; Lu, Y.H.; Wei, G.H. *Improvement of Cultivated Land Fertility and Efficient Utilization of Chemical Fertilizer Nutrients*; Science Press: Beijing, China, 2022.

43. State Administration for Market Regulation (SAMR). *Standardization Administration of the People's Republic of China (SAC). Well-Facilitated Farmland Construction-General Rules*; SAMR: Beijing, China, 2022.

44. Bryan, B.A.; Gao, L.; Ye, Y.; Sun, X.; Connor, J.D.; Crossman, N.D.; Stafford-Smith, M.; Wu, J.; He, C.; Yu, D.; et al. China's response to a national land-system sustainability emergency. *Nature* **2018**, *559*, 193–204. [CrossRef]

45. *Fertile Soil Project Phase II Construction Plan (2004–2010)*; Department of Crop Management, Ministry of Agriculture (DCM, MOA): Beijing, China, 2003.

46. The State Council of the People's Republic of China (SCPRC). National Plan for High Standard Farmland Construction. 2013. Available online: https://www.ndrc.gov.cn/fzggw/jgsj/njs/sjdt/201312/P020191101560721899254.pdf (accessed on 30 August 2021).

47. Ministry of Agriculture and Rural Affairs (MARA); Ministry of Finance of the People's Republic of China (MOF). Northeast Black Soil Conservation Tillage Action Plan (2020–2025). 2020. Available online: http://www.moa.gov.cn/gk/tzgg_1/tz/202003/P020200318408776059442.pdf (accessed on 30 April 2022).

48. Zhao, Y.; Wang, M.; Hu, S.; Zhang, X.; Ouyang, Z.; Zhang, G.; Huang, B.; Zhao, S.; Wu, J.; Xie, D.; et al. Economics- and policy-driven organic carbon input enhancement dominates soil organic carbon accumulation in Chinese croplands. *Proc. Natl. Acad. Sci. USA* **2018**, *115*, 4045–4050. [CrossRef]

49. Luo, Y. *Remediation Mechanism and Technological Development of Heavy Metal Polluted Soil*; Science Press: Beijing, China, 2016.

50. Chen, X.-X.; Zhong, H.-T.; Wang, Q.; Peng, C.-L.; Yan, S.-L.; Qin, Q. Application and Research Progress of "VIP+n" Remediation Measures for Heavy Metal Pollution of Farmland in Hunan Province. *Mod. Agric. Sci. Technol.* **2019**, *740*, 149–150, 152.

51. National Agricultural Technology Extension Service Center (NATEC). *Compilation of Dryland Water-Saving Agricultural Technology Models*; China Agricultural Press: Beijing, China, 2020.

52. China Agricultural Green Development Research Association (CAGDRA); Institute of Agricultural Resources and Regional Planning (IARRP); Chinese Academy of Agricultural Sciences (CAAS). *China Agricultural Green Development Report 2020*; China Agricultural Press: Beijing, China, 2021.

53. Ministry of Agriculture of the People's Republic of China (MOA). National Action Plan for Popularization of Soil Testing and Formula Fertilization in 2010. 2010. Available online: http://www.moa.gov.cn/govpublic/ZZYGLS/201006/t20100606_1534647.htm (accessed on 30 April 2022).

54. Chen, X.; Cui, Z.; Fan, M.; Vitousek, P.; Zhao, M.; Ma, W.; Wang, Z.; Zhang, W.; Yan, X.; Yang, J.; et al. Producing more grain with lower environmental costs. *Nature* **2014**, *514*, 486–489. [CrossRef]

55. Ministry of Agriculture of the People's Republic of China (MOA). National Action Plan for Replacing Chemical Fertilizer with Organic Fertilizer for Fruit, Vegetable and Tea. 2017. Available online: http://www.moa.gov.cn/nybgb/2017/derq/201712/t20171227_6130977.htm (accessed on 30 April 2022).

56. Guo, J.H.; Liu, X.J.; Zhang, Y.; Shen, J.L.; Han, W.X.; Zhang, W.F.; Christie, P.; Goulding, K.W.; Vitousek, P.M.; Zhang, F.S. Significant acidification in major Chinese croplands. *Science* **2010**, *327*, 1008–1010. [CrossRef]

57. Sun, B.; Shen, R.P.; Bouwman, A.F. Surface N balances in agricultural crop production systems in China for the period 1980-2015. *Pedosphere* **2008**, *18*, 135–143. [CrossRef]

58. Zhang, X.; Davidson, E.A.; Mauzerall, D.L.; Searchinger, T.D.; Dumas, P.; Shen, Y. Managing nitrogen for sustainable development. *Nature* **2015**, *528*, 51–59. [CrossRef]

59. Lehmann, J.; Bossio, D.A.; Kögel-Knabner, I.; Rillig, M.C. The concept and future prospects of soil health. *Nat. Rev. Earth Environ.* **2020**, *1*, 544–553. [CrossRef]

60. Honeycutt, C.W.; Morgan, C.L.S.; Elias, P.; Doane, M.; Mesko, J.; Myers, R.; Odom, L.; Moebius-Clune, B.; Nichols, R. Soil Health: Model programs in the USA. *Front. Agric. Sci. Eng.* **2020**, *7*, 356–361. [CrossRef]
61. Ministry of Agriculture and Rural Affairs of the People's Republic of China (MARA). National High Standard Farmland Construction Plan (2021–2030). 2021. Available online: http://www.ntjss.moa.gov.cn/zcfb/202109/t20210915_6376511.htm (accessed on 30 April 2022).

International Journal of
***Environmental Research
and Public Health***

Article

Quantitative Response of Gray-Level Co-Occurrence Matrix Texture Features to the Salinity of Cracked Soda Saline–Alkali Soil

Yue Zhao [1], Zhuopeng Zhang [1], Honglei Zhu [2] and Jianhua Ren [1,*]

[1] Heilongjiang Province Key Laboratory of Geographical Environment Monitoring and Spatial Information Service in Cold Regions, Harbin Normal University, Harbin 150025, China; zhaoyue_hrbnu@163.com (Y.Z.); zhuopeng_hrbnu@163.com (Z.Z.)

[2] College of Life Science, Henan Normal University, Xinxiang 453007, China; 2015111@htu.edu.cn

* Correspondence: renjianhua@hrbnu.edu.cn; Tel.: +86-431-88060524

Abstract: Desiccation cracking during water evaporation is a common phenomenon in soda saline–alkali soils and is mainly determined by soil salinity. Therefore, quantitative measurement of the surface cracking status of soda saline–alkali soils is highly significant in different applications. Texture features can help to determine the mechanical properties of soda saline–alkali soils, thus improving the understanding of the mechanism of desiccation cracking in saline–alkali soils. This study aims to provide a new standard describing the surface cracking conditions of soda saline–alkali soil on the basis of gray-level co-occurrence matrix (GLCM) texture analysis and to quantitatively study the responses of GLCM texture features to soil salinity. To achieve this, images of 200 field soil samples with different surface cracks were processed and calculated for GLCMs under different parameters, including directions, gray levels, and step sizes. Subsequently, correlation analysis was then conducted between texture features and electrical conductivity (EC) values. The results indicated that direction had little effect on the GLCM texture features, and that four selected texture features, contrast (CON), angular second moment (ASM), entropy (ENT), and homogeneity (HOM), were the most correlated with EC under a gray level of 2 and step size of 1 pixel. The results also showed that logarithmic models can be used to accurately describe the relationships between EC values and GLCM texture features of soda saline–alkali soils in the Songnen Plain of China, with calibration R^2 ranging from 0.88 to 0.92, and RMSE from 2.12×10^{-4} to 9.68×10^{-3}, respectively. This study can therefore enhance the understanding of desiccation cracking of salt-affected soil to a certain extent and can also help to improve the detection accuracy of soil salinity.

Keywords: GLCM; texture feature; soda saline–alkali soil; soil surface crack; Songnen Plain

Citation: Zhao, Y.; Zhang, Z.; Zhu, H.; Ren, J. Quantitative Response of Gray-Level Co-Occurrence Matrix Texture Features to the Salinity of Cracked Soda Saline–Alkali Soil. *Int. J. Environ. Res. Public Health* **2022**, *19*, 6556. https://doi.org/10.3390/ijerph19116556

Academic Editor: Paul B. Tchounwou

Received: 14 April 2022
Accepted: 26 May 2022
Published: 27 May 2022

Publisher's Note: MDPI stays neutral with regard to jurisdictional claims in published maps and institutional affiliations.

1. Introduction

Soil salinization is a very serious issue in China, with the total area of salt-affected soil being almost 9.91×10^7 ha [1], which is increasing due to the growing population and deteriorating ecological environment. This has caused great damage to China's social economy, natural environment, and ecosystem. As one of the three major distribution areas in China, the total area of saline soil in the Songnen Plain is over 3.73×10^6 ha [2]. The main soil salt minerals in this area are $NaHCO_3$ (sodium bicarbonate) and Na_2CO_3 (sodium carbonate), together with small amounts of sulfate and chloride, indicating that the salt-affected soils belong to a typical type of inland soda saline–alkali soil. Because of the high content of clay particles and adsorbable cations, shrinkage and cracking on the surface of soda saline–alkali soil are very common during water evaporation. Desiccation cracks indicate a surface state of saline–alkali soils and are commonly considered as a characterization of the salinity levels of salt-affected soils, which also indicate a mechanical state of the salt-affected soil [3]. Therefore, exploring the quantitative relationship between crack characteristics and salt content will help to further understand the cracking mechanism of the saline–alkali soil surface. In addition, effective crack characteristics can also be used to further predict soil salinity in order to provide important guidance for both ecological restoration and improvement of salinized soil in soil sciences, which will thus alleviate the conflicts between human and local land and will thus guarantee China's food security to a certain extent.

Many field and laboratory studies have focused on the relationship between soil salinity and the morphology of desiccation cracks. Amaya et al. [4] measured salinity of a saline cracking soil within the peds to a depth of 50 cm over a three-year period during reclamation drying in field, finding that although the initial EC within the interior of peds and below the cracking depth ranged from 22 to 35 ds/m, the soil salinity redistributes and decreases with the EC range from only 12 to 16 ds/m at the end of three years of desalinization. Lima and Grismer [5] conducted field measurements considering the effects of soil salinity on soil crack morphology at various times after irrigation and found that as salinity increased, soil crack width, island width, crack area, and crack volume tended to increase, whereas crack depth decreased. In order to present a field dataset to quantitatively evaluate the contribution of bypass flow to the leaching salts, Fujimaki and Baki [6] carried out soil sampling and monitoring of groundwater and discharge from a tile drain in farmland with a cracking soil in the Nile Delta, finding that the evidence for the occurrence of significant bypass flow through cracks was the salinity of the pore water. After conducting a field evaluation of the impact of soil cracking on irrigation, drainage, and soil salinity on a heavy clay soil in the Imperial Valley of California, Van der Tak and Grismer [7] found that water movement within soil cracks controls the water application uniformity, soil profile wetting, and salt leaching to irrigation. Ben-Hur and Assouline [8] selected a cotton field located in the Yizre'el Valley, Israel, as the experiment site to study the tillage effects on water and salt distribution in a vertisol during effluent irrigation and rainfall, with their results showing that the high infiltration of the runoff through cracks limited the effects of the runoff downhill flow on the water and salt distribution along the slope. After studying the effects of roots and salinity on law of development for farmland soil desiccation crack, Zhang et al. [9] found that when the moisture content is less than 27%, salt content will greatly increase the area density of soil cracks at steady state with the length density appearing as a rather opposite trend. However, most field experiments only deal with the qualitative relationship between salt content and desiccation cracks. With the development of image processing technology, crack feature can be extracted with a quite precise accuracy, which makes more scholars tend to carry out controllable experiments to quantitatively analyze the influence of salt content on soil desiccation cracking. From a laboratory dry test, Zhang et al. [10] investigated the development law of desiccation cracks on the soil surface under different salt content, with their results indicating that soil salinity can increase area density but lead to a decrease in length density of cracks. Zhang et al. [11] conducted desiccation tests in the laboratory on initially saturated slurry specimens with different NaCl (sodium chloride) content selected from Yar City, northwest China, and their results indicated that as the NaCl content increased, the intersection number, segment number, and total length of the cracks all decreased. After conducting a laboratory study to investigate the effects of four salt cations, namely, Na^+ (sodium), K^+ (potassium), Ca^{2+} (calcium), and Mg^{2+} (magnesium), on soil shrinkage and cracking during dehydration, Xing et al. [12] found that parameters including crack length, crack area, crack length density, and crack area density decreased with an increase in the concentrations of K^+, Na^+, and Ca^{2+}, but a reduction was found when the concentration of Mg^{2+} increased, indicating that the four crack parameters also increased with the content of HCO_3^- (bicarbonate), CO_3^{2-} (carbonate), and SO_4^{2-} (sulfate), while an opposite trend was found with the concentration of Cl-. After studying the effect of cation type on the process of shrinkage and desiccation cracking, Wang et al. [13] found that salt cations, including Na^+, K^+, Ca^{2+}, and Mg^{2+}, have a strong effect on the crack areas of salt-affected soils.

From the studies mentioned above, it can be seen that the relationships between salt content and the extent of desiccation cracks are quite different, indicating that the effect of salinity on the cracking patterns and dynamics of salt-affected soils still remains an open question. This is because of the complicated interaction between salt and soil particles during desiccation cracking is affected by the ion concentration and the valence state of ions, which are determined by the physical and chemical properties of the soil samples, indicating that the effect of salinity on the cracking of the cohesive soil surface remains partially unclear [14]. In addition, the response of soil cracking to salt content is inseparable from the selection of crack characterization indicators. Although many quantitative indexes in previous studies have been established to characterize the extent of shrinkage and desiccation cracks [15–18], there is still no unified soil crack description standard. In addition, most previous indicators have focused on geometric features, indicating that they usually have severe limitations for characterizing cracks and cannot reflect the spatial distributions of desiccation cracks with convergence in many applications. Moreover, these geometric indicators hardly reflect the direction of crack propagation, which usually depends on the local hydrological conditions and topographical fluctuations of salinized soil. Therefore, an effective characterization indicator of soil cracks is of great significance for analyzing the response of crack characteristics to soil salt content, determining the salt migration process of the saline–alkaline soil, monitoring the range of soil salinization, and ameliorating salt-affected soil.

Texture features often refer to visual characteristics that do not depend on the color or brightness of the image and can reflect the homogeneous phenomenon of the image and describe the pixel distribution in the neighborhood space [19–21]. For a special object within an image, texture features often contain important information about the surface structure arrangement and thus can reflect its connection with the surrounding environment [22,23]. Texture analysis aims to select a unique method to describe the underlying characteristics, which generally consist of four types: statistical, modeling, signal processing, and structural methods [24]. Among these types of texture analysis, GLCM is the most common [25–30] method because it can reflect a large amount of information within a grayscale image, such as the direction, interval, amplitude, and change ratio. GLCM texture features are commonly extracted for analyzing the local features and overall arrangement rules of an image and are widely used for pattern recognition, accurate classification, feature extraction, and image segmentation in many applications [31–39]. Because the propagation and development of desiccation cracks are rather random in statistics, texture features always contain important information about the crack arrangement and the pixel distributions within the crack patterns. Therefore, it is quite certain that GLCM texture analysis can aid in describing the complex structures and variation in the surface intensity of desiccation cracks in cohesive saline–alkali soil. This is because GLCM texture features can describe both the important arrangement of the surface structure and the distribution of the pixels of the soil crack image in the neighborhood space. In addition, GLCM texture features extracted from crack patterns can also reflect differences in the physical and chemical properties of saline–alkali soils, such as clay minerals and salt content. However, very few studies have focused on the relationship between soil salinity and GLCM texture features computed from crack images. However, research on the correlation between cracked soil surface texture features and soil salinity is still very rare. Although Ren et al. [40,41] studied the influence of salt content on the shrinkage and cracking process of soda saline–alkali soils on the basis of the theory of GLCM texture analysis, only one type of GLCM texture feature (corresponding to the contrast) was extracted from the binary crack image in their research, and their cracked soil samples were prepared in the laboratory, indicating that these samples cannot accurately reflect the real status of desiccation cracks generated in nature.

Surface cracking is a mechanical state of saline–alkali soil, which indicates that exploring the correlation between crack characteristics and salt content can therefore effectively improve the cognition level of cracking process of saline soil. To achieve this objective and develop an effective standard for crack characteristics, this study intended to quantify the ability of GLCM texture features extracted from crack patterns in characterizing the desiccation cracks generated on the surface of soda saline–alkali soil samples. In particular, the effects of parameters, including the gray level, step size, and direction, on the results of GLCM texture features were individually analyzed and compared. Subsequently, correlation analysis was also conducted between GLCM texture features and EC values of the soil samples to quantitatively study the response of GLCM texture features to soil salinity for further understanding the cracking process of soda saline–alkali soils in the Songnen Plain of China. Finally, the logarithmic regression model between EC values and several common texture features were developed under the optimal GLCM computing parameters, which both leads to a better recognition for desiccation cracking process of soda saline–alkali soils and also provides a possibility for effective detection of the characteristics of soda saline–alkali soils.

2. Materials and Methods

2.1. Study Area and Soil Sampling

The western part of the Songnen Plain is a typical salt-affected soil region in China with salt minerals mainly composed of $NaHCO_3$ and Na_2CO_3, which makes the soil a typical type of soda saline–alkaline soil. In this study, Baicheng City was selected as the study area, with an average annual precipitation of 400 to 500 mm, which is mainly concentrated in July and August; however, the average annual evaporation in Baicheng City is as high as 1500 to 1900 mm. This severely unbalanced evaporation-to-precipitation ratio, coupled with special hydrogeological conditions and over-irrigation in human agricultural production, makes the area heavily salinized. In addition, the desiccation cracks commonly occur on the soil surface since the soil can be classified as a texture of clay loam with high clay content. After considering the heterogeneity of soil salinity, 200 soil sample points with different extents of desiccation cracking were selected, with a small region ranging from 45°18′14″ N to 45°29′47″ N and 123°39′8″ E to 124°21′6″ E in November 2018 (Figure 1). This is because, after months of evaporation, the soil moisture content is very low in November, and the cracking process is complete on the soil surface. All soil samples were determined on the basis of the rule of plum blossom spots.

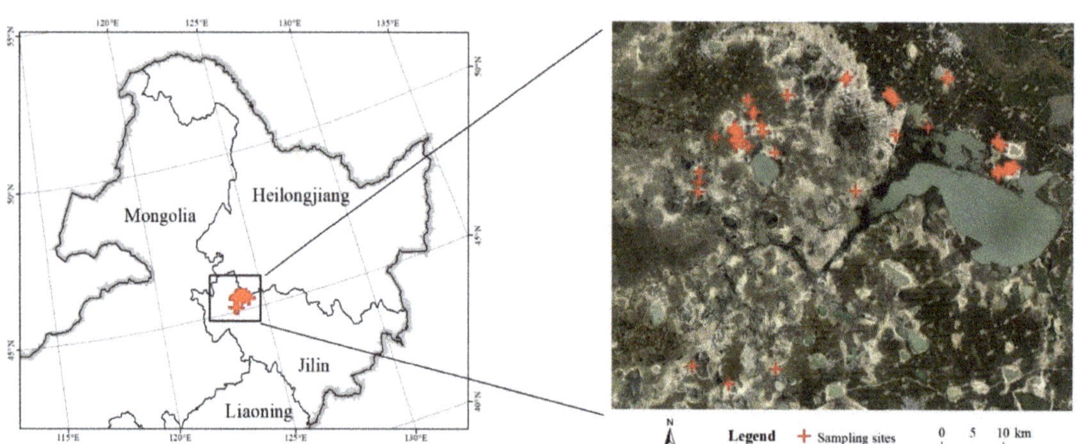

Figure 1. This study area and sampling points located on a Landsat-8 satellite image.

2.2. Preprocessing of Crack Images

According to Ren et al. [40], the GLCM contrast feature becomes stable at a scale of approximately 38 × 38 cm. After considering the effects of sample size on the degree of soil cracking, soil samples with a rectangular size of 50 × 50 cm were selected in this study, and the cracking patterns of all the soil samples were measured using the following standard. First, a digital camera was selected and fixed on a platform with the lens 1.5 m above the ground. Second, a rectangular wooden frame with an inner size of 50 × 50 cm was designed and placed on the ground to ensure that it coincided with the vertical projection of the digital camera lens. Third, the white balance processing of the camera was performed together with the aperture size and exposure time set to standardize the same lighting environment and camera parameters. Finally, the desiccation crack was photographed and the calibration image of a black and white grid plate with a size of 50 × 50 cm was taken for further geometric distortion corrections. When the images of the cracked soil samples were taken, a unified pre-processing operation was conducted. In detail, all crack patterns of the soil samples were geometrically corrected using the polynomial method, cropped to a standard size of 50 × 50 cm, and converted into grayscale images (Figure 2).

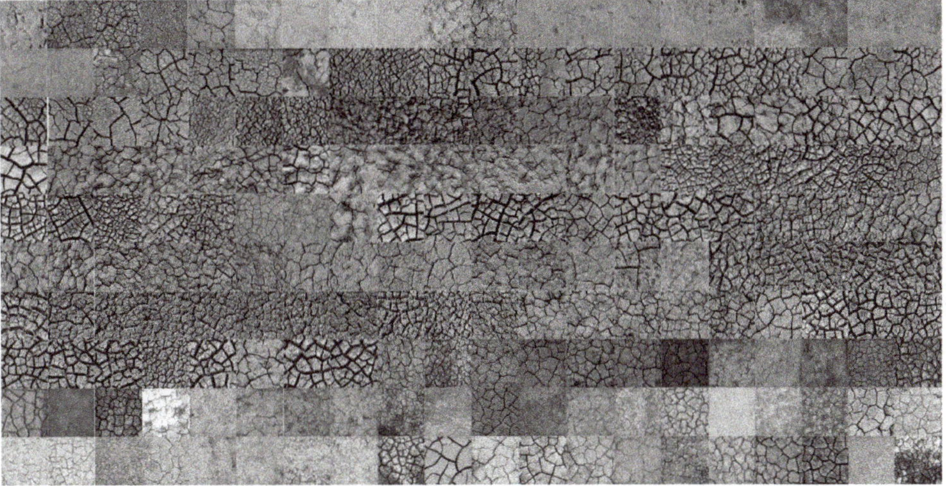

Figure 2. Grayscale images of cracked soil samples.

2.3. Soil Property Measurements

After the crack images were taken at the sampling points, soil samples were collected at a depth of 20 cm in the center of the wooden frame to determine the soil properties. This is because the kind of soda saline soil largely prevents salt moving downwards due to its bad infiltration capacity, which indicates the properties of soil from the top 20 cm soil layer are very stable [42]. The soil samples were weighed before and after they were evenly oven-dried for water contents, after which they were ground and passed through a 2 mm sieve for soil properties. Note that as the osmotic pressure of the soil solution increased by soil salinity is strictly proportional to the EC values under certain water conditions, the EC value was thus determined as the indicator of soil salinity in this study. In particular, soil suspensions for all soil samples were configured using CO_2-free distilled water (pH = 7), with a water–soil mass ratio of 5:1 [43–45]. After stirring with a glass rod and leaving for about half an hour, the pH and EC values of all the soil samples were measured using potentiometric and conductometric methods, respectively (Figure 3). To facilitate the analysis and comparison of different samples, the unit of the EC value was uniformly converted into ds/m. In addition, the particle size distributions of all soil samples were measured using an Mllvern-200 laser particle size analyzer.

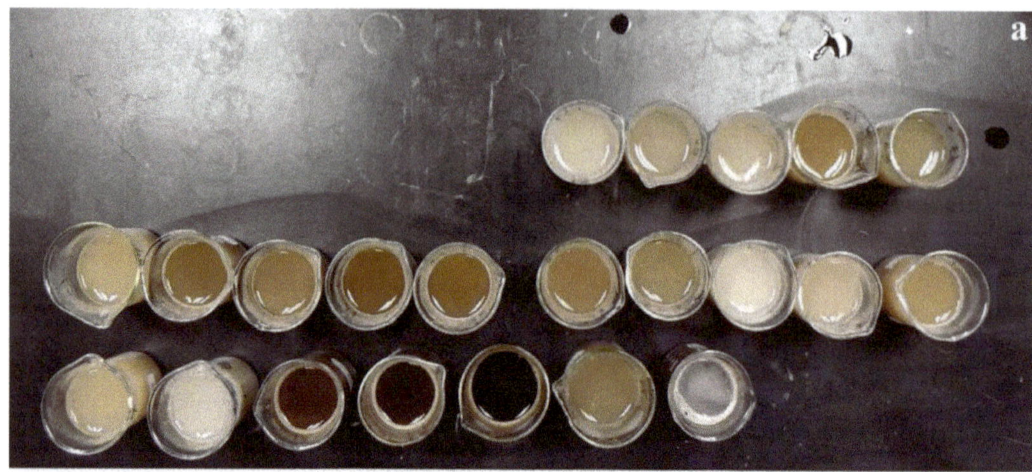

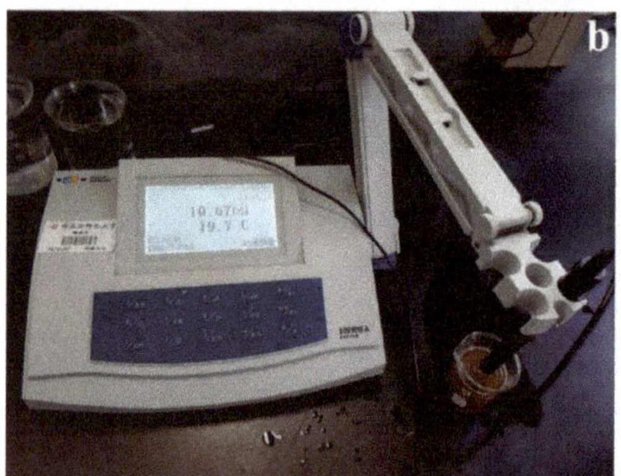

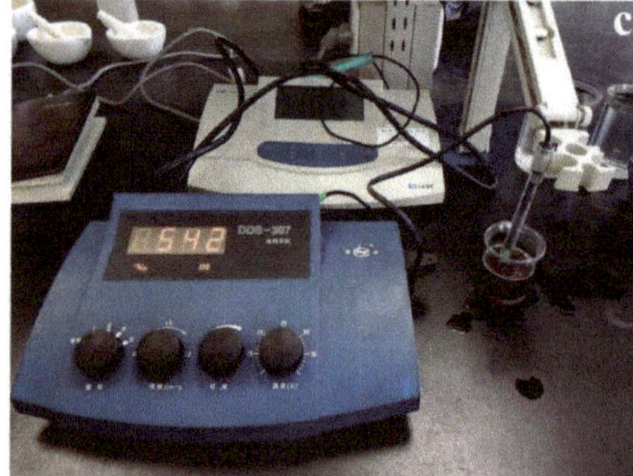

Figure 3. Measurements of main chemical soil properties. (**a**) Prepared soil suspensions; (**b**) pH measurement process; (**c**) EC measurement process.

2.4. Gray Level Co-Occurrence Matrix Texture Features

In this study, we selected the classic GLCM texture features as the statistical texture characteristics of the cracked soil surface to determine the relevance of different pixels by calculating the second-order combined conditional probability density between the image pixel gray levels at a certain distance and direction. For a known gray image $f(x, y)$ of a soil sample with surface cracks, the second-order combined conditional probability density can be calculated using the following equation:

$$p(i, j) = g\{(x_1, y_1), (x_2, y_2) \in m \times n | f(x_1, y_1) = i, f(x_2, y_2) = j\} \tag{1}$$

where i and j are the gray values of the gray image $f(x, y)$ at the (x_1, y_1) and (x_2, y_2) coordinate positions in Equation (1), respectively. In practical applications, a series of texture feature features must be extracted according to the calculation results of the gray-level co-occurrence matrix so that the images can be more intuitively extracted and target recognitions can be performed. After the GLCMs of all cracked soil samples were extracted, 13 common texture features were then computed using the formulas listed in Table 1 [46].

Table 1. GLCM texture feature calculation formulas.

Texture Features	Formulas
Contrast (CON)	$CON = \sum\limits_{n}^{Ng-1} n^2 \left\{ \sum\limits_{i=1}^{Ng} \sum\limits_{j=1}^{Ng} p(i,j) \right\}, |i-j| = n$
Angular second moment (ASM)	$ASM = \sum\limits_{I} \sum\limits_{J} \{P(i,j)\}^2$
Entropy (ENT)	$ENT = - \sum\limits_{i=1}^{Ng} \sum\limits_{j=1}^{Ng} p(i,j) \log(p(i,j))$
Homogeneity (HOM)	$HOM = \sum\limits_{i=1}^{Ng} \sum\limits_{j=1}^{Ng} \frac{1}{1+(i-j)^2} p(i,j)$
Correlation (COR)	$COR = \frac{\sum_{i=1}^{Ng} \sum_{j=1}^{Ng} (ij)p(i,j) - u_x u_y}{\sigma_x \sigma_y}$
Cluster shade (CS)	$CS = \sum\limits_{i=1}^{Ng} \sum\limits_{j=1}^{Ng} \left((i - u_i) + \left(j - u_j \right) \right)^3 p(i,j)$
Cluster prominence (CP)	$CP = \sum\limits_{i=1}^{Ng} \sum\limits_{j=1}^{Ng} \left((i - u_i) + \left(j - u_j \right) \right)^4 p(i,j)$
Max probability (MP)	$MP = max\{p(i,j)\}$
Sum average (SA)	$SA = \sum\limits_{1=2}^{2Ng} i p_{x+y}(i)$
Sum entropy (SE)	$SE = - \sum\limits_{1=2}^{2Ng} p_{x+y}(i) \log\{ p_{x+y}(i) \}$
Sum variance (SV)	$SV = \sum\limits_{1=2}^{2Ng} (i - SumEntropy)^2 p_{x+y}(i)$
Information of correlation (IC1)	$IC1 = \frac{HXY - HXY1}{max(HX, HY)}$
Information of correlation (IC2)	$IC2 = (1 - exp[-2^*(HXY2 - HXY)])^{\frac{1}{2}}$

Here, $p(i,j)$ is the value of the normalized GLCM at the (i,j) coordinate position; Ng is the gray level of the co-occurrence matrix; u_x and u_y are the mean of $p_x(i)$ and $p_y(j)$, respectively; and σ_x and σ_y are the variance of $p_x(i)$ and $p_y(j)$, respectively. The relevant intermediate parameters in the formulas for texture features can be calculated from the following Equations (2)–(7):

$$P_x(i) = \sum_{j=1}^{Ng} p(i,j) \tag{2}$$

$$P_y(j) = \sum_{i=1}^{Ng} p(i,j) \tag{3}$$

$$P_{x+y}(k) = \sum_{i=1}^{Ng} \sum_{j=1}^{Ng} p(i,j), i+j = k, k = 2,3,\ldots 2Ng \tag{4}$$

$$HXY = - \sum_{i=1}^{Ng} \sum_{j=1}^{Ng} p(i,j) log\{p(i,j)\} \tag{5}$$

$$HXY1 = - \sum_{i=1}^{Ng} \sum_{j=1}^{Ng} p(i,j) log\{p_x(i)p_y(j)\} \tag{6}$$

$$HXY2 = - \sum_{i=1}^{Ng} \sum_{j=1}^{Ng} p_x(i)p_y(j) log\{p_x(i)p_y(j)\} \tag{7}$$

To study the influence of GLCM calculation parameters on the results of texture features, the crack images of all soil samples were transformed into eight different gray levels (2, 4, 8, 16, 32, 64, 128, and 256, as shown in Figure 4), and the GLCMs of the transformed images with different gray levels were calculated at seven different pixel steps (1, 5, 10, 20, 40, 60, and 80 pixels) and four directions (0°, 45°, 90°, and 135°) for texture features. Correlation analysis was then performed between EC and texture features on the basis of different GLCM parameters.

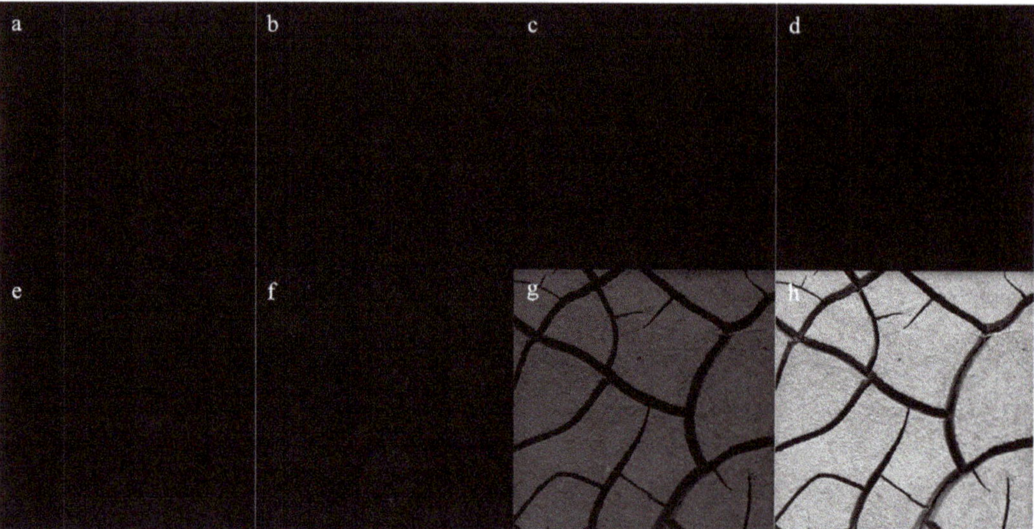

Figure 4. Extraction results of eight gray-scale images of a typical cracked soil sample. (**a**) gray level of 2; (**b**) gray level of 4; (**c**) gray level of 8; (**d**) gray level of 16; (**e**) gray level of 32; (**f**) gray level of 64; (**g**) gray level of 128; (**h**) gray level of 256.

Note that there were too many soil samples in this study, indicating that different types of measurement results cannot be listed individually. In order to express the characteristics of the data more clearly, statistical analysis was therefore performed on the measured physical and chemical properties of soil samples and the GLCM texture features extracted from crack images. Specially, the statistical parameters including the minimum, maximum, average, standard deviation, coefficient of variance (CV), skewness, and kurtosis values of soil properties and GLCM texture features were computed using the SPSS software.

3. Results

3.1. Chemical and Physical Properties

Table 2 lists the statistical indexes of soil properties for all the cracked soil samples. From the table, it can be seen that EC values of the samples ranged from 0.2 to 6.37 ds/m with a mean value of 0.95 ds/m, while pH ranged from 8.55 to 11.16 with a mean value of 10.06. In addition, measurement results from Li and Wang [47] showed the ESP and pH from soil samples of the western Songnen Plain (covering all the sampling points in this study) were higher than 20% and 8.5, respectively, indicating that all the soil samples exhibited intensive alkali characteristics according to the international classification standard proposed by USSLS [48]. The standard deviation of 0.915 ds/m and CV of 96.45% also showed that there was significant heterogeneity among the EC values of all the soil samples in the study. Table 2 also shows that the samples selected belong to a typical soil texture of clay loam, with the clay content of all soil samples varying in a narrow range from 25.01% to 30.99%; the standard deviation of 1.61% and CV of 5.74% also explained that the differences in clay content were not evident within the soil samples. As the soil samples were fully dried in natural conditions, the water content varied from 2.01% to 4.47%.

Table 2. Statistical description of soil properties of the soil samples.

Parameters	Min	Max	Mean	Standard	CV%	Skewness	Kurtosis
EC (ds/m)	0.20	6.37	0.95	0.915	96.45	3.04	10.9
pH	8.55	11.16	10.06	0.53	5.36	0.34	−0.85
Moisture (%)	2.01	4.47	2.95	0.58	19.32	−1.14	0.06
Clay (%)	25.01	30.99	27.88	1.61	5.74	−1.01	0.06
Silt (%)	30.06	41.95	35.98	3.51	9.77	−1.30	0.02
Sand (%)	28.19	39.38	33.86	3.39	10.01	−1.13	0.13

N = 200; CV, coefficient of variation.

From Table 2, it also can be seen that the clay content of the soil samples in this study covered a very small range of 25.01% to 30.99%. To quantify the effects of clay content on the extent of desiccation cracks, a complementary study was carried out between the clay content and the four selected GLCM texture features (Figure 5). Notably, the scatter diagrams in Figure 5 did not show clear regularity for the data points, indicating that the clay content of soda saline–alkali soils was not sensitive to GLCM texture features.

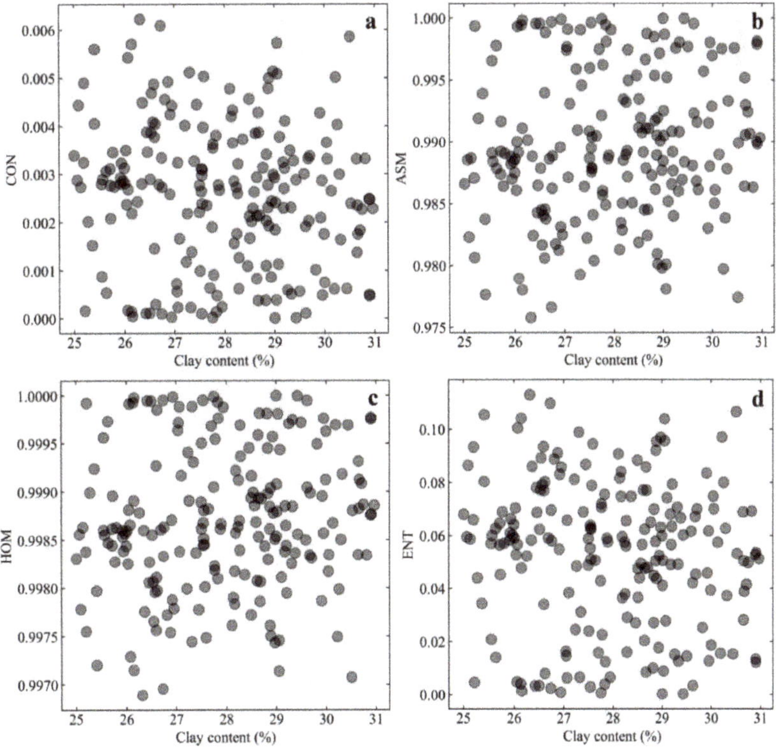

Figure 5. Scatter diagram between clay content and GLCM texture features. (**a**) for texture feature of CON; (**b**) for texture feature of ASM; (**c**) for texture feature of HOM; (**d**) for texture feature of ENT.

3.2. Optimal Texture Features

Table 3 shows the maximum correlation coefficients between EC values and the 13 GLCM texture features in the $0°$, $45°$, $90°$, and $135°$ directions under different gray levels and pixel steps. The table also indicates that the same trends appeared between each texture feature and EC values of all soil samples. Texture features such as CON, ASM, ENT,

HOM, CS, CP, MP, SA, and SE had a high correlation with EC, with correlation coefficients all above 0.7, while the other four texture features, namely, COR, SV, IC1, and IC2, had poor correlation with EC values with correlation coefficients less than 0.58. Therefore, four optimal texture features (CON, ASM, ENT, and HOM) were selected in this study to analyze the ability of GLCM texture features to characterize the surface cracks of soda saline–alkali soils, considering both their high relationship with EC and their common application in texture feature analysis. Particularly, CON returns the amount of local variation in an image, reflecting the sharpness of the image and the intensity of the texture. ASM (also referred to as energy) measures the sum of squared elements in the GLCM and ranges from 0 to 1. It describes the thickness of the image texture feature and the uniformity of pixel distributions. ENT measures the randomness of the intensity distribution in the image and represents the amount of information in the image. HOM usually represents a value that measures the distribution closeness of elements in the GLCM to the GLCM diagonal and ranges from 0 to 1.

Table 3. Maximum correlation coefficient between texture features and EC in four directions under different gray levels and step sizes.

Texture Features	0°	45°	90°	135°
CON	0.82	0.78	0.76	0.78
ASM	−0.77	−0.76	−0.75	−0.76
ENT	0.74	0.73	0.72	0.73
HOM	−0.82	−0.78	−0.76	−0.78
COR	−0.47	−0.29	−0.31	−0.39
CS	−0.75	−0.75	−0.76	−0.75
CP	0.75	0.76	0.77	0.76
MP	−0.77	−0.76	−0.76	−0.76
SA	0.74	0.73	0.72	0.73
SE	−0.74	−0.73	−0.72	−0.73
SV	0.37	0.44	0.37	0.34
IC1	0.57	0.58	0.61	0.58
IC2	−0.31	−0.43	−0.31	−0.43

N = 200; significance level α = 0.05; CON, contrast; ASM, angular second moment; ENT, entropy; HOM, homogeneity; COR, correlation; CS, cluster shade; CP, cluster prominence; MP, max probability; SA, sum average; SE, sum entropy; SV, sum variance; IC1 and IC2, information of correlation based on different equations.

3.3. Analysis of GLCM Parameters

3.3.1. Effects of Directions

Figure 6 shows the coefficients of variation (CV) of CON, ASM, ENT, and HOM extracted from the 0°, 45°, 90°, and 135° directions, which were computed from the GLCMs of a typical soil sample under different gray levels and pixel steps. As shown in Figure 6a, the CV of CON in the four directions reached its highest when the step size was 1 pixel, after which the CV decreased significantly with increasing step size and gradually stabilized until it reached 60 pixels. Figure 6b shows that when the gray level was larger than 16, the CV of ASM decreased with the steps and stabilized at a step size of 10 pixels. Figure 6b also indicates that CV was no longer affected by step size when the gray level was less than 8. Figure 6c shows that the CV of HOM had the same trend as ASM but with a larger difference at various gray levels. Figure 6d shows that the CV of ENT had a similar trend with CON at the same gray level, but the difference was larger than that of CON when the step sizes were 10, 20, and 40 pixels. Therefore, there were certain differences in the extraction results of texture features computed from different directions, gray levels, and step sizes, which also explained the differences in the maximum correlation coefficients between the texture features and EC values in Table 3. To effectively consider the workload and remove the effect of direction, the mean values of texture features in the directions of 0°, 45°, 90°, and 135° were selected to further analyze the effect of texture features on the surface cracking status of soda saline–alkali soils under different gray levels and step sizes.

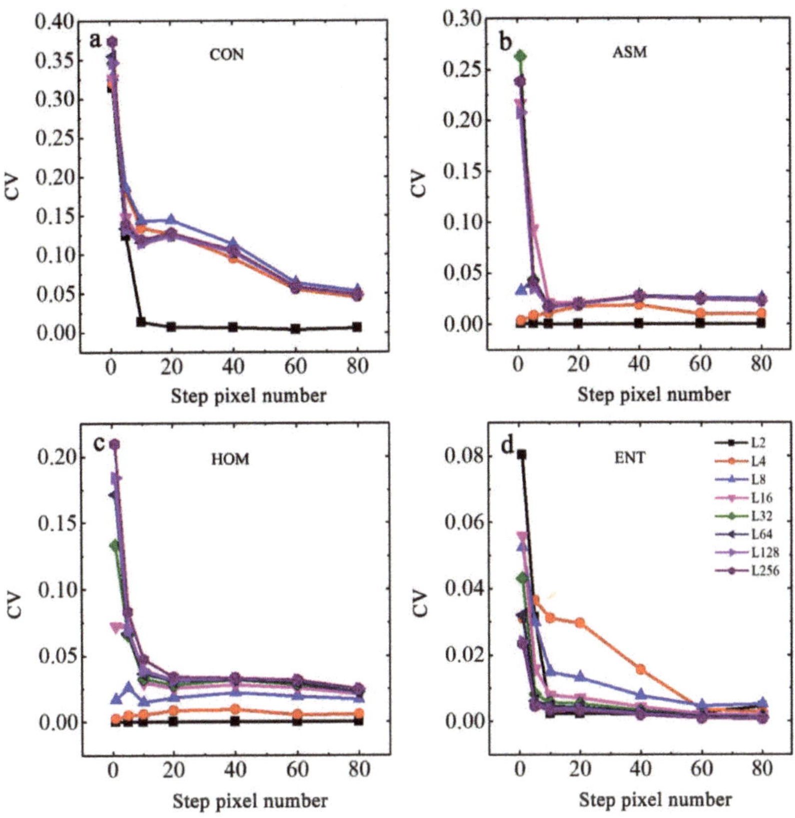

Figure 6. Coefficients of variation of four texture features in four directions of a typical soil sample. (**a**) for texture feature of CON; (**b**) for texture feature of ASM; (**c**) for texture feature of HOM; (**d**) for texture feature of ENT.

3.3.2. Effects of Gray Levels and Step Sizes

To quantify the response of GLCM texture features on the salinity of soda saline–alkali soil in the Songnen Plain of China, the mean values in four directions ($0°$, $45°$, $90°$, and $135°$) were extracted under different gray levels and step sizes for all selected texture features, including CON, ASM, ENT, and HOM. Subsequently, correlation analysis was performed between the EC values and the texture features derived from different GLCM parameters, as shown in Figure 7. It can be seen from Figure 7a that the correlation coefficient between CON and EC of the soil samples was significantly positive, with an initial decreasing and then increasing trend, which finally became stable at a step size of 16 pixels. Although the lowest correlation coefficient of CON for 1 pixel in the step size occurred at a gray level of 8, the lowest values in all correlation coefficient curves under other step sizes were found at the same gray level of 4. Figure 7b shows that the correlation coefficient between the ASM and EC of the soil samples decreased with increasing gray levels, while the step size had little effect on the results of the correlation analysis. Figure 7c shows that although the trend of the correlation coefficient curves between HOM and EC were basically the same as those of ASM, a significant difference was still found under various step sizes. Figure 7d indicates that the curve shapes of the positive correlation coefficients between ENT and EC decreased until a gray level of 8, after which the correlation curves gradually increased and became stable at a gray level of 32. In addition, Figure 7 also shows that when the gray level of the crack images was determined, the correlation coefficients of the four texture features and EC values gradually decreased with increasing step size, especially when the gray level was less than 8, and the difference in the correlation coefficients also showed a more evident tendency when the gray level increased.

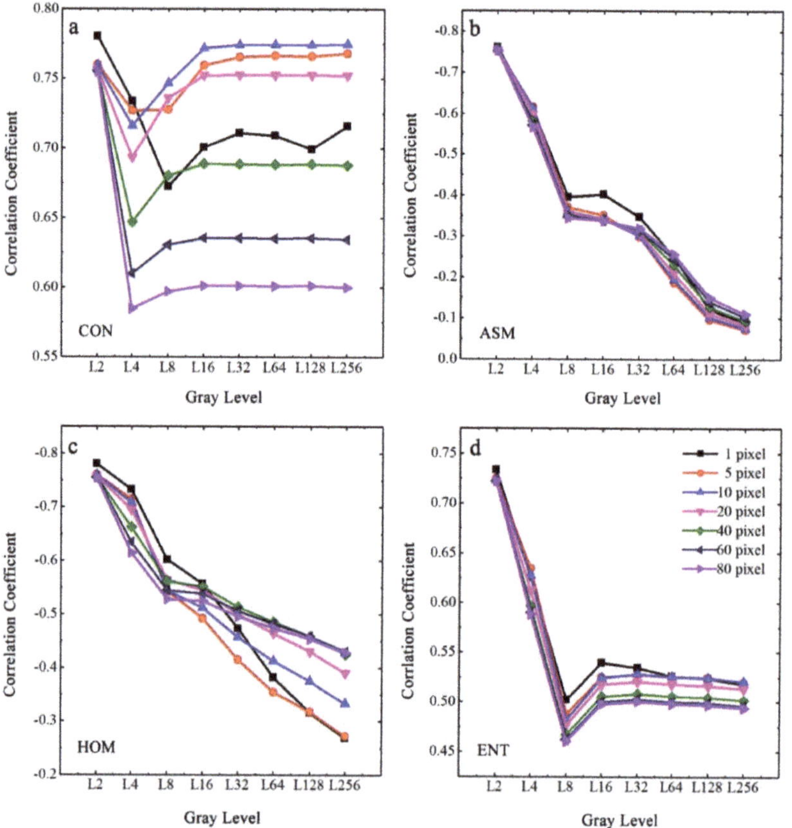

Figure 7. Correlation coefficients between four selected GLCM texture features and EC of the cracked soil samples. (**a**) for texture feature of CON; (**b**) for texture feature of ASM; (**c**) for texture feature of HOM; (**d**) for texture feature of ENT.

3.4. Cross-Correlation Analysis between Different Texture Features

As discussed in Section 3.3, the difference in the correlation between each texture feature and EC values of soil samples was quite small, given the step size of 1 pixel and gray level of 2. As the gray level and step size increased, the difference between different texture features and the EC values of soil samples increased. Figure 8 shows the analysis results of the cross-correlation among the four selected GLCM texture features of CON, ASM, HOM, and ENT under different step sizes and gray levels. We observed that when the gray level was 2 and the step size was 1 pixel, the two different texture features were highly correlated with a correlation coefficient close to 1. However, the correlation between any two texture features rapidly decreased as the gray level increased once the step size was determined. Figure 8 also indicates that among all combinations of two texture features, the intervals of cross-correlation curves under different gray levels usually increased until the largest difference appeared at a step size of five pixels. After that, the curves of cross-correlation became close and gradually stabilized with increasing step sizes, except for the combination between CON and ENT.

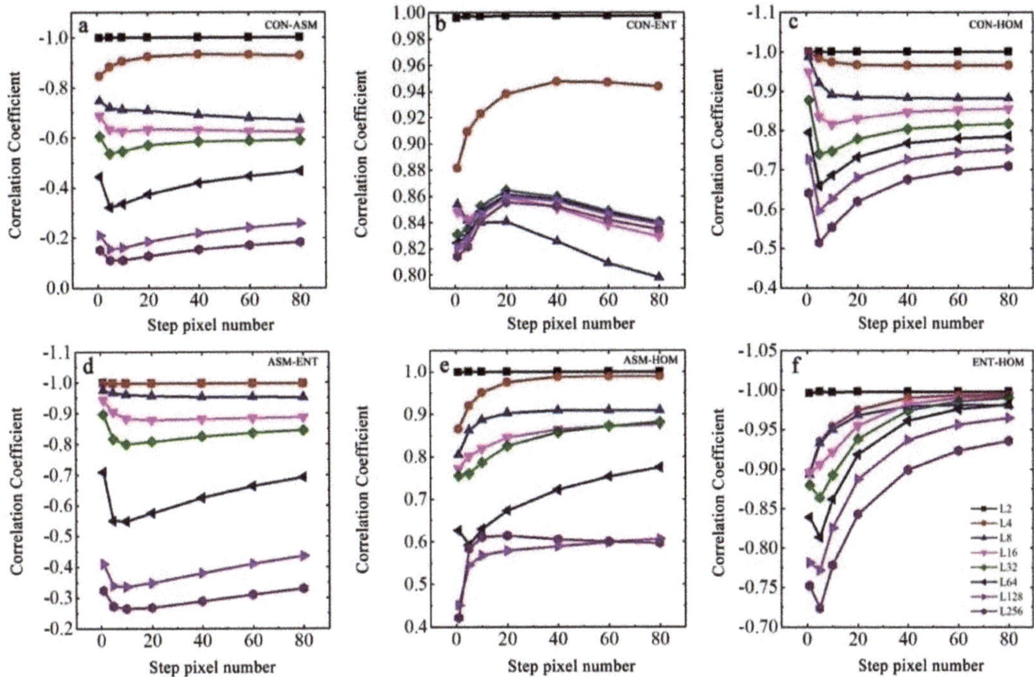

Figure 8. Cross−correlation coefficients of four GLCM texture features of the cracked soil samples. (**a**) between CON and ASM; (**b**) between CON and ENT; (**c**) between CON and HOM; (**d**) between ASM and ENT; (**e**) between ASM and HOM; (**f**) between ENT and HOM.

3.5. Logarithmic Regression Models between EC and Texture Features

Table 4 shows the statistical parameters of the mean values of the four texture features in four directions of 0°, 45°, 90°, and 135° for all soil samples, which were calculated under a gray level of 2 and step size of 1 pixel. Table 4 shows that the CON extraction results had the lowest values, ranging from 4.82×10^{-6} to 6.2×10^{-3}, followed by ENT varying from 1.92×10^{-4} to 1.13×10^{-1}, while ASM and HOM had relatively high values ranging from 9.76×10^{-1} to 9.99×10^{-1} and from 9.97×10^{-1} to 9.99×10^{-1}, respectively. In addition, the distributions of CON and HOM of different soil samples were relatively discrete, with CVs of 58.25% and 52.71%, respectively. The coefficient of variation (CV) values of 0.59% and 0.08% indicated that the extraction results of ASM and HOM texture features were relatively concentrated. Table 4 also shows that the kurtosis values of the four texture features were all less than 0, indicating that the overall distribution of the texture features was relatively flat, but the characteristic was not evident because the range was only from −0.70 to −0.61, whereas the skewness between −0.22 and 0.06 indicated that texture features of the soil samples, including CON, ASM, ENT, and HOM, basically conformed to the normal distribution.

Table 4. Statistical description of four GLCM texture features of cracked soil samples.

Parameters	Min	Max	Mean	Standard	CV%	Skewness	Kurtosis
CON	4.82×10^{-6}	6.2×10^{-3}	2.6×10^{-3}	1.5×10^{-3}	58.25	0.06	−0.61
ASM	9.76×10^{-1}	9.99×10^{-1}	9.89×10^{-1}	5.9×10^{-3}	0.59	0.02	−0.69
ENT	1.92×10^{-4}	1.13×10^{-1}	5.29×10^{-2}	2.79×10^{-2}	52.71	−0.22	−0.70
HOM	9.97×10^{-1}	9.99×10^{-1}	9.98×10^{-1}	7.1×10^{-4}	0.08	−0.06	−0.61

N = 200; CON, contrast; ASM, angular second moment; ENT, entropy; HOM, homogeneity; CV, coefficient of variation.

To further illustrate the response of the GLCM texture features to the salinity of soda saline–alkali soils in the Songnen Plain, regression models were developed between the EC values and the texture features of CON, ASM, ENT, and HOM of the cracked soil samples. Figure 9 shows the scatter points between the mean values in the 0°, 45°, 90°, and 135° directions of the four typical texture features mentioned above and the EC values of all cracked soil samples under the optimal GLCM parameters (gray level of 2 and step size of 1 pixel). As shown in Figure 9, all four texture features had a logarithmic relationship with the EC values of soil samples, where CON and ENT were positively correlated with EC values and ASM and HOM were inversely proportional to EC values.

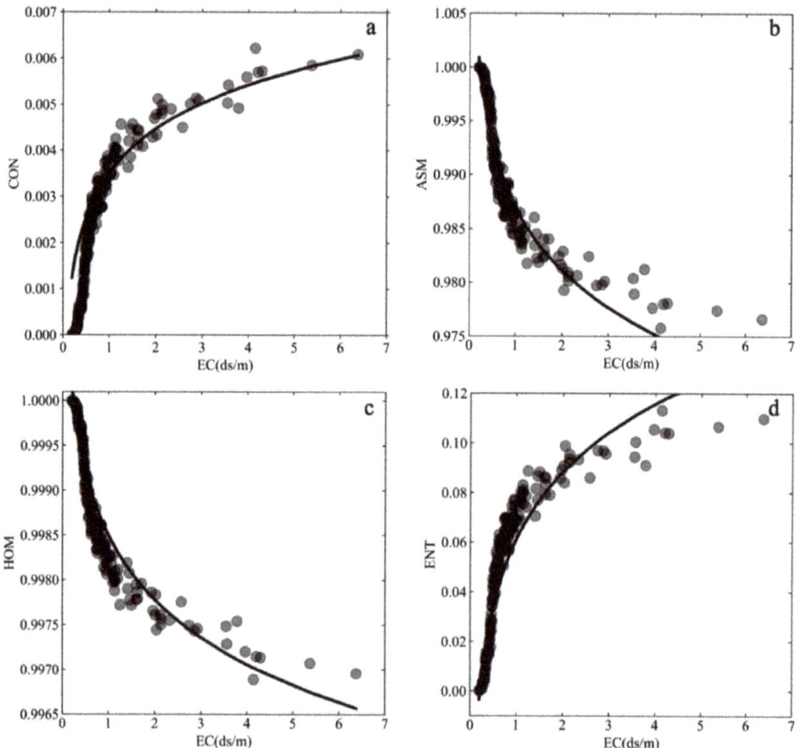

Figure 9. Logarithmic fitting results between EC and four GLCM texture features of the cracked soil samples. (**a**) between EC and CON; (**b**) between EC and ASM; (**c**) between EC and HOM; (**d**) between EC and ENT.

The established models in Table 5 show that both CON and HOM had the best logarithmic correlation with EC, both with R^2 of 0.92, followed by ASM and EC with an R^2 of 0.90. Although the weakest relationship was found between ENT and EC, R^2 was still high, with a value of 0.88. In addition, Table 5 shows that all logarithmic models had very low RMSE, ranging only from 2.12×10^{-4} to 9.68×10^{-3}, which suggested that texture features of GLCM could effectively characterize the salt content of soil samples. Thus, GLCM texture features can be considered as good indicators of the salt-determined crack status of the soda saline–alkali soil.

Table 5. Logarithmic regression models based on EC and texture features of soda saline–alkali soil samples with surface cracks.

Texture Features	Logarithmic Regression Models	R^2	RMSE
CON	$y = 0.0032 \times \lg(x) + 0.005$	0.92	4.24×10^{-4}
ASM	$y = -0.0196 \times \lg(x) + 0.987$	0.90	1.86×10^{-3}
ENT	$y = 0.091 \times \lg(x) + 0.065$	0.88	9.68×10^{-3}
HOM	$y = -0.0024 \times \lg(x) + 0.998$	0.92	2.12×10^{-4}

RMSE, root mean square error; RMSE $= \sqrt{\frac{\sum_{i=1}^{n}(y'-y)^2}{n}}$, n represents the number of soil samples, y stands for the measured texture feature, and y' stands for the observed texture feature based on models; CON, contrast; ASM, angular second moment; ENT, entropy; HOM, homogeneity.

4. Discussion

From Table 2, it can be seen that the clay content did not play an important role in desiccation cracking of soda saline–alkali soils in this study. This was because the original minerals of soda saline–alkali soils in the Songnen Plain are quartz and feldspar, and the secondary mineral relates to illite/smectite formation with an interlayer ratio above 0.5 according to the results of X-ray diffraction complete analysis measured by Zhang et al. [49] and Wang et al. [50]. This result indicates that the clay content and clay mineral composition did not play important roles in the shrinkage and cracking of the soil samples because of the activity index covering a very low range from 0.33 to 0.48. The soil salinity therefore can be considered as the main role in the desiccation cracking process of salt-affected in this study. Yang et al. [51] found that the soil salinity was highly correlated with both ESP and SAR of the salinized soils in Songnen Plain since the main salt minerals were $NaHCO_3$ and Na_2CO_3, which made Na^+ the dominant exchangeable cation; their results also indicated that this kind of salt mineral composition leads to an alkalinization reaction and thus disperses the cementation between the clay soil particles. In addition, according to the measurements results of different ions, Chi and Wang [52] found that Na^+ occupies an absolute dominant role in the cations of the saline–alkali soils in the Songnen Plain with content much larger than those of the cations including K^+, Ca^{2+}, and Mg^{2+}, indicating that Na^+ affects the desiccation cracking of soda saline–alkali soils in Songnen Plain, China, to a certain extent. Specifically, the previous studies from Zhang et al. [53] and Yu et al. [54] turned out that a kind of thick bound water film can be found forming around the soil particles, which is caused by the interaction between colloidal particles and adsorbent cations (especially for Na^+ with a hydrolysis radius compared with other cations within the type of soda saline–alkali soils in Songnen Plain). This kind of water film always in turn reduces the cohesion between soil particles and then results in a decrease in the soil strength [55]. Aksenov et al. [56] also found that the combined water film generated among soil particles could reduce the internal friction angle and shear strength of salinized soil samples, thus making the surface of saline–alkali soils more prone to shrinkage and cracking. Moreover, many studies have indicated that the diffuse double layer (DDL) also plays an important role in determining the shrinking and cracking processes of saline soils during water evaporation [57,58]. Specifically, water evaporation causes thinning of the DDL and a reduction in the distance between soil particles, which results in the propagation of desiccation cracks on the soil surface. Therefore, a higher salt content makes soil particles combine more tightly, which is manifested by more soil volume shrinkage and more complex soil cracking.

The selection of crack parameters usually shows great effects on the quantification of soil surface cracks. Although many previous studies have shown that geometric parameters (such as crack length, crack area, crack length density, and crack area density) have been commonly used, they still have certain defects in quantitatively characterizing the surface cracks compared with texture features. Specifically, crack length and crack area cannot show convergence [59,60], while crack length density and crack area density are unable to adequately describe the distributions of desiccation cracks generated from the random cracking locations of the soil samples with low repeatability [61,62]. In addition, the crack densities mentioned above also fail to provide an idea of the propagation of soil cracks in different directions [63]. Due to its strong self-correlation under natural conditions, GLCM texture analysis can thus offer an easy way to describe the cracking status of the soil surface quantitatively and effectively.

Under natural conditions, there are differences in the extraction results of GLCM texture features in various directions, including 0°, 45°, 90°, and 135°, which may be related to the presence of slight slopes on certain soil surfaces that cause the water to flow in a specific direction after precipitation. However, the difference in texture features of field soil samples due to the direction is not distinct because most of the process and direction of crack development are random. Moreover, the direction of the crack patterns of soil samples may not be accurately obtained in the photo process in the field, and there may be operations, such as rotation of the images in the post-processing stage, which make it difficult to ensure the same standard direction of the texture features for all soil samples. The characterization ability of GLCM texture features on the EC values generally showed a decreasing trend with the increasing step size. This was because when the step size is small, the extraction results of the texture features can effectively distinguish the grayscale of the crack region. However, when the step size is larger than the width of the soil crack, the calculation results of the GLCM texture features ignore the information of some crack regions, resulting in a decrease in the correlation with soil EC. Further, the increasing gray level reduced the correlation between texture features and soil salinity, which may be because when the gray level is relatively low, the gray value of the crack region is more distinct from the gray level of other surface areas of the soil samples. In contrast, with an increasing gray level, the uncertainty of the pixel distribution is enhanced, and the consistency of gray values is weakened [64–66], which rapidly decreases the ability of ASM and HOM to characterize crack conditions. In addition, although the amount of local variation in the cracked patterns of soil samples was the most notable in the lowest image information when the gray level was set to 2 (all the images only represent crack and non-crack regions), increasing the gray level gradually reduced the proportion of gray levels within the crack regions and thus made CON (returns the local variation of crack images) and ENT (returns the randomness and complexity of crack images) decrease and gradually become stable. Therefore, GLCM texture features were the best for characterizing the surface cracking status of soda saline–alkali soils under the optimal gray level of 2 and step size of 1 pixel.

From Section 3.5, it can be seen that the EC values of soda saline–alkali soils showed clear logarithmic relationships with different GLCM texture features, indicating that a new online measurement method of soil salinity can therefore be proposed using regression models based on the GLCM texture features, which are computed with optimal GLCM computing parameters. Specially, the procedures can be described as follows: firstly, taking crack patterns under field conditions using a unified photographic standard; secondly, performing geometric correction and preprocessing operations on all crack images; after that, extracting GLCM texture features under optimal computing parameters including gray level, direction, and step size; subsequently, developing regression models with a certain number of soil samples; thereafter, importing the texture features extracted from all other soil samples into the best regression model to predict the EC values of soil samples; and finally, calibrating all estimated EC values using the measured ones in laboratory for better accuracy. In practical applications, this potential method can thus be selected to measure the soil salinity non-destructively, effectively, and accurately. In addition, this kind of method can be further extended to aerial remote sensing for simultaneous measurement of soil salinity in a large range, which can provide important guidance for ecological restoration, agricultural production, and engineering construction in saline–alkali areas.

5. Conclusions

In this study, quantitative analyses were conducted to study the effects of GLCM computing parameters on the relationship between different texture features and the EC values of soda saline–alkali soil samples. Although the texture features in different directions were different, their influence was quite limited; as the gray level and step size increased, the correlation between texture features and soil salinity greatly decreased. Although the cross-correlation between various texture features decreased rapidly with the increase in gray level, it was weakly affected by step size. The soil salinity of the soda saline–alkali soils in the Songnen Plain was correlated with the surface crack status, indicating that they can be determined by the GLCM texture features. In further studies, aerial remote sensing for simultaneous prediction of soil salinity in a large range are required to provide guidance for ecological restoration, agricultural production, and engineering construction.

Author Contributions: Conceptualization, J.R.; data curation, Z.Z.; formal analysis, J.R., Y.Z. and Z.Z.; funding acquisition, J.R.; investigation, J.R.; methodology, J.R. and Y.Z.; project administration, H.Z.; resources, J.R.; software, Y.Z., Z.Z. and H.Z.; supervision, Z.Z. and H.Z.; validation, Y.Z. and Z.Z.; visualization, Z.Z. and Y.Z.; writing—original draft, J.R. and Y.Z.; writing—review and editing, J.R. and H.Z. All authors have read and agreed to the published version of the manuscript.

Funding: This research was founded by the Strategic Priority Research Program of the Chinese Academy of Sciences (No. XDA28110501), the University Nursing Program for Young Scholar with Creative Talents in Heilongjiang Province of China (No. UNPYSCT-2018180), the PhD startup Funds of Harbin Normal University (No. XKB201904), and the National Natural Science Foundation of China (No. 41601382).

Institutional Review Board Statement: Not applicable.

Data Availability Statement: Not applicable.

Acknowledgments: The authors appreciate the assistance in the soil property measurement and the image processing from the "Cold Region Ecological Security" Collaborative Innovation Center of Heilongjiang Province, China.

Conflicts of Interest: The authors declare no conflict of interest.

References

1. Rengasamy, P. World salinization with emphasis on Australia. *J. Exp. Bot.* **2006**, *57*, 1017–1023. [CrossRef] [PubMed]
2. Zhang, W.; Feng, Y. Physico-chemical properties and ecological recovery of saline-alkali soil in Songnen Plain. *Acta Pedol. Sin.* **2009**, *46*, 169–172, (In Chinese with English Abstract). [CrossRef]
3. Ren, J.; Li, X.; Zhao, K. Quantitative analysis of relationships between crack characteristics and properties of soda-saline in Songnen Plain, China. *Chin. Geogr. Sci.* **2015**, *25*, 591–601. [CrossRef]
4. Amaya, T.; Nagahori, K.; Takahashi, T. Lysimeter experiment on the progressive process of salt exclusion-Positive studies on the salt behavior and salt exclusion in Kasaoka Bay polder. *Trans. JPSIDER* **1983**, *104*, 1–8. (In Japanese) [CrossRef]
5. Lima, L.A.; Grismer, M.E. Soil crack morphology and soil-salinity. *Soil Sci.* **1992**, *153*, 149–153. [CrossRef]
6. Fujimaki, H.; Baki, H.M.F.A.E. Effect of bypass-flow on leaching of salts in a cracking soil in the Nile Delta. *Water* **2021**, *13*, 993. [CrossRef]
7. Van der Tak, L.D.; Grismer, M.E. Irrigation, drainage and soil salinity in cracking soils. Transactions of the ASAE. *Am. Soc. Agric. Eng.* **1987**, *30*, 740–744. [CrossRef]
8. Ben-Hur, M.; Assouline, S. Tillage effects on water and salt distribution in a vertisol during effluent irrigation and rainfall. *Agron. J.* **2002**, *94*, 1295–1304. [CrossRef]
9. Zhang, Z.; Zhu, W.; Zhu, L.; Wang, C.; Sheng, L.; Chen, Y. Effects of roots and salinity on law of development for farmland soil desiccation crack. *Trans. Chin. Soc. Agric. Eng.* **2014**, *30*, 83–89, (In Chinese with English Abstract). [CrossRef]
10. Zhang, Z.; Wang, C.; Zhu, C.; Zhu, W.; Wu, C. Geometric characteristics of shrinkage crack network in soil. *Earth Sci.* **2014**, *39*, 1465–1472. [CrossRef]
11. Zhang, Y.; Ye, W.M.; Chen, B.; Chen, Y.G.; Ye, B. Desiccation of NaCl-contaminated soil of earthen heritages in the Site of Yar City, northwest China. *Appl. Clay Sci.* **2016**, *124–125*, 1–10. [CrossRef]
12. Xing, X.; Kuan, D.; Ma, X. Differences in loam water retention and shrinkage behavior: Effects of various types and concentrations of salt ions. *Soil Tillage Res.* **2017**, *167*, 61–72. [CrossRef]
13. Wang, C.; Feng, G.; Zhang, Z.; Huang, M.; Qi, W.; Ma, L. Geometrical and statistical analysis of dynamic crack morphology in shrink-swell soils with addition of maize roots or salinity (NaCl). *Soil Tillage Res.* **2021**, *212*, 105057. [CrossRef]

14. Tang, C.; Zhu, C.; Cheng, Q.; Zeng, H.; Xu, J.; Tian, B. Desiccation cracking of soils: A review of investigation approaches underlying mechanisms, and influencing factors. *Earth Sci. Rev.* **2021**, *216*, 103586. [CrossRef]
15. Flowers, M.; Lal, R. Axle load and tillage effects on the shrinkage characteristics of a Mollic Ochraqualf in northwest Ohio. *Soil Tillage Res.* **1990**, *50*, 251–258. [CrossRef]
16. Tang, C.; Shi, B.; Liu, C.; Zhao, L.; Wang, B. Influencing factors of geometrical structure of surface shrinkage cracks in clayey soils. *Eng. Geol.* **2008**, *101*, 204–217. [CrossRef]
17. Vogel, H.J.; Hoffmann, H.; Roth, K. Studies of crack dynamics in clay soil-I. Experimental methods, results, and morphological quantification. *Geoderma* **2005**, *125*, 203–211. [CrossRef]
18. Yoshida, S.; Adachi, K. Effects of cropping and puddling practices on the cracking patterns in paddy fields. *Soil Sci. Plant Nutr.* **2001**, *47*, 519–532. [CrossRef]
19. Castellano, G.; Bonilha, L.; Li, L.M.; Cendes, F. Texture analysis of medical images. *Clin. Radiol.* **2004**, *59*, 1061–1069. [CrossRef]
20. Lan, R.; Zhong, S.; Liu, Z.; Shi, Z.; Luo, X. A simple texture feature for retrieval of medical images. *Multimed. Tools Appl.* **2018**, *77*, 10853–10866. [CrossRef]
21. Li, Y.; Lu, Z.; Li, J.; Deng, Y. Improving deep learning feature with facial texture feature for face recognition. *Wirel. Pers. Commun.* **2018**, *103*, 1195–1206. [CrossRef]
22. Akbal, E. An automated environmental sound classification methods based on statistical and textural feature. *Appl. Acoust.* **2020**, *167*, 107413. [CrossRef]
23. Yin, S.; Shao, Y.; Wu, A.; Wang, Y.; Gao, Z. Texture features analysis on micro-structure of paste backfill based on image analysis technology. *J. Cent. South Univ.* **2018**, *25*, 2360–2372. [CrossRef]
24. Liu, L.; Kuang, G. Overview of image textural feature extraction methods. *Int. J. Image Graph.* **2009**, *14*, 622–635. [CrossRef]
25. Aouat, S.; Ait-hammi, I.; Hamouchene, I. A new approach for texture segmentation based on the Gray Level Co-occurrence Matrix. *Multimed. Tools Appl.* **2021**, *80*, 24027–24052. [CrossRef]
26. Bakheet, S.; Al-Hamadi, A. Automatic detection of COVID-19 using pruned GLCM-Based texture features and LDCRF classification. *Comput. Biol. Med.* **2021**, *137*, 104781. [CrossRef]
27. Lian, M.; Huang, C. Texture feature extraction of gray-level co-occurrence matrix for metastatic cancer cells using scanned laser pico-projection images. *Laser Med. Sci.* **2019**, *34*, 1503–1508. [CrossRef]
28. Rafi, M.; Mukhopadhyay, S. Texture description using multi-scale morphological GLCM. *Multimed. Tools Appl.* **2018**, *77*, 30505–30532. [CrossRef]
29. Srivastava, D.; Rajitha, B.; Agarwal, S.; Singh, S. Pattern-based image retrieval using GLCM. *Neural Comput. Appl.* **2020**, *32*, 10819–10832. [CrossRef]
30. Vimal, S.; Robinson, Y.H.; Kaliappan, M.; Vijayalakshmi, K.; Seo, S. A method of progression detection for glaucoma using K-means and the GLCM algorithm toward smart medical prediction. *J. Supercomput.* **2021**, *77*, 11894–11910. [CrossRef]
31. Alvarado, F.A.P.; Hussein, M.A.; Becker, T.A. Vision System for Surface Homogeneity Analysis of Dough Based on the Grey Level Co-occurrence Matrix (GLCM) for Optimum Kneading Time Prediction. *J. Food Process Eng.* **2016**, *39*, 166–177. [CrossRef]
32. Li, W.; Jiang, X.; Sun, W.; Wang, S.; Liu, C.; Zhang, X.; Zhang, Y.; Zhou, W.; Miao, L. Gingivitis identification via multichannel gray-level co-occurrence matrix and particle swarm optimization neural network. *Int. J. Imaging Syst. Technol.* **2019**, *30*, 401–411. [CrossRef]
33. Lloyd, K.; Rosin, P.L.; Marshall, D.; Moore, S.C. Detecting violent and abnormal crowd activity using temporal analysis of grey level co-occurrence matrix (GLCM)-based texture measures. *Mach. Vision Appl.* **2017**, *28*, 361–371. [CrossRef]
34. Oghaz, M.M.; Maarof, M.A.; Rohani, M.F.; Zainal, A.; Shaid, S.Z.M. An optimized skin texture model using gray-level co-occurrence matrix. *Neural Comput. Appl.* **2019**, *31*, 1835–1853. [CrossRef]
35. Olaniyi, E.O.; Adekunle, A.A.; Odekuoye, T.; Khashman, A. Automatic system for grading banana using GLCM texture feature extraction and neural network arbitrations. *J. Food Process Eng.* **2017**, *40*, e12575. [CrossRef]
36. Raju, P.; Rao, V.M.; Rao, B.P. Optimal GLCM combined FCM segmentation algorithm for detection of kidney cysts and tumor. *Multimed. Tools Appl.* **2019**, *78*, 18419–18441. [CrossRef]
37. Singh, A.; Armstrong, R.T.; Regenauer-Lieb, K.; Mostaghimi, P. Rock Characterization Using Gray-Level Co-Occurrence Matrix: An Objective Perspective of Digital Rock Statistics. *Water Resour. Res.* **2018**, *55*, 1912–1927. [CrossRef]
38. Tahir, M.A.; Bouridane, A.; Kurugollu, K. An FPGA Based Coprocessor for GLCM and Haralick Texture Features and their Application in Prostate Cancer Classification. *Analog. Integr. Circuits Signal Process.* **2005**, *43*, 205–215. [CrossRef]
39. Varish, N.; Pal, A.K. A novel image retrieval scheme using gray level co-occurrence matrix descriptors of discrete cosine transform based residual image. *Appl. Intell.* **2018**, *48*, 2930–2953. [CrossRef]
40. Ren, J.; Li, X.; Zhao, K.; Fu, B.; Jiang, T. Study of an on-line measurement method for the salt parameters of soda-saline soils based on the texture features of cracks. *Geoderma* **2016**, *263*, 60–69. [CrossRef]
41. Ren, J.; Li, X.; Li, S.; Zhu, H.; Zhao, K. Quantitative analysis of spectral response to soda saline-alkali soil after cracking process: A laboratory procedure to improve soil property estimation. *Remote Sens.* **2019**, *11*, 1406. [CrossRef]
42. Hossain, M.S.; Rahman, G.K.M.M.; Solaiman, A.R.M.; Alam, M.S.; Mia, M.A.B. Estimating electrical conductivity for soil salinity monitoring using various soil-water ratios depending on soil texture. *Commun. Soil Sci. Plant Anal.* **2020**, *51*, 635–644. [CrossRef]
43. Zhang, X.; Huang, B.; Liang, Z.; Zhao, Y.; Sun, W.; Hu, W. Study on salinization characteristics of surface soil in Western Songnen Plain. *Soils.* **2013**, *45*, 332–338, (In Chinese with English Abstract). [CrossRef]

44. Ma, C. Quantitative retrieval of soil salt content in the Songnen Plain based on HJ1A-HIS images. *Arid Zone Res.* **2014**, *31*, 226–230, (In Chinese with English Abstract). [CrossRef]

45. Klaustermeier, A.; Tomlinson, H.; Daigh, A.L.M.; Limb, R.; DeSutter, T.; Sedivec, K. Comparison of soil-to-water suspension ratios for determining electrical conductivity of oil-production-water contaminated soils. *Can. J. Soil Sci.* **2016**, *96*, 233–243. [CrossRef]

46. Haralick, R.M.; Shanmugan, K.; Dinstein, I. Textural features for image classification. *IEEE Trans. Syst. Man Cybern.* **1973**, *3*, 610–621. [CrossRef]

47. Li, B.; Wang, Z. The alkalization parameters and their influential factors of saline-sodic soil in the Songnen Plain. *J. Arid Land Resour. Environ.* **2006**, *20*, 183–191. [CrossRef]

48. USSLS. *Diagnoses and Improvement of Saline and Alkali Soils*; US Government Printing Office: Washington, DC, USA, 1954. [CrossRef]

49. Zhang, G.; Yu, Q.; Wei, G.; Chen, B.; Yang, L.; Hu, C.; Li, J.; Chen, H. Study on the basic properties of the soda-saline soils in Songnen plain. *Hydrogeol. Eng. Geol.* **2007**, *2*, 37–40. [CrossRef]

50. Wang, W.; Wang, Q.; Zhang, J.; Chen, H. An experiment study of on the fundamental properties of the carbonate-saline soil in west of Jilin Province. *J. Beijing Univ. Technol.* **2011**, *37*, 217–224. [CrossRef]

51. Yang, F.; An, F.; Wang, Z.; Yang, H.; Zhao, C. Spatial variation of apparent electrical conductivity of saline-sodic paddy soils in the Songnen Plain. *Chin. J. Eco-Agric.* **2015**, *23*, 614–619, (In Chinese with English Abstract). [CrossRef]

52. Chi, C.; Wang, Z. Conversion relationship between the chemical parameters in saturated and in 1:5 soil water extracts of saline and alkali soils in Songnen Plain of Northeast China. *Chin. J. Ecol.* **2009**, *28*, 172–176, (In Chinese with English Abstract). [CrossRef]

53. Zhang, G.; Li, J.; Yu, Q.; Zhang, B.; Yang, R.; Chen, H. Influence of salt content on shearing strength of the carbonate saline soil in Songnen Plain. *Chin. J. Geol. Hazard Control.* **2008**, *19*, 128–131. [CrossRef]

54. Yu, Q.; Sun, W.; Chen, B.; Yang, L.; Zhang, G. Research on strength of Song-Nen Plain soda-saline soil. *Rock Soil Mech.* **2008**, *29*, 1793–1796. [CrossRef]

55. Jeong, S.; Locat, J.; Leroueil, S. The effects of salinity and shear history on the rheological characteristics of illite-rich and Na-montmorillonite-rich clays. *Clay. Clay Miner.* **2012**, *60*, 108–120. [CrossRef]

56. Aksenov, V.I.; Kal'Bergenov, R.G.; Leonov, A.R. Strength characteristics of frozen saline soils. *Soil Mech. Found. Eng.* **2003**, *40*, 55–59. [CrossRef]

57. DeCarlo, K.F.; Shokri, N. Salinity effects on cracking morphology and dynamics in 3-D desiccating clays. *Water Resour. Res.* **2014**, *50*, 3052–3072. [CrossRef]

58. Shokri, N.; Zhou, P.; Keshmiri, A. Patterns of desiccation cracks in saline bentonite layers. *Transport Porous Med.* **2015**, *110*, 333–344. [CrossRef]

59. Ren, J.; Zhao, K.; Wu, X.; Zheng, X.; Li, X. Comparative analysis of the spectral response to soil salinity of saline-sodic soils under different surface conditions. *Int. J. Environ. Res. Public Health* **2018**, *15*, 2721. [CrossRef]

60. Ren, J.; Li, X.; Zhao, K.; Zheng, X.; Jiang, T. Quantitative research on the relationship between salinity and crack length of soda saline-alkali soil. *Pol. J. Environ. Stud.* **2019**, *28*, 823–832. [CrossRef]

61. Lakshmikantha, M.R.; Prat, P.C.; Ledesma, A. Boundary Effects in the Desiccation of Soil Layers with Controlled Environmental Conditions. *Geotech. Test. J.* **2018**, *41*, 20170018. [CrossRef]

62. Ren, J.; Xie, R.; Zhu, H.; Zhao, Y.; Zhang, Z. Comparative study on the abilities of different crack parameters to estimate the salinity of soda saline-alkali soil in Songnen Plain, China. *Catena* **2022**, *213*, 106221. [CrossRef]

63. Bordoloi, S.; Ni, J.; Ng, C.W.W. Soil desiccation cracking and its characterization in vegetated soil: A perspective review. *Sci. Total Environ.* **2020**, *729*, 138760. [CrossRef] [PubMed]

64. Numbisi, F.N.; Coillie, F.M.B.V.; Wulf, R.D. Delineation of cocoa agroforests using multi-season sentinel-1 SAR images: Low grey level range reduces uncertainties in GLCM texture-based mapping. *ISPRS Int. J. Geo-Inf.* **2019**, *8*, 179. [CrossRef]

65. Eichkitz, C.G.; Amtmann, J. Impact of gray level transformation and chosen amplitude range on GLCM-based anisotropy estimation. *First Break.* **2019**, *37*, 67–74. [CrossRef]

66. Chaahat, G.; Kumar, G.N.; Kumar, L.P. Gray level co-occurrence matrix (GLCM) parameters analysis for pyoderma image variants. *J. Comput. Theor. Nanosci.* **2020**, *17*, 353–358. [CrossRef]

International Journal of
Environmental Research and Public Health

MDPI

Article

Effects of Biochar and Manure Co-Application on Aggregate Stability and Pore Size Distribution of Vertisols

Taiyi Cai [1,2], Zhigang Wang [1], Chengshi Guo [3], Huijuan Huang [1], Huabin Chai [1,*] and Congzhi Zhang [2,*]

[1] School of Surveying and Land Information Engineering, Henan Polytechnic University, Jiaozuo 454000, China
[2] State Experimental Station of Agro-Ecosystem in Fengqiu, State Key Laboratory of Soil and Sustainable Agriculture, Institute of Soil Science, Chinese Academy of Sciences, Nanjing 210008, China
[3] Farmland Irrigation Research Institute, Chinese Academy of Agricultural Sciences, Xinxiang 453002, China
* Correspondence: chaihb@hpu.edu.cn (H.C.); czzhang@issas.ac.cn (C.Z.); Tel.: +86-391-3987-662 (H.C.)

Abstract: Background: The combination of biochar and organic manure has substantial local impacts on soil properties, greenhouse gas emissions, and crop yield. However, the research on soil health or quality is still in its early stages. Four pot experiments were carried out: C (30 g biochar (kg soil)$^{-1}$), M (10 g manure (kg soil)$^{-1}$), CM (15 g biochar (kg soil)$^{-1}$ + 5 g manure (kg soil)$^{-1}$), and the control (without any amendments). Results: When compared to C and M treatments, the MWD of CM was reduced by 5.5% and increased by 4.9%, respectively, and the micropore volume (5–30 m) was increased by 17.6% and 89.6%. The structural equation model shows that soil structural parameters and physical properties regulate the distribution of micropores (5–30 μm) in amended soil. Conclusion: Our studies discovered that biochar mixed with poultry manure had antagonistic and synergistic effects on soil aggregate stability and micropore volume in vertisol, respectively, and thus enhanced crop yield by 71.1%, which might be used as a technological model for farmers in China's Huang-Huai-Hai region to improve low- and medium-yielding soil and maintain soil health.

Keywords: soil health; biochar; manure; soil structure; vertisol

Citation: Cai, T.; Wang, Z.; Guo, C.; Huang, H.; Chai, H.; Zhang, C. Effects of Biochar and Manure Co-Application on Aggregate Stability and Pore Size Distribution of Vertisols. *Int. J. Environ. Res. Public Health* **2022**, *19*, 11335. https://doi.org/10.3390/ijerph191811335

Academic Editor: Paul B. Tchounwou

Received: 18 August 2022
Accepted: 6 September 2022
Published: 9 September 2022

Publisher's Note: MDPI stays neutral with regard to jurisdictional claims in published maps and institutional affiliations.

1. Introduction

Soil health is defined as soil's ongoing ability to support vital ecosystem functions for plants, animals, and humans [1]. Soil health or quality has become a hot topic in research on sustainable agricultural development today, and it can be evaluated quantitatively and qualitatively through soil health indices (soil processes and traits) [2]. Soil aggregate stability and soil pore characteristics, for example, are important indicators for characterizing soil health and the sustainability of various agricultural management practices [3,4].

Aggregate stability refers to the ability of soil aggregates to withstand damage when subjected to external forces (such as rain) [5]. It describes the physical ability of soils to maintain their aggregation and structure in wet conditions (such as irrigation after severe rainfall or protracted drought). Aggregate stability is regarded as a reliable indicator of soil's biological and physical health. According to research, poorly structured soils, such as vertisol, frequently develop surface crusts or compaction, which can impair water transport and gas exchange and increase the risk of flooding and drought. Internal and extrinsic factors both influence soil aggregate stability [6]. Meteorological conditions (e.g., average annual precipitation and temperature) and agricultural management practices (e.g., fertilization and tillage) are examples of external influences, while intrinsic factors include soil clay content, organic matter content, and soil minerals [7].

Soil pores are the spaces between soil particles or aggregates. Pore properties are the "leading indicator" for determining many soil processes and functions, including water storage and transport, microbial activity, and soil mechanical resistance to root penetration [8]. Soil pore size distribution (PSD) has been shown in studies to be an important indicator for a thorough understanding of soil aggregate stability, water transport, and carbon sequestration [9]. Many studies have shown that fertilization, tillage, and land use can all have a significant impact on the total porosity, size distribution, and function of soil pores [4,10,11], thereby affecting soil quality. This implies that soil pore structure is highly susceptible to soil management practices and environmental changes. As a result, studying soil pore characteristics is vital for assessing soil health.

Organic manure is an efficient way to improve soil health and plant growth [12]. In comparison to inorganic fertilizers, it primarily promotes plant growth in the soil by slowly decomposing and releasing various nutrients. Many types of organic manure on the market are derived from biochar, animal manure, and wood waste compost, and their effects on promoting healthy soils have been thoroughly researched [13,14].

Biochar is a byproduct of high-temperature pyrolysis of natural organic materials that are rich in carbon and porous materials [15]. Currently, the primary application of biochar in agriculture is as a soil amendment [16]. In-depth research has demonstrated that biochar may improve soil fertility, detoxify the soil, promote soil microbial diversity, and improve plant health, and it is projected to become one of the main efforts to preserve soil health and sustainable agricultural development. However, biochar prices are high and cannot be offset by prospective economic gains based on greater average yields and current CO_2 pricing [17]. As a result, biochar has yet to be used on a broad basis in agricultural operations [18,19].

Livestock and poultry waste are the primary sources of animal manure. Poultry manure, in particular, is widely used as a soil improvement material due to its obvious advantages of high nitrogen content, low price, and high yield [20], which can increase the source of soil nutrients and organic matter and can improve the soil microbial population, affecting soil structure and soil physical properties, as well as chemical and biological changes in other parameters [21]. According to a literature study, the amount of manure generated by large-scale chicken dung in China is 3.83×10^9 t, with Henan Province producing the most, followed by Shandong, Hebei, Sichuan, and Hunan provinces [22]. This suggests that poultry manure has a high potential for organic manure generation in soil.

Vertisol is soil with a similar mineral composition and characteristics to its parent material [23]. It is commonly available in Australia, China, India, the United States, and other countries [24]. Deep cracks and sticking behavior are easily developed as a result of the continual expansion and shrinkage of clay-rich minerals, posing a possible threat to agricultural production. However, due to its potential high natural fertility, it has piqued the interest of many scholars both domestically and internationally. Vertisols cover approximately 4×10^6 ha in China, with the majority of them located in semi-arid northern China and belonging to low- and medium-yielding soils [4]. To enhance its poor physical structure, predecessors attempted to implement various technologies (organic fertilizer, fly ash, biochar, and amendment, among others) and achieved considerable progress [3,4].

Previous research has mostly focused on the single effect of organic manure or biochar on the structural improvement of vertisols, focusing little on the combined effect of the two. Several studies combining the two have shown promising results in terms of soil C and N cycling processes, soil biological indicators (soil microbial biomass, enzyme activity, and soil microbial diversity), and crop yield enhancement [14,25,26], but there has been a severe lack of studies on the effects of the two on soil structure. Due to the uniqueness and typicality of the vertisol soil structure, improving its poor physical structure has always been the primary goal of vertisol research.

As a result, it is critical to understand the impact of the two on the physical structure of the modified soil. Through pot experiments, the goal of this study is to investigate the effects of combining biochar and organic fertilizer on soil aggregate stability, soil pore structure and their combined response (crop yield and economic profit). When compared to biochar or organic manure treatments alone, we hypothesized that co-applying biochar and organic manure improves soil's aggregate stability and optimizes soil pore size distribution in vertisols, which in turn will increase crop yield and economic efficiency, thus maintaining soil health.

2. Materials and Methods

2.1. Potted Experiment Method

The experimental soil was originally taken from a typical vertisol of a 0–30 cm soil layer in Dancheng County, Zhoukou City, Henan Province, China (33°38 N, 115°23 E, elevation of 23 m). Large aggregates were crushed, and debris, such as rocks larger than 3 cm in diameter, was removed. The soil is classified as a typical vertisol by the Chinese soil classification method [27]. Before the test, the selected soil physicochemical characteristics were determined using the Zhang and Gong [28] test method, and the results are shown in Table 1.

Table 1. Selected physicochemical properties of vertisol.

Parameter	Vertisol
Sand (2–0.02 mm, %)	26.0
Silt (0.02–0.002 mm, %)	30.7
Clay (<0.002 mm, %)	43.3
Porosity (%)	37.7
Total carbon (g kg^{-1})	5.92
C/N	10.3
CEC (cmol (+) kg^{-1})	23.7
pH	7.50

The pH was determined using the ratio of solid material to water of 1:2.5; particle size distribution was determined using sieving and the pipette method; cation exchangeable capacity was determined using the ammonium saturation and distillation methods; total carbon was estimated through potassium dichromate oxidation and titration with ferrous sulfate.

The test pots were 52 cm tall, with an upper inner diameter of 44 cm and a lower diameter of 30 cm, and a water outlet hole at the bottom. Each pot was filled with 70 kg of drying soil with a filling capacity of 1.3 g cm^{-3}.

The experiment started in 2012. Given the characteristics of poor soil physical properties (soil viscosity, high bulk density, and poor ventilation and water permeability) of sand ginger black soil, a total of four treatments were set up in a completely randomized block design and repeated four times. The four treatments were as follows: C (30 g biochar (kg soil)$^{-1}$), M (10 g manure (kg soil)$^{-1}$, CM (15 g biochar (kg soil)$^{-1}$ + 5 g manure (kg soil)$^{-1}$), and the control (without any amendments). These amounts refer to the ranges generally applicable in the region and other countries [4,14]. After the crops were harvested, the soil for each treatment was removed from the pot and sieved, as biochar is inert and basically does not decompose, and the biochar treatment was no longer required. Each pot of organic manure treatment was added again according to the soil ratio of 1.04 g/kg, and it was used for potting experiments in the summer. The applied crop rotation method was winter wheat–summer corn.

Wheat (*Triticum aestivum* L.) was planted on 9 October and harvested on 1 June; maize (*Zea mays* L.) was planted on June 10 and harvested on 1 October.

Biochar was created in a factory-scale reactor by pyrolysis at 500–550 °C for 20 h, then sieved to 2 mm and analyzed (Table 2) according to the methodology recommended by Xie et al. [29]. Poultry manure was composted at 30–70 °C for 30 days and kept above 55 °C for 7 days before its physicochemical properties were determined using the method recommended by Xie et al. [29].

Table 2. Selected chemical characteristics of the biochar and manure used in the test.

Parameter	Biochar	Manure
Porosity (%)	78.4	-
pH	9.20	8.60
Total carbon (g kg^{-1})	647.2	442.5
CEC (cmol (+) kg^{-1})	41.7	36.2
Phosphorus, mg kg^{-1}	2.6	5.3
Potassium, cmol$_c$ kg^{-1}	4.5	4.9
Calcium, cmol$_c$ kg^{-1}	3.3	2.0
Nitrogen, %	-	3.1

The pH was determined by the ratio of solid material to water of 1:2.5; particle size distribution was determined by sieving and the pipette method; cation exchangeable capacity was determined using the ammonium saturation and distillation methods; total carbon was estimated by potassium dichromate oxidation and titration with ferrous sulfate. "-" indicates undetermined.

Wheat seeds were sown in each pot in four rows. Maize was sown in pots with 10 seeds per pot, and four plants were kept until harvest. To calculate yield, the aboveground parts of the plants (straw and grain) were harvested separately and dried at 70 °C for 72 h.

2.2. Laboratory Methods

2.2.1. Soil Samples

In October 2019, after the end of the experiment (after maize harvesting), 15 cm soil samples were collected from each pot, air-dried, and sieved (<2 mm), and physicochemical properties were analyzed according to traditional methods [28].

2.2.2. Determination of Physical Properties

(1) Coefficient of linear expansion (COLE)

The COLE of soil was determined on ground remolded soils according to Schafer and Singer [30]. The COLE was calculated using the formula below:

$$COLE = (Lm - Ld)/Ld$$

where Lm and Ld are the length of moist and dry soils, respectively.

(2) Distribution and stability of soil aggregate

Soil agglomeration distribution was performed using the wet sieve method. For the main steps, 50 g of soil samples was weighed and placed on the top of a set of sieves with diameters of 2, 1, 0.5, and 0.25 mm, and all the sieves were placed in an agglomerate shaker model (DM200-II) and shaken up and down at a frequency of 30 times per minute for 30 min. Finally, the remaining agglomerates of each sieve were dried in a drying oven at 105 °C for 24 h and weighed, and the percentage of soil water stability mass for each particle size was calculated. The calculation formula of soil aggregate stability (average weight diameter) is as follows [31]:

$$MWD = \sum_{i}^{n} X_i W_i$$

In the formula, MWD is the average weight diameter, X_i is the grain size average diameter (mm); W_i size X_i is the body weight percentage.

2.2.3. Pore Size Distribution (PSD)

Using a mercury porosimeter (MIP) model, PSD was calculated (Autopore IV9500, Micromeritics Inc., Norcross, GA, USA). The mercury porosimetry method is based on the hypothesis that the soil pores are small, irregularly shaped, and cylindrical. Each pore can extend to the sample's outer surface and come into direct contact with the mercury (theta is 140°). The Washburn formula is as follows:

$$r = \frac{2\gamma \cos \theta}{p}$$

The pores were assumed to be cylindrical, where r is the pore radius (μm), p is the pressure (kPa), γ is the Hg surface tension (0.47 N m^{-1}), and θ is the mercury–soil contact angle (140°). Under certain pressure, the mercury will infiltrate into pores of the corresponding size, and the amount of indented mercury represents the volume inside the pores; if the pressure is gradually increased, the amount of indented mercury can be calculated, and the volume distribution of soil pores can be measured [32]. To gain an in-depth understanding of soil pore size distribution, soil pores are divided into five levels: macropores (>60 μm), mesopores (60–30 μm), micropores (30–5 μm), ultra-micropores (5–0.1 μm), and crypto pores (0.1–0.01 μm and <0.01 μm) [33].

2.2.4. Determination of Chemical Properties

The pH was determined using a standard method, i.e., a soil-to-water ratio of 1:2.5, and the mixture was shaken uniformly after 60 min of settling; subsequently, the pH of the suspension was measured using a pH meter [28].

Total organic carbon was determined using the C oxidation method with potassium dichromate, followed by the titration of the remaining $Cr_2O_7^{2-}$ with ammonium iron (II) sulfate [34].

2.3. Statistical Analysis

The multivariate statistical analysis of soil pore types and soil physical and chemical properties was conducted using the SPSS 21.0 Statistical Package (Tsinghua University Press, Beijing, China). Structural equation modeling (SEM) is a multivariate statistical method that can perform hypothesis testing on complex path relationships between indicators [35]. Based on the relative contribution rates of biochar and organic manure to soil micropore volume and their interaction, we used AMOS 21.0 software (IBM, Armonk, NY, USA) for structural equation modeling and quantitative analysis. Drawing using Sigmaplot12.5.

3. Results

3.1. Selected Physical and Chemical Properties of Vertisol Soil

The physical and chemical properties of vertisol soil selection are shown in Table 3. The soil has typical alkaline vertisol characteristics (pH > 7.98). The pH values of C, M, and CM treatments were not significantly different, and the pH values of the three treatments were all lower than those of the control. Soil organic carbon (SOC) ranged from 22.47 to 27.33 g kg^{-1}. On average, the soil organic carbon (SOC) in C (27.33 g kg^{-1}), M (26.87 g kg^{-1}), and CM (26.61 g kg^{-1}) were significantly higher than in the control (22.47 g kg^{-1}), respectively. The COLE ranged from 0.09 to 0.12, with an average of 0.105. The COLE values of the C and CM treatments were significantly lower than that of the control and M treatments; however, there was no significant difference between the C and CM treatments.

Table 3. Selected physical and chemical properties.

Treatment	pH	SOC (g/kg)	COLE	Porosity (%)	Fractal Dimension
Control	8.18 ± 1.08 a	10.47 ± 0.61 c	0.12 ± 0.74 a	24.14 ± 0.09 b	2.79 ± 0.01 b
C	8.07 ± 0.31 b	13.14 ± 0.26 a	0.10 ± 0.26 c	40.44 ± 0.15 a	2.89 ± 0.01 a
M	8.01 ± 0.14 b	11.87 ± 0.16 b	0.13 ± 0.16 b	21.31 ± 0.08 c	2.80 ± 0.01 b
CM	7.98 ± 1.04 b	12.66 ± 0.63 a	0.09 ± 0.80 c	40.63 ± 0.15 a	2.91 ± 0.01 a

Mean ± SD; n = 4. Values followed by different letters within a column are significantly different ($p < 0.05$).

3.2. Soil Aggregates

Figure 1 shows the improved vertisol aggregate size distribution with biochar and poultry manure. Biochar and poultry manure changed the soil aggregate properties of vertisol. Compared with the control, the percentage of water-stable macroaggregates of 2–0.25 mm was significantly increased for the C, M, and CM treatments, while the percentage of microaggregates < 0.053 mm was significantly decreased; however, the percentage of soil aggregates > 2 and 0.25–0.053 mm did not differ among the three treatments. Meanwhile, the CM treatment (biochar and manure co-application) had the highest percentage of 2–0.25 mm aggregates (26.3%), while the percentage of <0.053 mm aggregates was the lowest (61.9%), which was increased by 28.7% compared with the control, which decreased by 9.4%. This indicates that the interaction of biochar and poultry manure upon aggregate formation was significant ($p < 0.05$).

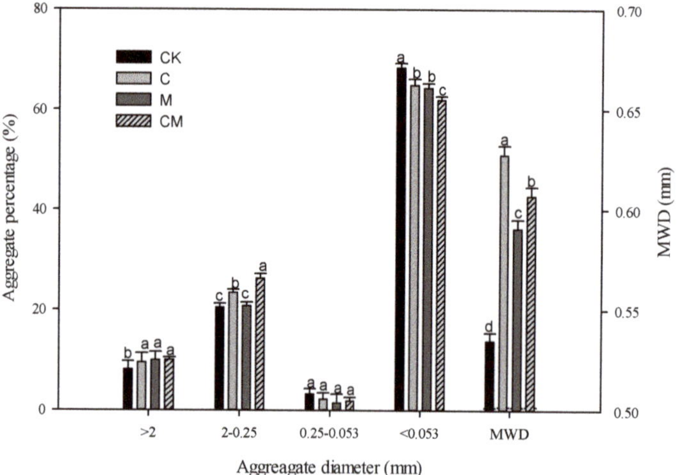

Figure 1. Effects of different treatments on the vertisol aggregate size distribution using wet sieving method and MWD, where error bars represent standard deviation and lowercase letters indicate significant differences between treatments ($p < 0.05$).

The mean weight diameter (MWD) plays an important role in assessing the stability of soil aggregates, with large MWD values indicating better water stability of aggregates [36]. Compared with the control, the MWD of the C, M, and CM treatments was significantly improved, and the CM treatment was the most effective (Figure 1). The MWD of CM was 4.9% higher than that of the M treatment, but 5.5% lower than that of the C treatment, which indicated that the combination of C and M would have obvious antagonistic effects, that is, reducing the positive effect of biochar treatment and enhancing the negative effect of poultry manure treatment.

3.3. Pore Characteristics and PSD of Soil

3.3.1. Pore Size Distribution

The cumulative pore volume and differential pore size distribution (PSD) curves of the vertisol soil show that (Figure 2) the PSD curve of vertisol has a unimodal distribution overall. The peak value (0.12–0.30 μm) was shifted to the right, and when the equivalent pore size was <0.01 μm, there was no significant difference between the two pore size distribution curves. The PSD changes in the C and CM treatments were similar. Compared with the control and M treatments, their PSD distribution curves shifted to the right as a whole, with peaks at 12–16 μm and 50–60 μm, respectively. The CM peak was significantly higher than that of the C treatment, which indicated that the CM treatment had a significant positive effect in increasing the pore volume with pore sizes of 5–30 μm; however, when the equivalent aperture was <5 μm, there was no significant difference in PSD distribution between the two treatments.

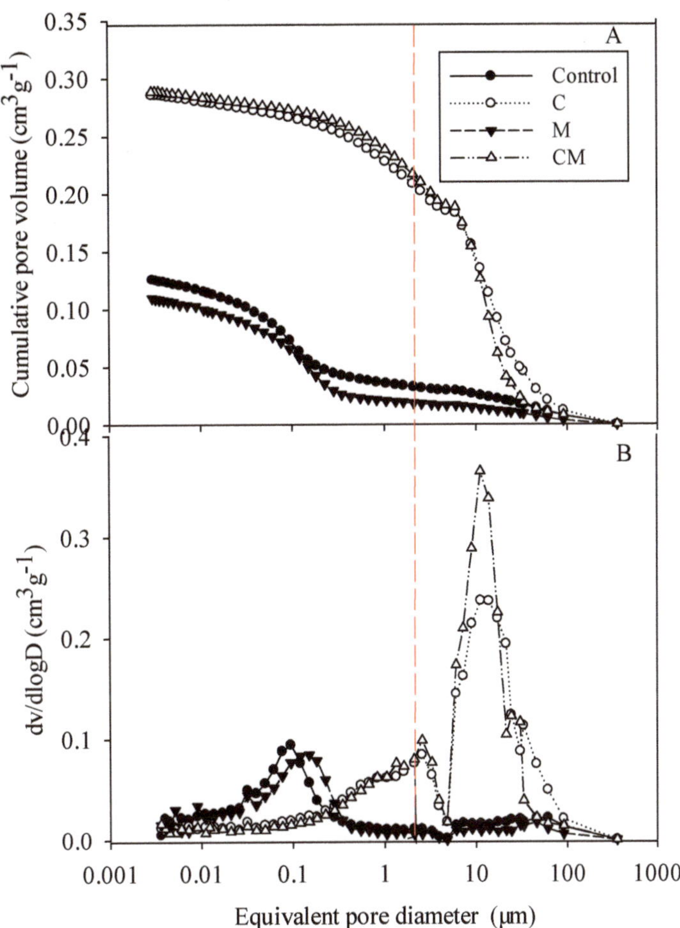

Figure 2. Cumulative (**A**) and differential (**B**) pore size distribution of different treatments determined by MIP in the range of 0.001–1000 μm. The red vertical line is the dividing line (2 μm) between the ineffective and effective pores.

3.3.2. Pore Volume Distribution

Pore size distribution (PSD) data can be used to characterize soil evolution and describe agricultural management effects [37,38]. The soil pore volume distribution and total pore volume determined based on the MIP method are shown in Figure 3A. The total pore volumes of soils treated with C and CM were 0.2859 and 0.2886 cm^3 g^{-1}, respectively, which were significantly higher than ($p < 0.05$) that of the control soils (0.1261 cm^3 g^{-1}), while the total pore volumes of soils treated with M (0.1014 cm^3 g^{-1}) were significantly lower than the control. Compared with the control, C, and M treatments, the total pore volume of the CM treatment was increased by 128.9%, 0.9%, and 184.6%, respectively, which indicated that the CM treatment reflected the synergistic effect of the mixing of biochar and poultry manure.

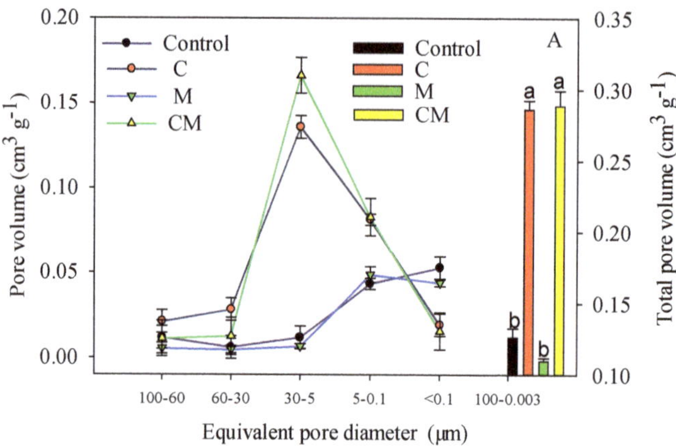

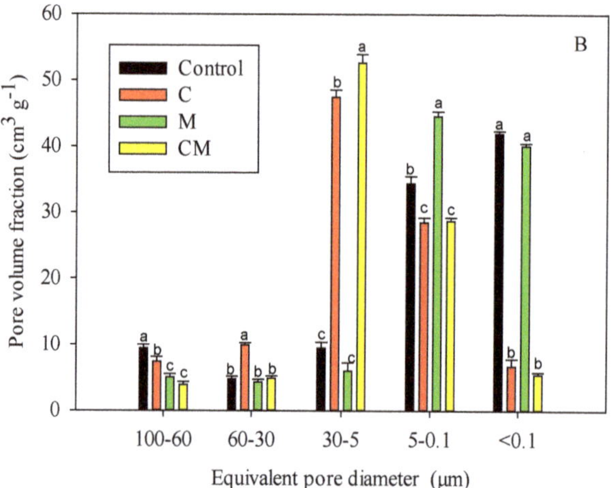

Figure 3. Determination of soil pore volume distribution through mercury intrusion (MIP) using the pore classification method of Cameron and Buchan [33,34]. (**A,B**) represent the pore volume and total pore volume at different treatment equivalent pore sizes and pore volume fraction, respectively. Values followed by different letters within horizontal row are significantly different ($p < 0.05$).

Figure 3B showed that the C, M, and CM treatments significantly reduced the volume percentage of macropores by 100–60 µm and significantly increased the volume percentage of micropores by 5–30 µm compared to the control, which further indicated the biochar and poultry manure in the pores. The improvement of the structure had a significant synergistic effect; however, the difference was that there was a positive synergistic effect on the 5–30 µm pore volume percentage, but a negative synergistic effect on the 100–60 µm pore volume percentage. Changes in the volume percentage of pores at other grades (60–30, 5–0.1, and <0.1 µm) were irregular.

3.4. Correlation between Soil Micropore Characteristics and Soil Physicochemical Properties

Pearson correlation coefficients were calculated to describe the correlation of micropore volume changes with soil properties and composition (Table 4). There was a significant positive correlation between the micropore volume (P5–30) and TPV, porosity, fractal dimension, and A2–0.25 ($p < 0.05$), but a significant negative correlation with P0.1–5, P0.01–0.1, $p < 0.01$, pH, COLE, and A < 0.053 ($p < 0.05$).

Table 4. Correlation between micropore volume and equivalent pore diameter (5–30 μm) and different varieties.

Variety	Correlation
A > 2	0.263
A2–0.25	0.601 *
A0.25–0.053	−0.115
A < 0.053	−0.636 *
MWD	0.432
SOC	0.497
pH	−0.866 **
COLE	−0.760 **
P > 60	−0.414
P30–60	0.23
P0.1–5	−0.844 **
P0.01–0.1	−0.986 **
$p < 0.01$	−0.983 **
TPV	0.985 **
Porosity	0.986 **
Fractal dimension	0.965 **

Percentage of A > 2 mm aggregate; percentage of A2–0.25 and 2–0.25 mm aggregates; percentage of A0.25–0.053 and 0.25–0.053 mm aggregates; percentage of A < 0.053 and <0.053 mm aggregates; MWD, mean weight diameter; SOC, soil organic carbon of bulk soil; pH, calculated using the ratio of solid material to water of 1:2.5; COLE, linear expansion coefficient, $p > 60$, macropore volume; P30–60, mesopore volume; P5–30, micropore volume; P0.1–5, ultramicropore volume; P0.01–0.1 and $p < 0.01$, cryptopore volume; TPV, total pore volume. ** Highly significant $p < 0.01$. * Significant at $p < 0.05$.

3.5. Structural Equation Modeling

Due to the strong correlation between the micropore volume (P5–30) and physico-chemical parameters, we introduced structural equation modeling to deeply explore the mechanism of biochar and livestock manure on the change in micropore volume (Figure 4). According to the three indicators of the model fitting effect ($\chi^2 = 6.437$, $df = 6$, and $p = 0.368$), the SEM model constructed in this study can better fit the internal relationship between the relevant indicators, assuming that causal variables explain 73.5% of the reason for the change in pore volume (5–30 μm).

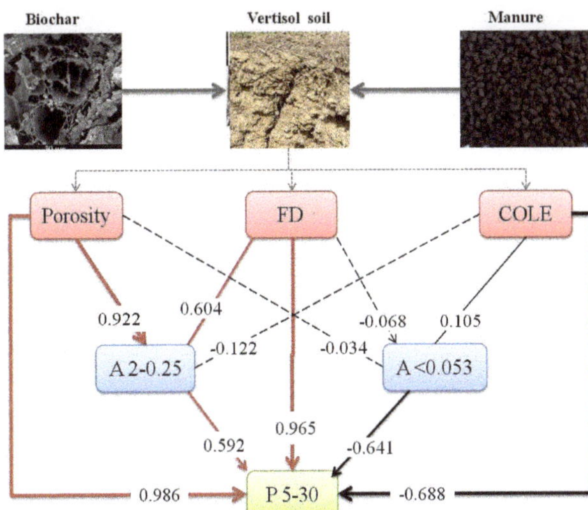

Figure 4. Structural equation modeling (SEM) analysis of the biochar and manure synergistic effects on micropore volume with equivalent pore diameter (5–30 μm).

The results of the optimal model fitting (Chi-square (χ^2) = 23.46; df = 12, p = 0.0481; comparative fit index (CFI) = 0.918; root square mean error of approximation (RMSEA) = 0.221). Square boxes denote variables included in the models. The solid arrows ($\rightarrow$) mean a single effect in the direction of the arrow, the lines (-) mean a cycle effect, and the thickness represents the magnitude of the path coefficients. Dashed arrows represent the directions, and the effects were non-significant (p > 0.05). (Porosity and total MIP porosity for the pore diameter range of 0.003–360 μm; FD, fractal dimension; COLE, linear expansion coefficient; Percentage of A2–0.25 and 2–0.25mm aggregates; Percentage of A < 0.053 and <0.053 mm aggregates; and P5–30 and 5–30 μm, soil pore volume.)

The structural equation model demonstrated that porosity, fractal dimension, and 2–0.25 mm aggregates had direct positive effects on the 5–30 μm pore volume, and porosity and fractal dimension had indirect positive effects on the 5–30 μm pore volume through 2–0.25 mm soil aggregates, respectively. COLE and <0.053 mm soil aggregates had a direct negative effect on the 5–30 μm pore volume, and COLE produced an indirect negative effect on the 5–30 μm pore volume through <0.053 mm soil aggregates.

3.6. Grain Yield and Economic Profit

Table 5 shows that the grain yields of C, M, and CM (0.37, 0.29, and 0.33 kg (Per Plant)$^{-1}$) were increased by 92.8%, 49.9%, and 71.1%, respectively, compared with the control (0.19 kg (Per Plant)$^{-1}$). Among them, the yield of CM was between C and M. This suggested that the combination of biochar and organic manure had an antagonistic effect on maize grain yield. In terms of economic Profit, the M treatment had the highest economic Profit, followed by the control treatment, and the C and CM treatments were the lowest.

Table 5. Effects of application of 20 t ha^{-1} biochar and 5 t ha^{-1} Poultry manure on summer maize yield and economic benefits (cost, income, and Profit).

Treatments	Grain Yield		Cost	Income	Profit
	kg (Per Plant)$^{-1}$	kg ha^{-1}		CNY	
Control	0.19	10,191.35	5100	15,287	10,187
C	0.37	19,652.48	23,100	29,479	6379
M	0.29	15,277.50	7600	22,916	15,316
CM	0.33	17,435.25	16,600	26,153	9553

4. Discussion

4.1. Amelioration of the Aggregate Stability

Soil aggregate stability is an indicator for assessing the effects of soil type and field management on soil quality [6]. The biochar-amended soil significantly increased the formation of 2–0.025 mm aggregates while decreasing the percentage of 0.053 mm microaggregates (Figure 1), indicating that the macroaggregates were formed through the combination of many microaggregates [31]. The biochar-enhanced carbon may act as a glue, concentrating microaggregates into macroaggregates. Our findings are consistent with those of Lu et al. [39], who discovered that rice husk biochar-amended soils have a significantly higher MWD than the control soils [36]. The improvement in aggregate stability was thought to be the result of two mechanisms. The first is that biochar improves soil properties through the physical meshing of carbon polymers or particles, increasing the internal cohesion of mineral particles to increase aggregate resistance to clay swelling [40], and the second is that biochar treatment promotes soil hydrophobicity properties, thereby reducing the degree of clay swelling and aggregate breakage [3]. The stability of aggregates in this study could be due to the interaction of these two mechanisms.

Several studies have suggested that adding poultry manure to soil improves MWD [41,42]. The use of poultry manure (2% by weight) increased wet aggregate stability (Figure 1). The MWD value of the poultry manure treatment was slightly higher than that of the control, but the difference was not significant, confirming a previously established effect of poultry manure on soil aggregation stability. This finding is consistent with that of Peng et al. [43], who concluded that adding swine manure increased dry aggregate stability and Peanut biomass in a Ultisol. Figure 1 also shows that biochar co-application with manure treatment significantly increased the percentage of 2–0.25 mm aggregate and MWD when compared to the C and M treatments, indicating a positive synergy effect between C and M. These findings are consistent with those of Sánchez et al. [41], who discovered that the addition of 3% biochar promoted the rapid degradation of organic matter, reduced the formation of large clumps, and accelerated stabilization and detoxification. One possible explanation for this is that biochar can help increase the carbon source for microorganisms or promote the rapid degradation of poultry manure [42].

4.2. Amelioration in the Soil Pore System

Soil pore characteristics (such as porosity, size distribution, and Shape of soil pore space) are important indicators of soil quality and are highly sensitive to soil management practices, such as biochar and manure application in the field [8,39]. Clay PSDs usually have more micropores (<0.01 μm). The 100–0.3 μm pore volume was significantly higher in the C and CM treatments than in the control and manure treatments (Figure 3). This variation could be attributed to the inherent properties of biochar (e.g., wider pore size distribution) [44]. Other authors have reported that organic waste-based biochar contains many macropores larger than 10 μm in diameter, which induce soil particle cohesion and thus improve the pore structure [45].

The use of poultry manure has been shown to improve soil structural stability [43,46]. In the current study, CM reduced soil porosity and was consistent with the changing trend of the control treatment; however, the pore volume fraction at a 0.5 m equivalent pore diameter was higher than that of other treatments (Figure 3B).

This finding contradicts previous research, which has found that organic manure treatment increases Porosity more than the control [14,47]. This result could be explained in two ways. On the one hand, the decrease in total pore porosity in soil may be due to an increase in ultramicropores (0.1–5 m) and crytopores (0.1–0.007 m), which have become the dominant component of the pore system. On the other hand, it is likely that changes in the composition and arrangement of soil particles (or aggregates) caused a decrease in porosity [46]. However, it is unknown how the composition and arrangement of soil particles (or aggregates) affect the decrease in total pore porosity; therefore, more research should be conducted.

Figures 3 and 4 show that the co-application of biochar with manure had a significant positive synergistic effect on microspore volumes of a 5–30 μm equivalent pore size. The synergistic effect was comparable to adding cow manure (2% by weight) and biochar (10% by weight) to calcareous soils, and it demonstrated a positive priming effect [48]. However, the findings of this study do not support those of Binh Thanh et al., [49] that the combination of biochar and cow manure had no positive synergistic effect. Although the underlying reasons for the opposite conclusion are yet to be determined, the dilution effect caused by the mixing of livestock manure and clay minerals could be one of them [46]. On the one hand, biochar's ion adsorption creates a favorable solution environment. Meanwhile, biochar accelerates the mineralization of livestock manure [42], Providing more ion adsorption for biochar, but biochar pores typically exceed 10 mm [45]. As biochar can only selectively adsorb parts of ions, it is the collaboration of biochar and manure that produces these synergistic effects. However, currently, this effect mechanism is only speculative, and more research is needed to confirm it.

4.3. MicroPore Change Based on Structural Equation Model

Interaction effects between different factors are frequently masked by simple bivariate correlations, and SEM can answer questions involving multiple regression analyses of factors [35]. We conducted a structural equation simulation analysis of soil pore structure parameters and soil micropore volume to reveal the pathway of biochar and organic manure affecting soil micropores (5–30 μm) via changes in soil aggregates and pore structure parameters. In this study, we hypothesized that soil micropores (5–30 μm) were affected by three pathways (Figure 4): the soil pore structure (represented by porosity and fractal dimension); soil aggregates (represented by the number of aggregates between 2–0.25 mm and 0.053 mm); and soil physical properties (represented by COLE), and that any difference in these parameters could explain soil micropore (5–30 μm) volume variation.

Soil porosity and fractal dimension contributed the most to the soil micropore (5–30 μm) volume in the first approach, with contributions of 0.986 and 0.965, respectively, and in the second pathway, the contribution rates of 2–0.25 mm and 0.053 mm aggregates were 0.592 and 0.641, respectively. COLE's contribution to the soil micropore (5–30 μm) volume was −0.688 in the third approach. It is worth noting that biochar and organic manure had a direct and indirect effect on the volume change in soil micropores (5–30 μm) via porosity, fractal dimension, and COLE, which was consistent with the findings of Guo et al. [50].

According to Table 1, biochar treatment enhanced porosity and fractal dimension while decreasing COLE. The use of organic manure reduced porosity while increasing the fractal dimension and COLE. CM treatment, on the other hand, enhanced porosity and fractal dimension while decreasing COLE. When these findings are combined with the structural equation model, it is obvious that the volume change in soil micropores (5–30 μm) is the result of the synergistic and antagonistic effects produced by the combination of biochar and organic fertilizer. However, given Present technological Capabilities, it is still impossible to quantify the synergistic and antagonistic effects. To this aim, the best microscopic technology must be developed, as well as advanced spectroscopy (NMR) and isotope labeling (^{13}C and ^{15}N) technologies.

4.4. Economic Profit

The combination of biochar and organic manure needs to be verified by comprehensive experiments with cost–benefit methods [18,20]. The most significant improvement effect of biochar and organic manure was an increase in crop yield (Table 5). As this was a pot experiment, only the corn yield was used as an example to facilitate the comparison with the field experiment. The field yield conversion method was as follows: convert the corn yield per plant in the pot experiment into the field yield based on 3500 plants (667 m^2)$^{-1}$, which is approximately 10,191 kg ha^{-1}. The use of biochar in combination with livestock manure increased production by 71.1%. When combined with the current situation of low local farmer income, high biochar costs, and abundant Poultry manure, this combination is recommended as a feasible alternative to increase maize yield in this region while considering appropriate economic benefits.

In general, vertisol improvement in China's Huang-Huai-Hai plain requires approximately 5 tons ha^{-1} of organic manure, with a total value of approximately CNY 500 [51], and with a local maize price of CNY 1.5/kg^{-1} [52], the application of organic manure would yield CNY 15316. Given that the cost of biochar varies and that the price in the region is approximately CNY 1800/t^{-1} [17], the benefit of 20 t biochar was approximately CNY 6379, which was lower than the benefit of organic manure alone and the control. However, as biochar is a major source of inert carbon with a long average retention time (decades to centuries), it may have a long-term impact on soil quality [20].

5. Conclusions

The effects of biochar, organic manure, and a combination of the two on soil properties and crop growth were significant. The combination (15 g biochar (kg soil)$^{-1}$ + 5 g manure (kg soil)$^{-1}$) demonstrated antagonistic effects on aggregate stability and crop yield, as well as synergistic effects on soil macropore and Micropore volume distribution, with the former being negative and the latter being positive.

Combining biochar and organic manure application can be a viable management practice for smallholder farmers in China's Huang-Huai-Hai Plain region, as it not only improves aggregate stability and the distribution of soil micropores (5–30 μm) corresponding to plant available water, but also reduces COLE and soil improvement costs, increases crop yields, and thus maintains soil health when compared to applying biochar or organic manure alone. Further research will be required to accurately match the ratio of biochar to organic manure to apply this improved technology more widely.

Author Contributions: T.C. and Z.W. designed the study and wrote the main body of the manuscript; C.Z. and H.C., conception of the study and final approval of the version to be published; H.H. and Z.W., data collection and analysis; C.G. revised the manuscript and figures; T.C. acquired the funding. All authors have read and agreed to the published version of the manuscript.

Funding: This research was funded by the National Natural Science Foundation of China (grant No: 41671225, 41877020, U1810203, U1261206).

Institutional Review Board Statement: Not applicable.

Informed Consent Statement: Not applicable.

Data Availability Statement: The datasets generated and/or analyzed during the current. Study are available from the corresponding author on reasonable request.

Acknowledgments: We are thankful to the anonymous reviewers and editors for their helpful comments and suggestions.

Conflicts of Interest: The authors declare no conflict of interest.

References

1. Lehmann, J.; Bossio, D.A.; Koegel-Knabner, I.; Rillig, M.C. The concept and future prospects of soil health. *Nat. Rev. Earth Environ.* **2020**, *1*, 544–553. [CrossRef]
2. Parr, J.F.; Papendick, R.I.; Hornick, S.B.; Meyer, R.E. Soil quality: Attributes and relationship to alternative and sustainable agriculture. *Am. J. Altern. Agric.* **2009**, *7*, 5–11. [CrossRef]
3. Sun, F.; Lu, S. Biochars improve aggregate stability, water retention, and pore-space properties of clayey soil. *J. Plant Nutr. Soil Sci.* **2014**, *177*, 26–33. [CrossRef]
4. Lu, S.G.; Sun, F.F.; Zong, Y.T. Effect of rice husk biochar and coal fly ash on some physical properties of expansive clayey soil (Vertisol). *Catena* **2014**, *114*, 37–44. [CrossRef]
5. Amézketa, E. Soil Aggregate Stability: A Review. *J. Sustain. Agric.* **1999**, *14*, 83–151. [CrossRef]
6. Rabbi, S.M.F.; Wilson, B.R.; Lockwood, P.V.; Daniel, H.; Young, I.M. Aggregate hierarchy and carbon mineralization in two Oxisols of New South Wales, Australia. *Soil Tillage Res.* **2015**, *146*, 193–203. [CrossRef]
7. Annabi, M.; Le Bissonnais, Y.; Le Villio-Poitrenaud, M.; Houot, S. Improvement of soil aggregate stability by repeated applications of organic amendments to a cultivated silty loam soil. *Agric. Ecosyst. Environ.* **2011**, *144*, 382–389. [CrossRef]
8. Kravchenko, A.N.; Guber, A.K. Soil pores and their contributions to soil carbon processes. *Geoderma* **2017**, *287*, 31–39. [CrossRef]
9. Zaffar, M.; Lu, S.G. Pore Size Distribution of Clayey Soils and Its Correlation with Soil Organic Matter. *Pedosphere* **2015**, *25*, 240–249. [CrossRef]
10. Pituello, C.; Dal Ferro, N.; Simonetti, G.; Berti, A.; Morari, F. Nano to macro pore structure changes induced by long-term residue management in three different soils. *Agric. Ecosyst. Environ.* **2016**, *217*, 49–58. [CrossRef]
11. Rabot, E.; Wiesmeier, M.; Schlüter, S.; Vogel, H.J. Soil structure as an indicator of soil functions: A review. *Geoderma* **2018**, *314*, 122–137. [CrossRef]
12. Pan, M.; Yau, P.C.; Lee, K.C.; Zhang, H.; Lee, V.; Lai, C.Y.; Fan, H.J. Nutrient Accumulation and Environmental Risks of Biosolids and Different Fertilizers on Horticultural Plants. *Water Air Soil Pollut.* **2021**, *232*, 480. [CrossRef]
13. Dias, B.O.; Silva, C.A.; Higashikawa, F.S.; Roig, A.; Sanchez-Monedero, M.A. Use of biochar as bulking agent for the composting of poultry manure: Effect on organic matter degradation and humification. *Bioresour. Technol.* **2010**, *101*, 1239–1246. [CrossRef] [PubMed]
14. Wang, Y.; Villamil, M.B.; Davidson, P.C.; Akdeniz, N. A quantitative understanding of the role of co-composted biochar in plant growth using meta-analysis. *Sci. Total Environ.* **2019**, *685*, 741–752. [CrossRef] [PubMed]
15. Sohi, S.P. Carbon Storage with Benefits. *Science* **2012**, *338*, 1034–1035. [CrossRef]
16. Sohi, S.P.; Krull, E.; Lopez-Capel, E.; Bol, R. A review of biochar and its use and function in soil. *Adv. Agron.* **2010**, *105*, 47–82.
17. Liu, X.; Zhang, A.; Ji, C.; Joseph, S.; Bian, R.; Li, L.; Pan, G.; Paz-Ferreiro, J. Biochar's effect on crop productivity and the dependence on experimental conditions—A meta-analysis of literature data. *Plant Soil* **2013**, *373*, 583–594. [CrossRef]
18. Sanchez-Monedero, M.A.; Cayuela, M.L.; Roig, A.; Jindo, K.; Mondini, C.; Bolan, N. Role of biochar as an additive in organic waste composting. *Bioresour. Technol.* **2018**, *247*, 1155–1164. [CrossRef]
19. Lorenz, K.; Lal, R. Biochar application to soil for climate change mitigation by soil organic carbon sequestration. *J. Plant Nutr. Soil Sci.* **2014**, *177*, 651–670. [CrossRef]

20. de Sousa Lima, J.R.; Cavalcanti de Goes, M.d.C.; Hammecker, C.; Dantas Antonino, A.C.; de Medeiros, E.V.; de Sa Barretto Sampaio, E.V.; de Barros Silva Leite, M.C.; da Silva, V.P.; de Souza, E.S.; Souza, R. Effects of Poultry Manure and Biochar on Acrisol Soil Properties and Yield of Common Bean. A Short-Term Field Experiment. *Agriculture* **2021**, *11*, 290. [CrossRef]

21. Albiach, R.; Canet, R.; Pomares, F.; Ingelmo, F. Microbial biomass content and enzymatic activities after the application of organic amendments to a horticultural soil. *Bioresour. Technol.* **2000**, *75*, 43–48. [CrossRef]

22. Li, H.; Feng, W.-T.; He, X.-H.; Zhu, P.; Gao, H.-J.; Sun, N.; Xu, M.-G. Chemical fertilizers could be completely replaced by manure to maintain high maize yield and soil organic carbon (SOC) when SOC reaches a threshold in the Northeast China Plain. *J. Integr. Agric.* **2017**, *16*, 937–946. [CrossRef]

23. Pal, D.K.; Wani, S.P.; Sahrawat, K.L. Vertisols of tropical Indian environments: Pedology and edaphology. *Geoderma* **2012**, *189*, 28–49. [CrossRef]

24. Bhattacharyya, T.; Pal, D.K.; Mandal, C.; Chandran, P.; Ray, S.K.; Sarkar, D.; Velmourougane, K.; Srivastava, A.; Sidhu, G.S.; Singh, R.S.; et al. Soils of India: Historical perspective, classification and recent advances. *Curr. Sci.* **2013**, *104*, 1308–1323.

25. Busscher, W.J.; Novak, J.M.; Evans, D.E.; Watts, D.W.; Niandou, M.A.S.; Ahmedna, M. Influence of Pecan Biochar on Physical Properties of a Norfolk Loamy Sand. *Soil Sci.* **2010**, *175*, 10–14. [CrossRef]

26. Yuan, Y.H.; Chen, H.H.; Yuan, W.Q.; Williams, D.; Walker, J.T.; Shi, W. Is biochar-manure co-compost a better solution for soil health improvement and N2O emissions mitigation? *Soil Biol. Biochem.* **2017**, *113*, 14–25. [CrossRef]

27. Li, D.C.; Zhang, G.L.; Gong, Z.T. On taxonomy of Shajiang black soils in China. *Soils* **2011**, *43*, 623–629.

28. Zhang, G.L.; Gong, Z.T. *Soil Survey Laboratory Methods*; Science Press: Beijing, China, 2012.

29. Xie, Z.; Xu, Y.; Liu, G.; Liu, Q.; Zhu, J.; Tu, C.; Amonette, J.E.; Cadisch, G.; Yong, J.W.H.; Hu, S. Impact of biochar application on nitrogen nutrition of rice, greenhouse-gas emissions and soil organic carbon dynamics in two paddy soils of China. *Plant Soil* **2013**, *370*, 527–540. [CrossRef]

30. Schafer, W.M.; Singer, M.J. A new method of measuring shrink-swell potential using soil pastes. *Soil Sci. Soc. Am. J.* **1976**, *40*, 805–806. [CrossRef]

31. LeBissonnais, Y. Aggregate stability and assessment of soil crustability and erodibility. 1. Theory and methodology. *Eur. J. Soil Sci.* **1996**, *47*, 425–437. [CrossRef]

32. Giesche, H. Mercury porosimetry: A general (practical) overview. *Part. Part. Syst. Charact.* **2006**, *23*, 9–19. [CrossRef]

33. Cameron, K.C.; Buchan, G.D. Porosity and pore size distribution. In *Encyclopedia of Soil Science*; Lal, R., Ed.; CRC Press: Boca Raton, FL, USA, 2006; pp. 1350–1353.

34. Nelson, D.W.; Sommers, L.E. Total carbon, organic carbon and organic matter. In *Methods of Soil Analysis. Part 2: Chemical and Microbial Properties*; Page, A.L., Miller, R.H., Keeney, D.R., Eds.; ASA and SSSA: Madison, WI, USA, 1982; pp. 539–579.

35. Jodie, B.U.; Peter, M.B. Structural Equation Modeling. Chapter 23, Volume 2. Research Methods in Psychology. In *Handbook of Psychology*, 2nd ed.; Wiley Online Library: Hoboken, NJ, USA, 2012.

36. Spohn, M.; Giani, L. Water-stable aggregates, glomalin-related soil protein, and carbohydrates in a chronosequence of sandy hydromorphic soils. *Soil Biol. Biochem.* **2010**, *42*, 1505–1511. [CrossRef]

37. Liu, J.; Yang, P.; Li, L.; Zhang, T. Characterizing the pore size distribution of a chloride silt soil during freeze-thaw processes via nuclear magnetic resonance relaxometry. *Soil Sci. Soc. Am. J.* **2020**, *84*, 1577–1591. [CrossRef]

38. Hajnos, M.; Lipiec, J.; Świeboda, R.; Sokołowska, Z.; Witkowska-Walczak, B. Complete characterization of pore size distribution of tilled and orchard soil using water retention curve, mercury porosimetry, nitrogen adsorption, and water desorption methods. *Geoderma* **2006**, *135*, 307–314. [CrossRef]

39. Lu, S.G.; Malik, Z.; Chen, D.P.; Wu, C.F. Porosity and pore size distribution of Ultisols and correlations to soil iron oxides. *Catena* **2014**, *123*, 79–87. [CrossRef]

40. Khademalrasoul, A.; Naveed, M.; Heckrath, G.; Kumari, K.; de Jonge, L.W.; Elsgaard, L.; Vogel, H.J.; Iversen, B.V. Biochar Effects on Soil Aggregate Properties Under No-Till Maize. *Soil Sci.* **2014**, *179*, 273–283. [CrossRef]

41. Sánchez-García, M.; Alburquerqu, J.A.; Sánchez-Monedero, M.A.; Roig, A.; Cayuela, M.L. Biochar accelerates organic matter degradation and enhances N mineralisation during composting of poultry manure without a relevant impact on gas emissions. *Bioresour. Technol.* **2015**, *192*, 272–279. [CrossRef]

42. Akdeniz, N. A systematic review of biochar use in animal waste composting. *Waste Manag.* **2019**, *88*, 291–300. [CrossRef]

43. Peng, X.H.; Zhu, Q.H.; Xie, Z.B.; Darboux, F.; Holden, N.M. The impact of manure, straw and biochar amendments on aggregation and erosion in a hillslope Ultisol. *Catena* **2016**, *138*, 30–37. [CrossRef]

44. Hayes, M.H.B. Biochar and biofuels for a brighter future. *Nature* **2006**, *443*, 144. [CrossRef]

45. Brodowski, S.; Amelung, W.; Haumaier, L.; Abetz, C.; Zech, W. Morphological and chemical properties of black carbon in physical soil fractions as revealed by scanning electron microscopy and energy-dispersive X-ray spectroscopy. *Geoderma* **2005**, *128*, 116–129. [CrossRef]

46. Haynes, R.J.; Naidu, R. Influence of lime, fertilizer and manure applications on soil organic matter content and soil physical conditions: A review. *Nutr. Cycl. Agroecosystems* **1998**, *51*, 123–137. [CrossRef]

47. Zhou, H.; Fang, H.; Mooney, S.J.; Peng, X. Effects of long-term inorganic and organic fertilizations on the soil micro and macro structures of rice paddies. *Geoderma* **2016**, *266*, 66–74. [CrossRef]

48. Ippolito, J.A.; Stromberger, M.E.; Lent, R.D.; Dungan, R.S. Hardwood biochar and manure co-application to a calcareous soil. *Chemosphere* **2016**, *142*, 84–91. [CrossRef] [PubMed]

49. Binh Thanh, N.; Nam Ngoc, T.; Chau Minh Thi, L.; Trang Thuy, N.; Thanh Van, T.; Binh Vu, T.; Tan Van, L. The interactive effects of biochar and cow manure on rice growth and selected properties of salt-affected soil. *Arch. Agron. Soil Sci.* **2018**, *64*, 1744–1758.
50. Guo, Y.; Fan, R.; Zhang, X.; Zhang, Y.; Wu, D.; McLaughlin, N.; Zhang, S.; Chen, X.; Jia, S.; Liang, A. Tillage-induced effects on SOC through changes in aggregate stability and soil pore structure. *Sci. Total Environ.* **2020**, *703*, 134617. [CrossRef] [PubMed]
51. Gurbuz, I.B.; Ozkan, G. A holistic approach in explaining farmers' intentional behaviour on manure waste utilization. *New Medit* **2021**, *20*, 83–99.
52. Liu, W.; He, X. Effects of maize policy reform in northeast china. *Manag. Theory Stud. Rural. Bus. Infrastruct. Dev.* **2018**, *40*, 348–360. [CrossRef]

International Journal of
Environmental Research and Public Health

Article

The Influence of Soil Erodibility and Saturated Hydraulic Conductivity on Soil Nutrients in the Pingshuo Opencast Coalmine, China

Mingjie Qian [1,*], Wenxiang Zhou [1], Shufei Wang [1], Yuting Li [1] and Yingui Cao [1,2]

[1] School of Land Science and Technology, China University of Geosciences (Beijing), Beijing 100083, China; zhouwenxiang@email.cugb.edu.cn (W.Z.); 3012200010@cugb.edu.cn (S.W.); 2012190028@cugb.edu.cn (Y.L.); caoyingui@cugb.edu.cn (Y.C.)
[2] Key Lab of Land Consolidation, Ministry of Natural Resources of the People's Republic of China, Beijing 100035, China
* Correspondence: qianmingjie@cugb.edu.cn

Abstract: Soil erodibility (K factor) and saturated hydraulic conductivity (Ks) are essential indicators for the estimation of erosion intensity and can potentially influence soil nutrient losses, making them essential parameters for the evaluation of land reclamation quality. In this study, 132 soil samples from 22 soil profiles were collected to measure soil physicochemical properties (e.g., particle size distribution, bulk density and soil nutrient content) and calculate the K factor and Ks of reclaimed soils across the South Dump of the Pingshuo opencast coalmine in the Loess Plateau, China. Geostatistical analysis and the kriging interpolation were employed to quantify the spatial variations in the K factor and Ks in different layers. The results show that the K factor at 0–10 cm is obviously lower than that of other soil layers due to the external input of organic matter, while the Ks tends to decrease along with soil depth. Horizontally, the K factor at 0–10 cm and 50–60 cm shows a decreasing tendency from west to east, while that of other soil layers seems not to show any spatial distribution pattern along latitude or longitude. Meanwhile, the Ks at 0–10 cm presents a striped distribution pattern, while that of other soil layers shows a patchy pattern. On the other hand, the independent-sample t-test and Spearman's correlation analysis were carried out to determine the effects of soil erodibility on total nitrogen (TN), soil organic matter (SOM), available phosphorus (AP) and potassium (AK). Overall, the K factor is negatively correlated with TN (r = −0.362, $p < 0.01$) and SOM contents (r = −0.380, $p < 0.01$), while AP and AK contents are mainly controlled by Ks. This study provides insight on the optimization of reclamation measures and the conservation of soil nutrients in reclaimed land of similar ecosystems.

Keywords: soil erodibility; saturated hydraulic conductivity; soil nutrients; opencast coalmine; Loess Plateau

Citation: Qian, M.; Zhou, W.; Wang, S.; Li, Y.; Cao, Y. The Influence of Soil Erodibility and Saturated Hydraulic Conductivity on Soil Nutrients in the Pingshuo Opencast Coalmine, China. *Int. J. Environ. Res. Public Health* **2022**, *19*, 4762. https://doi.org/10.3390/ijerph19084762

Academic Editors: Bo Sun, Ming Liu, Yan Chen and Paul B. Tchounwou

Received: 14 March 2022
Accepted: 11 April 2022
Published: 14 April 2022

1. Introduction

Opencast coal mining is a preferentially adopted exploitation method due to its cost-effective nature, despite the fact that it causes serious damage to original landforms through large-scale and high-intensity mining activities (e.g., stripping, excavation and transportation) [1]. Numerous studies report that the surface vegetation is thoroughly removed and the subsurface geological structures are largely reshaped during opencast mining, which triggers severe soil erosion in mining areas [2,3]. Moreover, Chinese opencast coalmines are primarily distributed in ecologically fragile regions (e.g., Inner Mongolia, Shanxi Province, Ningxia Province, Shaanxi Province and Gansu Province), where local ecosystems are increasingly threatened by soil and water losses [2]. The practice of land reclamation has been widely employed and developed all over the world as the most effective pathway for the ecological restoration of opencast coalmines; it is mainly characterized by soil reconstruction, landform remodeling and revegetation [4]. The optimal soil physicochemical and biological properties should be constructed for the optimal productivity of reconstructed soils. As such, it is vital for the implementation of land reclamation measures to understand and decipher the spatial distribution patterns of reconstructed soil properties [5]. For instance, reconstructed soils are always deficient in soil nutrients due to intensive mining activities and the removal of surface vegetation, making soil fertility an essential indicator for the quality of land reclamation. Moreover, large amounts of soil nutrients may be lost during the erosion process, highlighting the necessity of exploring the spatial distribution characteristics of soil erosion in the reclaimed land of opencast coalmines.

Characterized by the detachment and transportation of soil substances, soil erosion is regarded as a severe environmental problem that triggers serious soil degradation and threatens the service and function of ecosystems [6–10]. Soil erodibility (K factor) and saturated hydraulic conductivity (Ks) are essential parameters in influencing erosion intensity and have been widely employed to estimate soil loss rates during the erosion process [11–15]. The K factor reflects the inherent soil susceptibility to external erosivity forces and is reported to be mainly regulated by the structural stability of soil aggregates, while Ks can exert a non-negligible influence on the soil erosion process by regulating the infiltration and drainage of surface runoff [16–18]. Previous studies have reported that the K factor and Ks are largely influenced by surface vegetation, SOM and soil physicochemical properties (e.g., soil texture, bulk density and total porosity) [13,19–24]. Comprehensive and localized land management requires accurate spatial distribution information concerning soil properties, which highlights the importance of spatial analyses of soil erosion indicators. For instance, Wang et al. [20] quantified how vegetation restoration strategies influenced the distribution of the K factor in the Loess Plateau, and they concluded that the variations in K factor could be regulated by various factors, such as SOM content, plant litter density, soil bulk density (SBD) and biological crust thickness. Geostatistical methods can be effectively employed to estimate soil properties at unsampled locations based on a set of statistical tools [25]. Over the years, numerous interpolation techniques have been developed to quantitatively estimate the spatial distribution patterns of soil properties, such as pedo-transfer functions, inverse distance weighting (IDW), ordinary kriging (OK) and artificial neural networks [26,27]. The OK has been the most frequently adopted spatial interpolation method for soil properties because OK is considered the best linear unbiased predictor and can be easily conducted with high precision and accuracy [28–30]. Bonilla and Johnson [22] employed the OK interpolation to construct soil erodibility maps, and they found that soil erodibility was positively correlated with silt content but not correlated with SOM content in Central Chile. However, few studies have employed geostatistical methods to investigate the spatial variability of the K factor and Ks in reclaimed lands, despite the fact that their variability can reflect the overall quality of land reclamation efforts and guide the implementation of reclamation measures.

Soil nutrients are essential for the revegetation and ecological restoration of reclaimed lands; the dissolution and adsorption/desorption of these nutrients are strongly regulated by surface runoff, reclamation practices and the physicochemical properties of reclaimed soils [27,31,32]. Guan et al. [32] analyzed the spatiotemporal variations in soil nutrients under different land use types in reclaimed land of the Pingshuo opencast coalmine and concluded that the adopted reclamation measures significantly influence the evolution of reclaimed soil nutrients. Huo et al. [33] explored the interaction mechanisms among precipitation, surface runoff and soil nutrient losses, through which they found that high-intensity rainfall and runoff accelerate soil nutrient losses by increasing soil erosion. However, the effects of soil erodibility on soil nutrients have been rarely investigated, and several related studies concentrated only on the total nutrient content instead of the available nutrients that better reflect soil fertility [34,35]. Although Wang et al. [11] carried out laboratory simulation experiments to determine how soil erodibility affects the loss of available phosphorus and nitrogen, relevant research is still scarce, and research examining the effects of soil erodibility under natural conditions is required. Land reclamation measures can be greatly improved in order to reduce soil nutrient losses by optimizing the intrinsic soil properties if the interaction mechanisms between soil erodibility and soil nutrients can be better understood. Such an advance would provide scientific guidance for global land reclamation practices in similar ecosystems.

Overall, this study aims to: (1) analyze the spatial distribution characteristics of the K factor and Ks in different soil layers based on geostatistical analysis and kriging interpolation; (2) decipher the potential influencing factors for the spatial distribution patterns of the K factor and Ks; (3) explore how the K factor and Ks affect the accumulation and migration of total nitrogen (TN), SOM, available phosphorus (AP) and potassium (AK) in reclaimed soil profiles; (4) provide scientific guidance for land reclamation in opencast coalmine regions of similar ecosystems.

2. Materials and Methods

2.1. Study Area

The largest Chinese opencast coalmine is the Pingshuo opencast coalmine ($112°11'$–$113°30'$ E, $39°23'$–$39°37'$ N). It is located in the east of the Loess Plateau and the north of Shanxi Province, where the soil erosion intensity is one of the highest in the world and the erosion rates vary from 5000 to 10,000 Mg km^{-2} yr^{-1} as a combined result of natural factors (e.g., intensive rainfall, erodible loess soil and a low vegetation coverage rate) and anthropogenic factors (e.g., mining, cultivation and overgrazing) [20,36]. Moreover, the ecosystem of the opencast coalmine was severely damaged due to intensive mining activities that hinder the restoration of vegetation and land reclamation in this region [2,37]. The present research was carried out in the South Dump of the Pingshuo opencast coalmine, which was employed as the open dump of the Antaibao opencast coalmine from 1985 to 1989 [38]. The study area is dominated by a semi-arid continental monsoon climate with a mean annual precipitation of around 450 mm, most (>65%) of which is concentrated from June to September [39]. Based on Chinese Soil Taxonomic Classification, the local soils are mainly composed of chestnut cinnamon soil and chestnut soil, which are of poor soil structure and low organic matter content and have limited ability to resist wind/water erosion [38]. The study area has suffered severe soil erosion due to the concentrated rainfall and loose soil structure; as a result, large amounts of soil nutrients essential to plants have been lost. According to previous studies and local long-term records, the soil erosion rate of the dump areas without reclamation (15,060 t·km^{-2}·a^{-1}) is much higher than that of the original Loess Plateau (10,120 t·km^{-2}·a^{-1}), while the erosion rate of the dump areas that have experienced 10-year revegetation is reduced to 3438 t·km^{-2}·a^{-1} [2].

The South Dump is the earliest reclaimed region of the whole opencast coalmine and has experienced around 31 years of land reclamation, due to which the local ecological environment has been greatly improved [2]. The adopted land reclamation measures mainly comprise revegetation and soil reconstruction in the South Dump. The main plant configurations used for revegetation are *Hippophae rhamnoides*, *Pinus tabulaeformis*, *Ulmus pumila* and *Robinia pseudoacacia*. During the soil reconstruction process, various topsoil substitute materials (e.g., natural soils, coal gangue and rock fragments) were mixed to improve the soil productivity and landscape stability of the reclaimed land [38]. Different from natural soils, reconstructed soils are always characterized by high heterogeneity and bulk density.

2.2. Sampling and Analysis

A total of 132 soil samples were collected from 22 sample sites in May and August 2018 (Figure 1). These sample sites were selected because they represent the overall conditions in the South Dump (e.g., topography, vegetation types and soil conditions). Due to the exposure of hard rocks, only soil samples at 0–60 cm were collected and, based on our field observations, the sampling interval was determined as 10 cm at all sites. The quadrat was 10 m × 10 m on the platform in the first field sampling and 5 m × 5 m in the second field sampling. The size of the quadrat on slopes was adjusted to guarantee that the vertical projection area of the quadrat reached 10 m × 10 m or 5 m × 5 m [38].

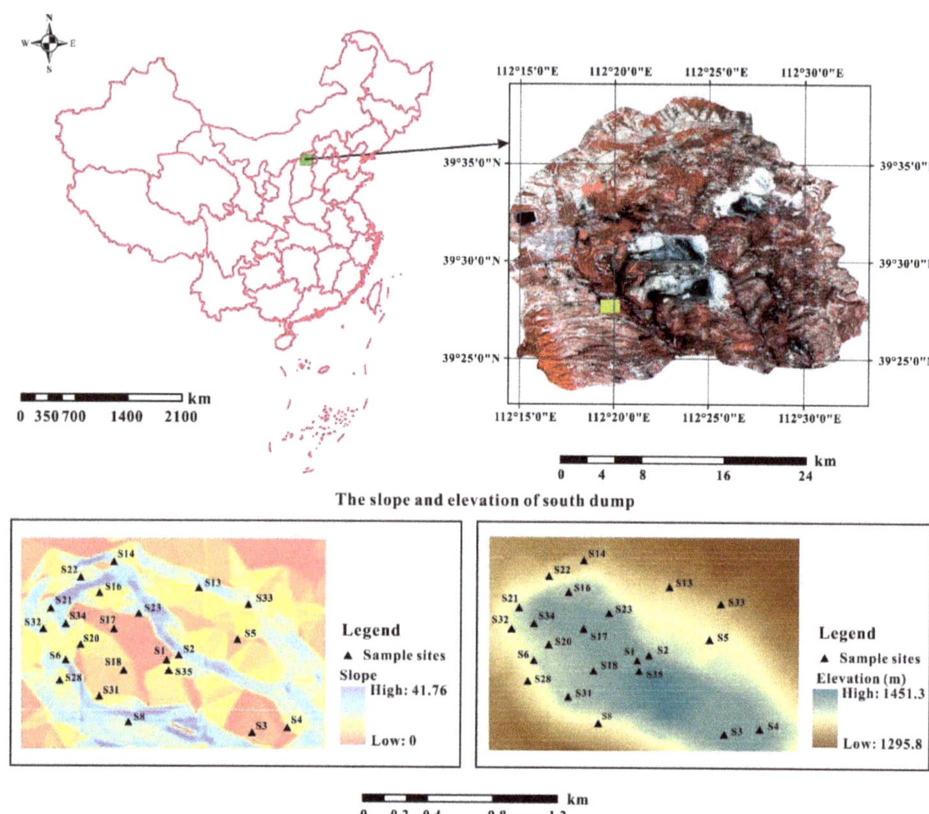

Figure 1. The location, slope, elevation and sample sites of the South Dump in the Pingshuo opencast coalmine, China.

After the collected soil samples were air-dried sufficiently, they were passed through a 2-mm sieve to remove unwanted materials (e.g., stones and plant residues) for subsequent laboratory analyses [40]. The contents of total nitrogen (TN) and soil organic matter (SOM) were measured based on the Kjeldahl and Walkley–Black methods [41,42]. Soil AK was extracted by NH_4OAc solutions and determined by atomic absorption spectroscopic analyses, while AP was extracted by $NaHCO_3$ solutions and measured using Olsen's bicarbonate method [43,44]. After the pretreatment, particle size distribution (PSD) was determined by the laser particle size analyzer "Longbench Mastersizer 2000" (Malvern Instrument, Malvern, England) with a precision of 1%; the detailed experimental procedure we employed has been described by Liu and Han [45]. To obtain the herbaceous biomass, the plants on 1 m × 1 m quadrat at each sample site were randomly collected and weighted after they were sufficiently dried to a constant weight at 65 °C. These soil physicochemical properties were measured in the Beijing Academy of Agriculture and Forestry Sciences. Additionally, the data and experimental procedures for determining soil bulk density have been reported by Huang et al. [37].

2.3. Statistical Analysis

2.3.1. Data Acquisition of the K Factor and Ks

The soil erodibility factor (K factor) has been widely employed to quantify the ability of soils to resist erosion processes and is mainly dominated by soil properties (e.g., organic matter, PSD and permeability) [46]. The K factor can be calculated by various models, such as the USLE model (Universal Soil Loss Equation) and the EPIC model (erosion productivity impact calculator) [47–50]. In this study, the K factor is calculated according to the EPIC model, the formula for which is listed as follows [51]:

$$K = \left(0.3e^{[-0.0256Sa(1-\frac{Si}{100})]} + 0.2\right) \times \left(\frac{Si}{Si+Cl}\right)^{0.3} \times \left[1 - \frac{0.25SOC}{SOC + e^{(3.72-2.95SOC)}}\right] \times \left[1 - \frac{0.7Sa'}{Sa' + e^{(22.9Sa'-5.51)}}\right] \quad (1)$$

where Sa, Si and Cl (%) represent the contents of sand, silt and clay, respectively. Sa' (%) = $1 - Sa/100$, while SOC (%) refers to the content of soil organic carbon that is derived from the measured SOM content (i.e., SOC = SOM × van Bemmelen factor) [52]. The van Bemmelen factor is a widely used conversion factor (0.58) to calculate SOC content based on measured SOM [53]. For the benefit of using an international unit (t ha h (ha MJ mm)$^{-1}$), the calculated K factors are multiplied by a conversion factor (0.1317).

The saturated hydraulic conductivity (Ks) is an essential soil property for the prediction of soil water movement, which always triggers the redistribution of elements or nutrients in soil profiles [23]. The direct measurement of Ks is very time-consuming and expensive due to the strong nonlinearity of the unsaturated hydraulic conductivity function and water retention [54,55]. Moreover, the reconstructed soils in the South Dump are highly compacted and heterogeneous, which means that accurate measurement of Ks is very difficult for reclaimed land, and it is more appropriate to calculate Ks values based on simulation models. Various pedotransfer functions (PTFs) have been developed for the accurate prediction of soil hydraulic parameters, of which the Rosetta model based mainly on artificial neural network analyses (ANN) and the bootstrap resampling method are two of the most frequently employed models [56]. Rosetta1 was first proposed by Schaap et al. [57]; its reliability has been validated by various studies comparing measured and estimated Ks over the past two decades [11,58,59]. Moreover, the Rosetta model has been directly employed to calculate Ks and other soil parameters in the Loess Plateau [60], indicating that the model is qualified for the calculation of Ks in this study. The optimized Rosetta3 model was adopted to calculate the Ks (m/day) values based on the test data of PSD (percentages of clay, silt and sand) and soil bulk density by running the corresponding code in Python.

2.3.2. Geostatistical Analysis

Geostatistical methods have been widely employed to describe the spatial variability of various soil physicochemical properties [61,62]. Previous studies have confirmed that the kriging interpolation method should be preferentially considered to predict the values of soil physicochemical properties at unsampled locations [23,48,63]. In this study, the ordinary kriging (OK) was adopted as the interpolation method to present the spatial distribution characteristics of the K factor and Ks, the optimal input parameters for which were obtained by the application of a semivariogram. The formula of the semivariogram is shown as follows:

$$\gamma(h) = \frac{1}{2N(h)} \sum_{i=1}^{N(h)} [F(x_i) - F(x_i + h)]^2 \quad (2)$$

where $\gamma(h)$ refers to the semivariance for the given lag distance h, $N(h)$ represents the total data pair number between sample sites separated by h, and $F(x_i)$ is the value of the variable F at x_i.

2.3.3. Data Validation

The cross-validation method was carried out to assess the quality of the OK interpolation in this study. This method picks out one datapoint from the sample pool each time and estimates this datapoint based on the model derived from the remaining data [48,63,64]. The errors between the predicted and actual values for all sample sites are calculated to assess the performance of the corresponding model (e.g., linear model, spherical model, exponential model and Gaussian model). The mean error (ME), absolute mean error (AME) and root mean square error (RMSE) are calculated as the evaluation indices for model stability. These indices can be calculated according to the following equations:

$$\text{ME} = \frac{1}{n}\sum_{i=1}^{n}(E_i - A_i) \tag{3}$$

$$\text{AME} = \frac{1}{n}\sum_{i=1}^{n}|E_i - A_i| \tag{4}$$

$$\text{RMSE} = \sqrt{\frac{1}{n}\sum_{i=1}^{n}(E_i - A_i)^2} \tag{5}$$

where n is the number of sample sites, E_i refers to the estimated value and A_i represents the actual observation value. Generally, ME, AME and RMSE values closer to zero denote better performance of the corresponding interpolation model. The geostatistical analysis and kriging interpolation in our study were carried out in GS$^+$ (Gamma Design Software, version 9.0).

3. Results and Discussion

3.1. The Statistical Characteristics of the K Factor, Ks and Herbaceous Biomass

3.1.1. Descriptive Statistics at Different Sample Sites

The average K factor at 0–10 cm and 10–20 cm, the surface herbaceous biomass and the vertical distribution characteristics of Ks at different sample sites are shown in Figure 2. The average K factor in topsoil (0–20 cm) ranges from 0.0252 to 0.0519 t ha h (ha MJ mm)$^{-1}$ with an average value of 0.0403 t ha h (ha MJ mm)$^{-1}$. The K factor is unevenly distributed among different sample sites, and the K factor values at sites S6, S20, S21 and S32 are much higher than those at S1, S2, S8 and S28 (Figure 2). As shown in Figure 2, the Ks values vary dramatically (0.006 m/day–2.278 m/day) and tend to decrease along with soil depth in most soil profiles. Overall, the Ks of topsoil is obviously higher than that of other soil layers at most sample sites, which results from the fact that deep soils are more compacted relative to topsoil. However, a vertical distribution tendency is not obvious at other sample sites (e.g., S16, S18, S20, S22 and S31), and the average Ks values are obviously different at different sample sites (Figure 2), which can be attributed to the high heterogeneity of reconstructed soils. Meanwhile, the herbaceous biomass varies dramatically (10.99–357.72 g m^{-2}) at different sample sites and presents somewhat opposite distribution characteristics relative to the K factor in topsoil (Figure 2). For instance, a relatively high K factor always corresponds to an extremely low level of herbaceous biomass (e.g., sites S6, S21 and S32). On the one hand, higher erodibility triggers massive soil nutrient losses and reduces biomass in soils; on the other hand, SOM can act as an essential bonding agent that benefits the formation of soil aggregates and reduces soil erodibility [2,65,66]. Overall, herbaceous biomass is an important source of external organic matter input for local soils, and higher soil erodibility is adverse for the accumulation of SOM. A greater slope means higher runoff velocity and intensity, which accelerates the loss of fine soil particles and further decreases the K factor and increases the Ks in topsoil. To identify the potential influence of slope, the independent-sample t-test was employed to compare the difference between the K factor and Ks in the topsoil on platform and slope. The results show that there is no significant difference in the average K factor ($p = 0.920$) or Ks ($p = 0.805$) on the platform ($K_{mean} = 0.0386$ t ha h (ha MJ mm)$^{-1}$, $Ks_{mean} = 0.716$ m/day; N = 14) and the slope ($K_{mean} = 0.0382$ t ha h (ha MJ mm)$^{-1}$, $Ks_{mean} = 0.642$ m/day; N = 8), indicating that the influence of slope on the K factor and Ks can be ignored in the study area.

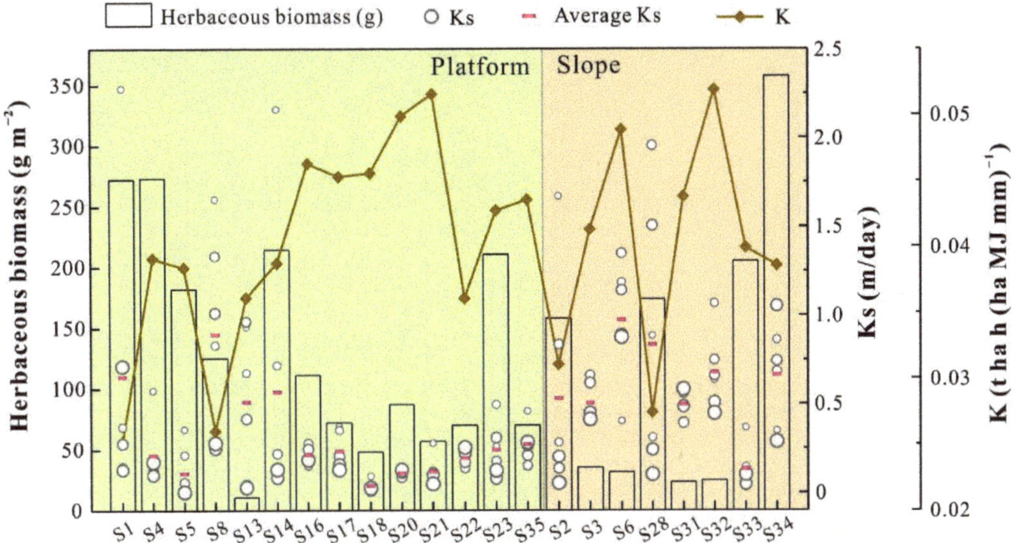

Figure 2. The distribution of K factor, Ks and herbaceous biomass at different sample sites. The K factor refers to the average value at 0–10 cm and 10–20 cm. The vertical distribution characteristics of Ks are presented by the bubble chart, where the larger bubbles represent deeper soil samples.

3.1.2. Descriptive Statistics in Different Soil Layers

The descriptive statistics for the K factor and Ks in different soil layers are displayed in Table 1. The results of the Kolmogorov–Smirnov test indicate that both the K factor and Ks conform to a normal distribution (Table 1). The average K factor does not show a certain distribution pattern among different soil layers, although the average K factor at 0–10 cm is lower than that of other soil layers. Soil erodibility (K factor) is an essential indicator of the vulnerability of soil to erosion processes, and a greater K factor means a higher possibility of soil erosion [21,67,68]. The low K factor value of topsoil is most likely attributable to the fact that topsoil receives more external organic matter input, which can enhance water-stable aggregate contents [69,70]. Meanwhile, the average Ks value tends to decrease along with soil depth, which results from the fact that deep soil is more compacted and mainly characterized by low soil porosity and high bulk density. The coefficient of variation (CV) is frequently employed to describe the dispersion degree of variables, which categorizes variable types into weak variability (CV < 10%), moderate variability (10% < CV < 100%) and strong variability (CV > 100%) [32]. The evidence from CV indicates that the K factor presents moderate variability (16.42% < CV < 24.21%) while the Ks shows moderate (0–10 cm, 10–20 cm and 50–60 cm, 76.90 < CV < 94.75) to strong (20–30 cm, 30–40 cm and 40–50 cm, 104.18 < CV < 106.35) variability.

Table 1. Statistical characteristics of the K factor and Ks in different soil layers.

Soil Depth (cm)	Mean	Median	Maximum	Minimum	CV (%)	K-S
K (t ha h (ha MJ mm)$^{-1}$)						
0–10	0.0384	0.0368	0.0543	0.0203	24.20	0.95
10–20	0.0423	0.0441	0.0540	0.0279	17.65	0.81
20–30	0.0421	0.0452	0.0571	0.0228	24.21	0.66
30–40	0.0412	0.0409	0.0554	0.0255	21.68	0.98
40–50	0.0445	0.0453	0.0566	0.0270	16.42	0.60
50–60	0.0423	0.0417	0.0531	0.0213	18.46	0.96

Table 1. *Cont.*

Soil Depth (cm)	Mean	Median	Maximum	Minimum	CV (%)	K-S
Ks (m/day)						
0–10	0.689	0.392	2.278	0.094	94.24	0.07
10–20	0.430	0.331	1.184	0.046	76.90	0.62
20–30	0.496	0.204	1.955	0.032	105.13	0.12
30–40	0.387	0.227	1.506	0.014	104.18	0.27
40–50	0.256	0.169	1.051	0.010	106.35	0.12
50–60	0.245	0.154	0.876	0.006	94.75	0.34

3.2. Geostatistical Analyses of K Factor and Ks

3.2.1. Semivariogram Analyses

Semivariograms are frequently applied to investigate the spatial variability of soil physicochemical properties [64,71], the parameters of which are shown in Table 2 in this study. The Gaussian model is the best-fit model for the K factor in all soil layers except for 50–60 cm, for which the best-fit model is the linear model (Table 2). Similarly, the best-fit model for Ks is the Gaussian model at 0–10 cm, 30–40 cm and 40–50 cm, while it is the linear model at 10–20 cm and 50–60 cm and the exponential model at 20–30 cm. The determination coefficient (R^2) ranges from 0.61 to 0.94, indicating that these optimal models are qualified for accurately describing the spatial variability of the K factor and Ks in different soil layers. The nugget (C_0) and sill ($C_0 + C$) of these models were obtained to describe the spatial structure of the K factor and Ks [64]. The ratio of $C_0/(C_0 + C)$ is an essential indicator for the spatial dependence degree of soil properties, which categorizes the degree into strong spatial dependence (<25%), moderate spatial dependence (25–75%) and weak spatial dependence (>75%) [64,71]. According to the classification criterion, the K factors at 10–20 cm, 20–30 cm, 30–40 cm and 40–50 cm have strong spatial dependence, while the K factors at 0–10 cm and 50–60 cm are characterized by moderate to weak spatial dependence, indicating that soil erodibility in the middle soil layers is mainly regulated by intrinsic factors, while extrinsic factors exert a non-negligible influence on soil erodibility in the surface (0–10 cm) and bottom (50–60 cm) soil layers. Meanwhile, it can also be concluded that the levels of saturated hydraulic conductivity at 10–20 cm and 50–60 cm are mostly regulated by extrinsic factors, and the Ks levels in the remaining soil layers are mainly determined by intrinsic factors. The evaluation indices of model stability (ME, MAE and RMSE) are shown in Table 3. The ME, MAE and RMSE values for the K factor and Ks in different soil layers are all close to zero and remain stable (K factor: $-2.90 \times 10^{-4} <$ ME $< 6.70 \times 10^{-4}$, $5.86 \times 10^{-3} <$ MAE $< 9.45 \times 10^{-3}$, $6.75 \times 10^{-3} <$ RMSE $< 1.11 \times 10^{-2}$; Ks: $-9.81 \times 10^{-3} <$ ME $< 1.24 \times 10^{-2}$, $1.81 \times 10^{-1} <$ MAE $< 5.51 \times 10^{-1}$, $2.34 \times 10^{-1} <$ RMSE $< 7.43 \times 10^{-1}$), which indicates that the kriging interpolation results are credible for the K factor and Ks in different soil layers.

Table 2. Parameters for the semivariogram analyses of the K factor and Ks in different soil layers.

Soil Depth (cm)	Optimal Model	$C_0/(C + C_0)$ (%) [a]	R^2 [b]	RSS [c]	A_0/m [d]
K (t ha h (ha MJ mm)$^{-1}$)					
0–10	Gaussian	36.78	0.80	9.07×10^{-10}	1143.15
10–20	Gaussian	0.17	0.92	3.47×10^{-10}	211.31
20–30	Gaussian	0.08	0.85	7.63×10^{-10}	145.49
30–40	Gaussian	0.12	0.64	1.95×10^{-9}	148.96
40–50	Gaussian	0.17	0.65	5.16×10^{-10}	143.76
50–60	Linear	100	0.66	1.96×10^{-10}	649.43

Table 2. *Cont.*

Soil Depth (cm)	Optimal Model	$C_0/(C + C_0)$ (%) [a]	R^2 [b]	RSS [c]	A_0/m [d]
Ks (m/day)					
0–10	Gaussian	13.82	0.85	1.40×10^{-2}	2793.80
10–20	Linear	100	0.80	3.60×10^{-3}	651.73
20–30	Exponential	17.12	0.94	2.41×10^{-3}	795.00
30–40	Gaussian	0.06	0.83	4.14×10^{-3}	226.90
40–50	Gaussian	0.13	0.61	7.02×10^{-4}	164.54
50–60	Linear	100	0.81	1.24×10^{-3}	631.17

[a] Ratio of spatial heterogeneity, where C_0 is Nugget and $C + C_0$ is Sill; [b] Determination coefficient; [c] Residual sum of squares; [d] Range.

Table 3. Stability evaluation for different models of the ordinary kriging interpolation.

Soil Depth (cm)	Optimal Model	ME	AME	RMSE
K (t ha h (ha MJ mm)$^{-1}$)				
0–10	Gaussian	4.86×10^{-4}	7.15×10^{-3}	8.94×10^{-3}
10–20	Gaussian	-2.90×10^{-4}	6.37×10^{-3}	7.44×10^{-3}
20–30	Gaussian	-1.67×10^{-4}	9.45×10^{-3}	1.11×10^{-2}
30–40	Gaussian	-1.36×10^{-4}	7.96×10^{-3}	9.30×10^{-3}
40–50	Gaussian	-1.18×10^{-4}	5.86×10^{-3}	7.57×10^{-3}
50–60	Linear	6.70×10^{-4}	5.86×10^{-3}	6.75×10^{-3}
Ks (m/day)				
0–10	Gaussian	-3.45×10^{-2}	5.51×10^{-1}	7.43×10^{-1}
10–20	Linear	3.03×10^{-3}	2.82×10^{-1}	3.35×10^{-1}
20–30	Exponential	-9.81×10^{-3}	4.08×10^{-1}	4.95×10^{-1}
30–40	Gaussian	3.64×10^{-5}	3.11×10^{-1}	3.95×10^{-1}
40–50	Gaussian	1.24×10^{-2}	2.05×10^{-1}	2.72×10^{-1}
50–60	Linear	1.08×10^{-2}	1.81×10^{-1}	2.34×10^{-1}

3.2.2. Spatial Distribution Maps of K factor and Ks

The spatial interpolation maps of the K factor and Ks in different soil layers of the South Dump are shown in Figures 3 and 4. The K factor in topsoil (0–10 cm) is more variable than that of other soil layers, which might result from the uneven input of external organic matter due to the variations in surface vegetation. For instance, the areas dominated by woody plants (e.g., *Robinia pseudoacacia*, *Pinus tabuliformis* and *Caragana korshinsk*) are characterized by lower K factor values compared with the areas dominated by herbaceous plants (e.g., *Incarvillea sinensis*, *Elymus dahuricus* and *Saussurea japonica*), indicating that the plant configuration modes can exert an essential influence on soil erodibility. The K factor generally presents a decreasing tendency from west to east at 0–10 cm and 50–60 cm, while the K factor seems not to show any spatial distribution pattern along latitude or longitude for other soil layers. Meanwhile, many high-K factor or low-K factor "points" are sporadically distributed across the study area, which results from the strong heterogeneity of the reconstructed soils triggered by the irregular mixing of various substances (e.g., rock fragments, soil particles and plant residues) [72]. Vertically, the K factor in topsoil (0–10 cm) is lower than that of other soil layers in most areas of the South Dump, which might be attributed to the relatively higher SOM content in topsoil. Previous studies have confirmed that the existence of SOM benefits the formation of anti-erosion soil aggregates and the increase in soil infiltration, which can reduce soil erosion and surface runoff [2,73,74]. Saturated hydraulic conductivity is regarded as one of the most important indicators for the mobility of solute in soils, and understanding the spatial distribution characteristics of hydraulic properties is necessary for land reclamation and revegetation [75]. Overall, the Ks presents a striped distribution pattern at 0–10 cm while showing a patchy pattern in other soil layers. Moreover, the Ks in topsoil is obviously higher than that in deeper soil layers. Interestingly, the spatial distribution patterns of Ks are opposite to those of soil bulk density (SBD) reported by Huang et al. [37]. Meanwhile, the K factor also shows opposite spatial distribution characteristics relative to the reported SBD at 50–60 cm. Generally, high SBD can lower soil permeability and increase the ability of soils to resist erosion processes, which might account for the opposite spatial distributions between K factor, Ks and SBD [23]. Moreover, the effects of soil properties on K factor and Ks can be further supported by Spearman's correlation analyses. The results show that clay contents are positively correlated with the K factor ($r = 0.332$, $p < 0.01$) and negatively correlated with Ks ($r = -0.615$, $p < 0.01$), indicating that fine soil particles can improve soil erodibility and reduce saturated hydraulic conductivity. The spatial distribution patterns of the K factor and Ks in the South Dump highlight the significance of plant configuration modes and reconstructed soil properties (e.g., SBD, clay and SOM contents) on regulating soil erodibility, which can provide insight for vegetation configuration and soil management in similar ecosystems.

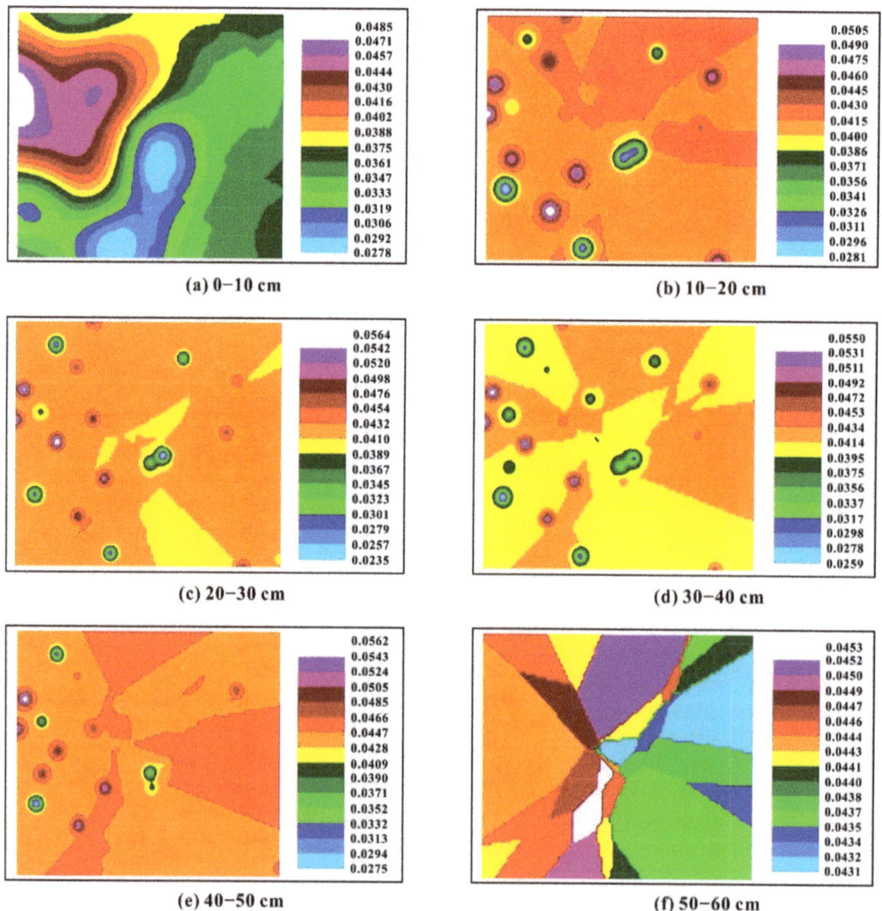

Figure 3. Spatial distribution maps of the K factor (t ha h (ha MJ mm)$^{-1}$) in the South Dump based on the ordinary kriging interpolation. (**a**–**f**) represent the soil layers at different depths (0–10 cm, 10–20 cm, 20–30 cm, 30–40 cm, 40–50 cm and 50–60 cm).

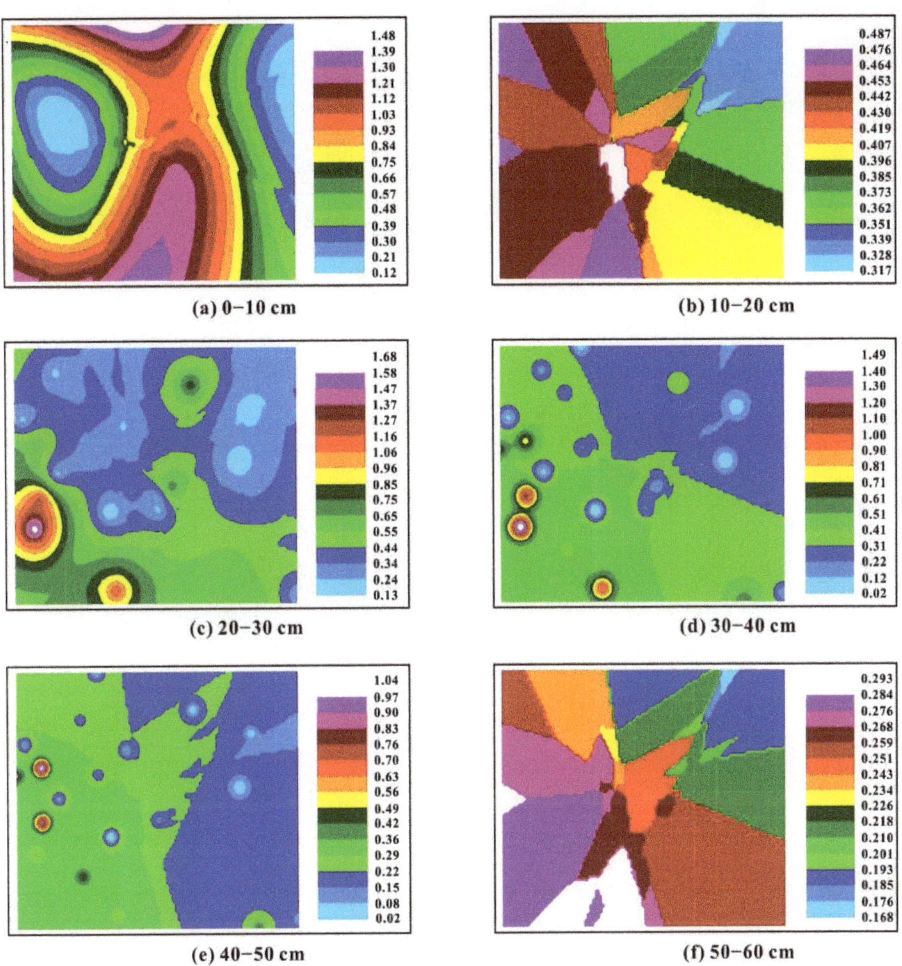

Figure 4. Spatial distribution maps of the Ks (m/day) in the South Dump based on the ordinary kriging interpolation. (**a**–**f**) represent the soil layers at different depths (0–10 cm, 10–20 cm, 20–30 cm, 30–40 cm, 40–50 cm and 50–60 cm).

3.3. Effects of Soil Erodibility and Saturated Hydraulic Conductivity on Soil Nutrients

To explore the effects of soil erodibility and saturated hydraulic conductivity on the distribution of soil nutrients (i.e., TN, SOM, AP and AK), soil samples were categorized into two groups (i.e., group 1 and group 2), and the independent-sample *t*-test was employed to compare the difference in soil nutrient contents between different groups. Group 1 comprises the samples whose K factor values (or Ks values) are lower than the average K factor (or average Ks), while group 2 is composed of the remaining soil samples (Figures 5 and 6). The average TN and SOM contents in group 1 are obviously higher than those in group 2 in all soil layers, and the difference is significant ($p < 0.01$) at 0–10 cm (Figure 5a,b). Meanwhile, the average AP and AK contents in group 1 are obviously lower than those in group 2 in all soil layers (except for 0–10 cm) (Figure 5c,d). For the Ks classification scheme, the average AP content in group 1 is higher than that in group 2 at 0–30 cm, and the average AK content in group 1 is higher than that in group 2 at 10–60 cm (Figure 6c,d).

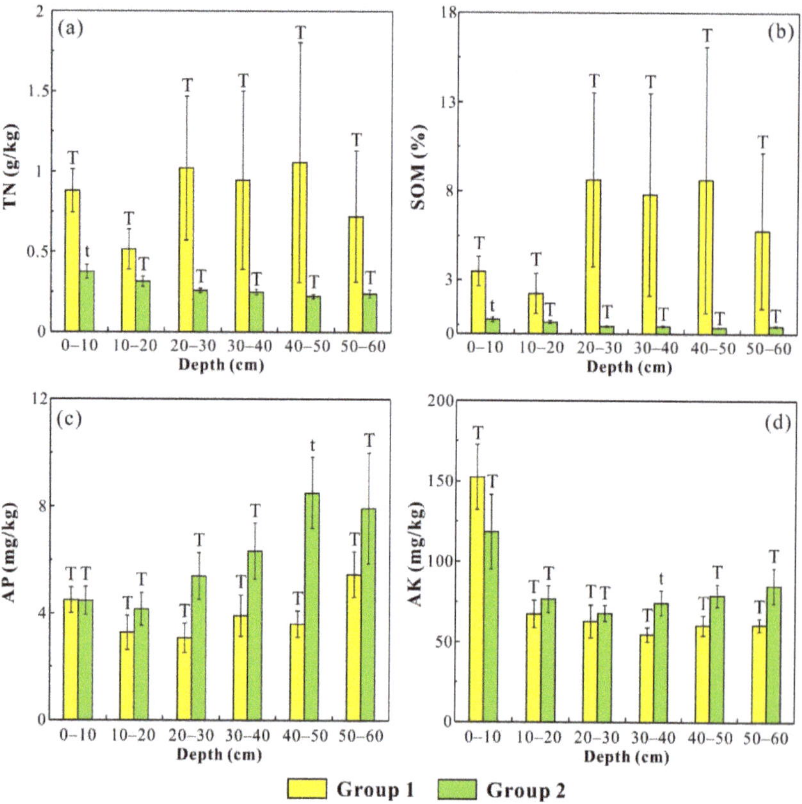

Figure 5. The average contents of TN (**a**), SOM (**b**), AP (**c**) and AK (**d**) in group 1 and group 2 (the classification of the two groups is based on the K factor values). The error bar represents the standard error and different forms of T denote significant difference.

Generally, soil nutrients are always dissolved in soil solution or bound by soil particles, due to which they can be transported in dissolved and particulate form during soil erosion process [11,76]. Theoretically, dissolved fractions of soil nutrients can move both horizontally and vertically, and the vertical mobility of dissolved soil nutrients is mainly regulated by the leaching process that is largely determined by saturated hydraulic conductivity. Although available soil nutrients only account for a small proportion of the total nutrient content, their mobility and accumulation in soil profiles are of great significance because they are mostly dissolved in soil solution and can be easily utilized by plant roots [11]. By comprehensively analyzing the distribution patterns of soil nutrients among the two groups, it is reasonable to conclude that lower soil erodibility benefits the accumulation of TN and SOM in reconstructed soils, while AP and AK contents are largely regulated by saturated hydraulic conductivity. This conclusion can be further supported by Spearman's correlation analysis, which denotes that the K factor is negatively correlated with TN ($r = -0.362$, $p < 0.01$) and SOM ($r = -0.380$, $p < 0.01$) while Ks is negatively associated with AP ($r = -0.229$, $p < 0.01$) and AK ($r = -0.180$, $p < 0.05$). The statistical results reflect that the mobility of TN and SOM is more controlled by surface lateral runoff during soil erosion processes, while the migration of AP and AK is mainly determined by the vertical leaching process in soil profiles.

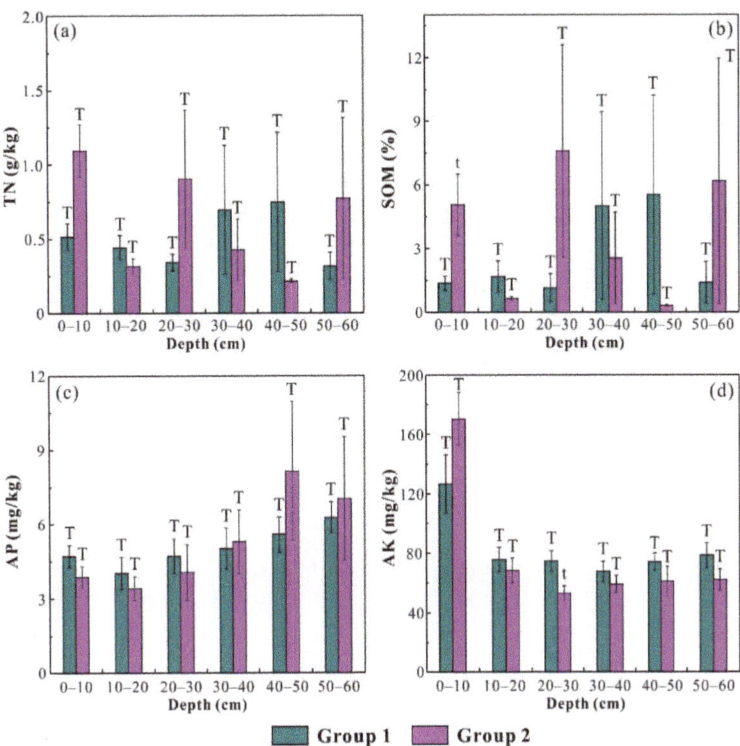

Figure 6. The average contents of TN (**a**), SOM (**b**), AP (**c**) and AK (**d**) in group 1 and group 2 (the classification of the two groups is based on the Ks values). The error bar represents the standard error and different forms of T denote significant difference.

3.4. Implications for Soil Management and Land Reclamation in Opencast Coalmine Areas

The kriging interpolation maps of the K factor and Ks can provide helpful guidance for the management practices of reconstructed soils and the spatial planning of land use during land reclamation in similar ecosystems. In the South Dump, the vegetation configuration modes should be optimized to improve the ability of reconstructed soils to resist erosion processes. Anti-erosion vegetation (e.g., black locust, shrub sophora and korshinsk peashrub) should be preferentially planted, and more exogenous organic materials (e.g., biochar and local organic fertilizer) should be exerted to reduce soil erodibility in high-K factor regions of the study area [23,31]. Moreover, it is quite important to sustain the soil fertility of reconstructed soils because the ability to maintain fertility is an essential indicator in evaluating the quality of land reclamation. The results of this study indicate that TN and SOM contents are mainly affected by the K factor, while AP and AK contents are more regulated by Ks. The influence of soil erodibility on TN and SOM can be mainly attributed to the soil erosion process, while there are many pathways through which saturated hydraulic conductivity affects the mobility of AP and AK. High Ks causes rapid water drainage and infiltration, which amplifies the leaching process of elements in soil profiles. Meanwhile, it also determines the abundance of air-filled porosity, which further exerts an important influence on the uptake of soil nutrients by plants [77]. This study demonstrates that different reclamation measures should be adopted to obtain good soil conditions with appropriate soil erodibility and saturated hydraulic conductivity for the conservation of different soil nutrients. This study can also benefit the improvement of the Chinese "completion standards on land reclamation quality" that have not yet considered soil erodibility or saturated hydraulic conductivity as indicators for land reclamation quality.

4. Conclusions

The effects of soil erodibility and saturated hydraulic conductivity on TN, SOM, AP and AK were determined, and spatial distribution maps of the K factor and Ks in different soil layers were presented based on geostatistical analyses. The main conclusions can be summarized as follows:

(1) The K factor in topsoil is obviously lower than that in other soil layers due to the external input of organic matter, while Ks generally tends to decrease along with soil depth. The K factor in middle soil layers has a strong spatial dependence and is mainly controlled by intrinsic factors, while Ks levels at 10–20 cm and 50–60 cm are mostly regulated by extrinsic factors based on the $C_0/(C_0 + C)$ values.

(2) The statistical results indicate that K factor is negatively correlated with SOM (r = −0.380, $p < 0.01$) and TN (r = −0.362, $p < 0.01$), while Ks is negatively associated with AP (r = −0.229, $p < 0.01$) and AK (r = −0.180, $p < 0.05$). The phenomenon should be attributed to the fact that TN and SOM mainly exist in particulate form, while AP and AK are always dissolved in soil solutions, and their mobility is largely influenced by the leaching process in soil profiles. The intriguing results indicate that the conservation of different soil nutrients requires different reclamation measures.

(3) Based on the spatial information for the K factor and Ks, anti-erosion vegetation should be preferentially planted in high-K factor regions to reduce the soil erodibility in the South Dump. The results of this study can guide the spatial planning of land use and the implementation of land reclamation measures in the reclaimed land of similar ecosystems.

Author Contributions: Conceptualization, M.Q., Y.C. and W.Z.; methodology, M.Q., Y.C. and W.Z.; software, W.Z. and Y.C.; validation, W.Z., M.Q. and Y.C.; formal analysis, W.Z.; investigation, S.W., Y.L., M.Q., Y.C. and W.Z.; resources, M.Q. and Y.C.; data curation, W.Z., S.W., Y.L. and M.Q.; writing—original draft preparation, M.Q. and W.Z.; writing—review and editing, M.Q., W.Z. and Y.C.; visualization, W.Z., M.Q., S.W., Y.L. and Y.C.; supervision, M.Q. and Y.C.; project administration, M.Q. and Y.C.; funding acquisition, Y.C. All authors have read and agreed to the published version of the manuscript.

Funding: The research and APC were funded by the National Natural Science Foundation of China (U1810107 and 41701607) and the Fundamental Research Funds for the Central Universities of China (2-9-2018-025 and 2-9-2019-307).

Data Availability Statement: Not applicable.

Acknowledgments: The authors gratefully acknowledge Gubai Luo and Xinyu Kuang for their assistance with field sampling and laboratory experiments.

Conflicts of Interest: The authors declare no conflict of interest.

References

1. Guan, Y.; Zhou, W.; Bai, Z.; Cao, Y.; Wang, J. Delimitation of supervision zones based on the soil property characteristics in a reclaimed opencast coal mine dump on the Loess Plateau, China. *Sci. Total Environ.* **2021**, *772*, 145006. [CrossRef] [PubMed]
2. Zhang, L.; Wang, J.; Bai, Z.; Lv, C. Effects of vegetation on runoff and soil erosion on reclaimed land in an opencast coal-mine dump in a loess area. *Catena* **2015**, *128*, 44–53. [CrossRef]
3. Zhao, Z.Q.; Shahrour, I.; Bai, Z.K.; Fan, W.X.; Feng, L.R.; Li, H.F. Soils development in opencast coal mine spoils reclaimed for 1–13 years in the West-Northern Loess Plateau of China. *Eur. J. Soil Biol.* **2013**, *55*, 40–46. [CrossRef]
4. Hu, Z.Q.; Atkinson, K. Principle and method of soil reconstruction for coal mine land reclamation. *J. China Coal Soc.* **1998**, *6*, 761–768.
5. Wang, J.M.; Yang, R.X.; Feng, Y. Spatial variability of reconstructed soil properties and the optimization of sampling number for reclaimed land monitoring in an opencast coal mine. *Arab. J. Geosci.* **2017**, *10*, 1–13. [CrossRef]
6. Bennett, H.; Chapline, W. *Soil Erosion: A National Menace*; United States Department of Agriculture (USDA): Washington, DC, USA, 1928.

7. Morgan, R.P.C. *Soil Erosion and Conservation*; John Wiley & Sons: Hoboken, NJ, USA, 2009.
8. Cerdà, A.; Lucas Borja, M.E.; Úbeda, X.; Martínez-Murillo, J.F.; Keesstra, S. *Pinus halepensis* M. versus *Quercus ilex* subsp. *Rotundifolia* L. runoff and soil erosion at pedon scale under natural rainfall in Eastern Spain three decades after a forest fire. *For. Ecol. Manag.* **2017**, *400*, 447–456. [CrossRef]
9. Poesen, J. Soil erosion in the Anthropocene: Research needs. *Earth Surf. Process. Landf.* **2018**, *43*, 64–84. [CrossRef]
10. Borrelli, P.; Alewell, C.; Alvarez, P.; Anache, J.A.A.; Baartman, J.; Ballabio, C.; Bezak, N.; Biddoccu, M.; Cerda, A.; Chalise, D.; et al. Soil erosion modelling: A global review and statistical analysis. *Sci. Total Environ.* **2021**, *780*, 146494. [CrossRef]
11. Wang, G.; Wu, B.; Zhang, L.; Jiang, H.; Xu, Z. Role of soil erodibility in affecting available nitrogen and phosphorus losses under simulated rainfall. *J. Hydrol.* **2014**, *514*, 180–191. [CrossRef]
12. Shipitalo, M.J.; Owens, L.B.; Bonta, J.V.; Edwards, W.M. Effect of no-till and extended rotation on nutrient losses in surface runoff. *Soil Sci. Soc. Am. J.* **2013**, *77*, 1329–1337. [CrossRef]
13. Yusof, M.F.; Azamathulla, H.M.; Abdullah, R. Prediction of soil erodibility factor for Peninsular Malaysia soil series using ANN. *Neural Comput. Appl.* **2014**, *24*, 383–389. [CrossRef]
14. Iaaich, H.; Moussadek, R.; Baghdad, B.; Mrabet, R.; Douaik, A.; Abdelkrim, D.; Bouabdli, A. Soil erodibility mapping using three approaches in the Tangiers province—Northern Morocco. *Int. Soil Water Conserv. Res.* **2016**, *4*, 159–167. [CrossRef]
15. Mallants, D.; Mohanty, B.P.; Vervoort, A.; Feyen, J. Spatial analysis of saturated hydraulic conductivity in a soil with macropores. *Soil Technol.* **1997**, *10*, 115–131. [CrossRef]
16. Tanner, S.; Katra, I.; Argaman, E.; Ben-Hur, M. Erodibility of waste (Loess) soils from construction sites underwater and wind erosional forces. *Sci. Total Environ.* **2018**, *616*, 1524–1532. [CrossRef] [PubMed]
17. Abel, S.; Peters, A.; Trinks, S.; Schonsky, H.; Facklam, M.; Wessolek, G. Impact of biochar and hydrochar addition on water retention and water repellency of sandy soil. *Geoderma* **2013**, *202*, 183–191. [CrossRef]
18. Bouma, J. Field measurement of soil hydraulic properties characterizing water movement through swelling clay soils. *J. Hydrol.* **1980**, *45*, 149–158. [CrossRef]
19. Zhu, B.B.; Li, Z.B.; Li, P.; Liu, G.B.; Xue, S. Soil erodibility, microbial biomass, and physical-chemical property changes during long-term natural vegetation restoration: A case study in the Loess Plateau, China. *Ecol. Res.* **2010**, *25*, 531–541. [CrossRef]
20. Wang, H.; Zhang, G.-H.; Li, N.-N.; Zhang, B.-J.; Yang, H.-Y. Soil erodibility as impacted by vegetation restoration strategies on the Loess Plateau of China. *Earth Surf. Process. Landf.* **2019**, *44*, 796–807. [CrossRef]
21. Deng, Y.; Shen, X.; Xia, D.; Cai, C.; Ding, S.; Wang, T. Soil erodibility and physicochemical properties of collapsing gully alluvial fans in southern China. *Pedosphere* **2019**, *29*, 102–113. [CrossRef]
22. Bonilla, C.A.; Johnson, O.I. Soil erodibility mapping and its correlation with soil properties in central Chile. *Geoderma* **2012**, *189*, 116–123. [CrossRef]
23. Usowicz, B.; Lipiec, J. Spatial variability of saturated hydraulic conductivity and its links with other soil properties at the regional scale. *Sci. Rep.* **2021**, *11*, 8293. [CrossRef]
24. Lim, H.; Yang, H.; Chun, K.W.; Choi, H.T. Development of pedo-transfer functions for the saturated hydraulic conductivity of forest soil in South Korea considering forest stand and site characteristics. *Water* **2020**, *12*, 2217. [CrossRef]
25. Nourzadeh, M.; Mahdian, M.H.; Malakouti, M.J.; Khavazi, K. Investigation and prediction spatial variability in chemical properties of agricultural soil using geostatistics. *Arch. Agron. Soil Sci.* **2012**, *58*, 461–475. [CrossRef]
26. Veronesi, F.; Corstanje, R.; Mayr, T. Landscape scale estimation of soil carbon stock using 3D modelling. *Sci. Total Environ.* **2014**, *487*, 578–586. [CrossRef] [PubMed]
27. Guan, F.; Xia, M.; Tang, X.; Fan, S. Spatial variability of soil nitrogen, phosphorus and potassium contents in Moso bamboo forests in Yong'an City, China. *Catena* **2017**, *150*, 161–172. [CrossRef]
28. Candela, L.; Olea, R.A.; Custodio, E. Lognormal kriging for the assessment of reliability in groundwater quality-control observation networks. *J. Hydrol.* **1988**, *103*, 67–84. [CrossRef]
29. Kravchenko, A.N. Influence of spatial structure on accuracy of interpolation methods. *Soil Sci. Soc. Am. J.* **2003**, *67*, 1564–1571. [CrossRef]
30. Matheron, G. Principles of geostatistics. *Econ. Geol.* **1963**, *58*, 1246–1266. [CrossRef]
31. Zhang, Q.; Liu, D.; Cheng, S.; Huang, X. Combined effects of runoff and soil erodibility on available nitrogen losses from sloping farmland affected by agricultural practices. *Agric. Water Manag.* **2016**, *176*, 1–8. [CrossRef]
32. Guan, Y.; Zhou, W.; Bai, Z.; Cao, Y.; Huang, Y.; Huang, H. Soil nutrient variations among different land use types after reclamation in the Pingshuo opencast coal mine on the Loess Plateau, China. *Catena* **2020**, *188*, 104427. [CrossRef]
33. Huo, J.Y.; Liu, C.J.; Yu, X.X.; Chen, L.H.; Zheng, W.G.; Yang, Y.H.; Yin, C.W. Direct and indirect effects of rainfall and vegetation coverage on runoff, soil loss, and nutrient loss in a semi-humid climate. *Hydrol. Process.* **2021**, *35*, e13985. [CrossRef]
34. Gilley, J.E.; Vogel, J.R.; Eigenberg, R.A.; Marx, D.B.; Woodbury, B.L. Nutrient losses in runoff from feedlot surfaces as affected by unconsolidated surface materials. *J. Soil Water Conserv.* **2012**, *67*, 211–217. [CrossRef]
35. Zhang, G.H.; Liu, G.B.; Wang, G.L.; Wang, Y.X. Effects of vegetation cover and rainfall intensity on sediment-associated nitrogen and phosphorus losses and particle size composition on the Loess Plateau. *J. Soil Water Conserv.* **2011**, *66*, 192–200. [CrossRef]
36. Wang, B.; Zhang, G.H.; Shi, Y.Y.; Li, Z.W.; Shan, Z.J. Effects of near soil surface characteristics on the soil detachment process in a chronological series of vegetation restoration. *Soil Sci. Soc. Am. J.* **2015**, *79*, 1213–1222. [CrossRef]

37. Huang, Y.H.; Cao, Y.G.; Pietrzykowski, M.; Zhou, W.; Bai, Z.K. Spatial distribution characteristics of reconstructed soil bulk density of opencast coal-mine in the loess area of China. *Catena* **2021**, *199*, 10. [CrossRef]

38. Cao, Y.; Wang, J.; Bai, Z.; Zhou, W.; Zhao, Z.; Ding, X.; Li, Y. Differentiation and mechanisms on physical properties of reconstructed soils on open-cast mine dump of loess area. *Environ. Earth Sci.* **2015**, *74*, 6367–6380. [CrossRef]

39. Zhou, W.; Yang, K.; Bai, Z.; Cheng, H.; Liu, F. The development of topsoil properties under different reclaimed land uses in the Pingshuo opencast coalmine of Loess Plateau of China. *Ecol. Eng.* **2017**, *100*, 237–245. [CrossRef]

40. Han, G.L.; Tang, Y.; Liu, M.; van Zwieten, L.; Yang, X.M.; Yu, C.X.; Wang, H.L.; Song, Z.L. Carbon-nitrogen isotope coupling of soil organic matter in a karst region under land use change, Southwest China. *Agric. Ecosyst. Environ.* **2020**, *301*, 107027. [CrossRef]

41. Parkinson, J.A.; Allen, S.E. A wet oxidation procedure suitable for the determination of nitrogen and mineral nutrients in biological material. *Commun. Soil Sci. Plant Anal.* **1975**, *6*, 1–11. [CrossRef]

42. Nelson, D.W.; Sommers, L.; Page, A.L.; Miller, R.H.; Keeney, D.R. Total carbon, organic carbon, and organic matter. *Methods Soil Anal.* **1982**, *9*, 539–552.

43. Sumner, M.; Miller, W. Cation exchange capacity and exchange coefficients. *Methods Soil Anal. Part 3 Chem. Methods* **1996**, *5*, 1201–1229.

44. Long, O.H.; Seatz, L.F. Correlation of soil tests for available phosphorus and potassium with crop yield responses to fertilization. *Soil Sci. Soc. Am. J.* **1953**, *17*, 258–262. [CrossRef]

45. Liu, M.; Han, G.L. Assessing soil degradation under land-use change: Insight from soil erosion and soil aggregate stability in a small karst catchment in southwest China. *PeerJ* **2020**, *8*, 19. [CrossRef] [PubMed]

46. Zeng, C.; Wang, S.J.; Bai, X.Y.; Li, Y.B.; Tian, Y.C.; Li, Y.; Wu, L.H.; Luo, G.J. Soil erosion evolution and spatial correlation analysis in a typical karst geomorphology using RUSLE with GIS. *Solid Earth* **2017**, *8*, 721–736. [CrossRef]

47. Wang, B.; Zheng, F.; Guan, Y. Improved USLE-K factor prediction: A case study on water erosion areas in China. *Int. Soil Water Conserv. Res.* **2016**, *4*, 168–176. [CrossRef]

48. Addis, H.K.; Klik, A. Predicting the spatial distribution of soil erodibility factor using USLE nomograph in an agricultural watershed, Ethiopia. *Int. Soil Water Conserv. Res.* **2015**, *3*, 282–290. [CrossRef]

49. Sharpley, A.N. *EPIC—Erosion/Productivity Impact Calculator: 1. Model Documentation*; Sharpley, A.N., Williams, J.R., Eds.; Technical Bulletin No. 1768; US Department of Agriculture: Washington, DC, USA, 1990.

50. Liu, M.; Han, G.L.; Li, X.Q. Comparative analysis of soil nutrients under different land-use types in the Mun River basin of Northeast Thailand. *J. Soils Sediments* **2021**, *21*, 1136–1150. [CrossRef]

51. Williams, J.R. Chapter 25: The EPIC model. In *Computer Models of Watershed Hydrology*; Singh, V.P., Ed.; Water Resources Publications: Littleton, CO, USA, 1995.

52. Pribyl, D.W. A critical review of the conventional SOC to SOM conversion factor. *Geoderma* **2010**, *156*, 75–83. [CrossRef]

53. Minasny, B.; McBratney, A.B.; Wadoux, A.M.J.C.; Akoeb, E.N.; Sabrina, T. Precocious 19th century soil carbon science. *Geoderma Reg.* **2020**, *22*, e00306. [CrossRef]

54. Sudicky, E.A. A natural gradient experiment on solute transport in a sand aquifer—Spatial variability of hydraulic conductivity and its role in the dispersion process. *Water Resour. Res.* **1986**, *22*, 2069–2082. [CrossRef]

55. Neuman, S.P. Trends, prospects and challenges in quantifying flow and transport through fractured rocks. *Hydrogeol. J.* **2005**, *13*, 124–147. [CrossRef]

56. Zhang, Y.G.; Schaap, M.G. Weighted recalibration of the Rosetta pedotransfer model with improved estimates of hydraulic parameter distributions and summary statistics (Rosetta3). *J. Hydrol.* **2017**, *547*, 39–53. [CrossRef]

57. Schaap, M.G.; Leij, F.J.; van Genuchten, M.T. Rosetta: A computer program for estimating soil hydraulic parameters with hierarchical pedotransfer functions. *J. Hydrol.* **2001**, *251*, 163–176. [CrossRef]

58. Qu, W.; Bogena, H.; Huisman, J.; Vanderborght, J.; Schuh, M.; Priesack, E.; Vereecken, H. Predicting sub-grid variability of soil water content from basic soil information. *Geophys. Res. Lett.* **2015**, *42*, 789–796. [CrossRef]

59. Cai, Y.; Wu, P.; Zhang, L.; Zhu, D.; Chen, J.; Wu, S.; Zhao, X. Simulation of soil water movement under subsurface irrigation with porous ceramic emitter. *Agric. Water Manag.* **2017**, *192*, 244–256. [CrossRef]

60. Turkeltaub, T.; Jia, X.; Zhu, Y.; Shao, M.-A.; Binley, A. Recharge and nitrate transport through the deep vadose zone of the Loess Plateau: A regional-scale model investigation. *Water Resour. Res.* **2018**, *54*, 4332–4346. [CrossRef]

61. Burgos, P.; Madejon, E.; Perez-De-Mora, A.; Cabrera, F. Spatial variability of the chemical characteristics of a trace-element-contaminated soil before and after remediation. *Geoderma* **2006**, *130*, 157–175. [CrossRef]

62. Marchant, B.P.; Lark, R.M. Robust estimation of the variogram by residual maximum likelihood. *Geoderma* **2007**, *140*, 62–72. [CrossRef]

63. Bogunovic, I.; Mesic, M.; Zgorelec, Z.; Jurisic, A.; Bilandzija, D. Spatial variation of soil nutrients on sandy-loam soil. *Soil Tillage Res.* **2014**, *144*, 174–183. [CrossRef]

64. Yao, X.; Yu, K.; Deng, Y.; Zeng, Q.; Lai, Z.; Liu, J. Spatial distribution of soil organic carbon stocks in Masson pine (*Pinus massoniana*) forests in subtropical China. *Catena* **2019**, *178*, 189–198. [CrossRef]

65. Liu, M.; Han, G.L.; Li, X.Q. Contributions of soil erosion and decomposition to SOC loss during a short-term paddy land abandonment in northeast Thailand. *Agric. Ecosyst. Environ.* **2021**, *321*, 107629. [CrossRef]

66. Huang, X.F.; Lin, L.R.; Ding, S.W.; Tian, Z.C.; Zhu, X.Y.; Wu, K.R.; Zhao, Y.Z. Characteristics of soil erodibility k value and its influencing factors in the Changyan watershed, southwest Hubei, China. *Land* **2022**, *11*, 134. [CrossRef]

67. Liu, M.; Han, G.L.; Li, X.Q. Using stable nitrogen isotope to indicate soil nitrogen dynamics under agricultural soil erosion in the Mun River basin, northeast Thailand. *Ecol. Indic.* **2021**, *128*, 107814. [CrossRef]
68. Veihe, A. The spatial variability of erodibility and its relation to soil types: A study from northern Ghana. *Geoderma* **2002**, *106*, 101–120. [CrossRef]
69. Polyakov, V.; Lal, R. Modeling soil organic matter dynamics as affected by soil water erosion. *Environ. Int.* **2004**, *30*, 547–556. [CrossRef]
70. Six, J.; Bossuyt, H.; Degryze, S.; Denef, K. A history of research on the link between (micro)aggregates, soil biota, and soil organic matter dynamics. *Soil Tillage Res.* **2004**, *79*, 7–31. [CrossRef]
71. Cambardella, C.A.; Moorman, T.B.; Novak, J.M.; Parkin, T.B.; Karlen, D.L.; Turco, R.F.; Konopka, A.E. Field-scale variability of soil properties in central Iowa soils. *Soil Sci. Soc. Am. J.* **1994**, *58*, 1501–1511. [CrossRef]
72. Pietrzykowski, M. Soil quality index as a tool for Scots pine (*Pinus sylvestris*) monoculture conversion planning on afforested, reclaimed mine land. *J. For. Res.* **2014**, *25*, 63–74. [CrossRef]
73. Zheng, F.L. Effect of vegetation changes on soil erosion on the Loess Plateau. *Pedosphere* **2006**, *16*, 420–427. [CrossRef]
74. Rodríguez, A.R.; Arbelo, C.D.; Guerra, J.A.; Mora, J.L.; Notario, J.S.; Armas, C.M. Organic carbon stocks and soil erodibility in Canary Islands Andosols. *Catena* **2006**, *66*, 228–235. [CrossRef]
75. Alletto, L.; Coquet, Y. Temporal and spatial variability of soil bulk density and near-saturated hydraulic conductivity under two contrasted tillage management systems. *Geoderma* **2009**, *152*, 85–94. [CrossRef]
76. Correll, D.L. The role of phosphorus in the eutrophication of receiving waters: A review. *J. Environ. Qual.* **1998**, *27*, 261–266. [CrossRef]
77. Reynolds, W.D.; Bowman, B.T.; Brunke, R.R.; Drury, C.F.; Tan, C.S. Comparison of tension infiltrometer, pressure infiltrometer, and soil core estimates of saturated hydraulic conductivity. *Soil Sci. Soc. Am. J.* **2000**, *64*, 478–484. [CrossRef]

International Journal of
***Environmental Research
and Public Health***

Article

Soil Contamination with Metals in Mountainous: A Case Study of Jaworzyna Krynicka in the Beskidy Mountains (Poland)

Sławomir Dorocki and Joanna Korzeniowska *

Institute of Geography, Pedagogical University of Krakow, 30-084 Krakow, Poland
* Correspondence: joanna.korzeniowska@up.krakow.pl; Tel.: +48-12-662-62-61

Abstract: The paper presents the content of six metals (Cd, Cr, Cu, Ni, Pb, and Zn) in the soils of the southern slope of Jaworzyna Krynicka in Poland. Soil samples were collected in polygons, starting from an altitude of 500 m above sea level and ending at an altitude of 1100 m above sea level. Ten soil samples were collected in each polygon. The polygons were set at every 100 m of absolute altitude. The selected research area is an important natural area. The fertile mountain beech forests located there are the most important forest communities in the mountain areas of Poland. They are valuable habitats for plants and animals (especially for large predatory mammals). Every year, numerous tourists and health resort patients visit this place. The results of the research showed that soil contamination in the study area is not high, in particular for altitudes of 500 and 900 m above sea level. At these altitudes, the contents of Cd, Cr, Cu, Ni, Pb, and Zn were similar to the concentrations of these metals in uncontaminated soils. The tests carried out showed very low cadmium content for all absolute altitudes. Zinc, the concentrations of which exceeded natural values, showed the highest content in the tested soils. All the metals tested showed a common tendency of increases in their content in the soils of Jaworzyna Krynicka up to 800 m above sea level. From an altitude of 900 m above sea level, the content of these metals decreased, except for Pb. Only Pb concentrations in Jaworzyna Krynicka soils also increased with the increasing altitude. The research significance of this work is that it is important for assessing the ecological balance in the selected area.

Keywords: heavy metals; soil; absolute altitude; Jaworzyna Krynicka; contamination

Citation: Dorocki, S.; Korzeniowska, J. Soil Contamination with Metals in Mountainous: A Case Study of Jaworzyna Krynicka in the Beskidy Mountains (Poland). *Int. J. Environ. Res. Public Health* **2023**, *20*, 5150. https://doi.org/10.3390/ijerph20065150

Academic Editors: Bo Sun, Ming Liu and Yan Chen

Received: 19 January 2023
Revised: 6 March 2023
Accepted: 13 March 2023
Published: 15 March 2023

1. Introduction

Heavy metals are very common in the environment. Metals present in the form of salt solutions, which are by-products or wastes of various industries, pose a threat to the environment [1–6].

The condition for the sustainable existence of any ecosystem is the balance of the main metabolic processes, i.e., homeostasis. In natural ecosystems, biogeochemical cycles are usually characterized by a certain regularity. However, as a result of human economic activity, they may be disturbed owing to the excessive discharge of one or more components. The constant increase in the consumption of trace elements leads to changes in the proportion between their activation and introduction into the biological environment and their re-deposition in geological formations [3,7].

Dusts contained in the atmosphere, in which heavy metals are present, get into the soil and fall on the above-ground parts of plants. As a result, the concentration of heavy metals in these elements of the environment increases. The natural content of heavy metals in soils is closely related to the type, kind, and grade of soil [3,4,7]. The intensity of metal deposition depends on the emission volume, physical properties of dusts, meteorological conditions, and soil characteristics.

Five metals (Cd, Cr, Cu, Pb, and Zn) from the group of elements with a very high degree of threat to the natural environment and one (Ni) with an average degree of threat were selected for the study. Introduced into the soil, these metals accumulate in it and remain there for a long time [8–10]. Heavy metals are often transported over long distances from emission sources [11–13]. This phenomenon is related to the long-term dustiness of the atmosphere and meteorological conditions. The residence time of particulate matter in the atmosphere also depends on the size of the particles and the configuration of the terrain onto which they fall. Low pressure and strong winds favor the spread of pollutants over long distances [14,15].

The aim of this study was to determine the content of Cd, Cr, Cu, Ni, Pb, and Zn in the soil, at various absolute altitudes, of an environmentally important area, the Krynica-Zdrój and Muszyna health resorts located in the Poprad Landscape Park. Krynica-Zdrój, as well as Muszyna, which borders it, are towns with the status of health resorts, which are visited by many tourists and health resort patients [16,17]. These towns are so popular owing to their healing waters, climatic conditions, and extensive tourist and skiing infrastructure [18,19]. It is an area located in the Beskid Sądecki. The object of the research was the area of the southern slope of Jaworzyna Krynicka, reaching an altitude of 1114 m above sea level. This peak is located on the border of the two mentioned health resorts; however, the samples were taken from the area administratively belonging to the Muszyna commune. The research work carried out is important for determining the impact of long-range emissions on the natural environment of mountain areas. The increased content of heavy metals in the soils of environmentally valuable areas clearly indicates an anthropogenic source of pollution.

2. Materials and Methods

2.1. Study Area

Soil samples were collected on the southern slope of Jaworzyna Krynicka, which is the highest peak (1114 m above sea level) in the Jaworzyna Krynicka range in the Beskid Sądecki, Poland. The Jaworzyna Krynicka range occurs within the Magura Nappe. The spatially dominant geological formations within the range are thick-bedded Magura sandstones [20]. The sandstones are represented by psamite–pelite deposits, conglomerated with a clay or lime–clay binder. They result in a rock mantle with a mechanical composition of light and medium loams, sometimes dusty, with a skeleton of various sizes, occurring at a depth of 40–50 cm, and sometimes shallower [21].

Climatic conditions in the mountainous area of the Carpathians are difficult to determine. Their parameters depend on numerous factors, such as altitude above sea level, exposure, lie of the land, etc. The climate of the study area is closely related to the orographic location and massiveness of the culminating part of the Jaworzyna Krynicka range. The conditions prevailing there correspond to the habitat types of forests, in particular to the growth and development of mountain forests and mountain mixed forests. In the case of advective weather, situations with air inflow from the west (18.8%) as well as from the south and north-west prevailed, on average, in 11.0% of all cases. These situations generally prevailed in all months, with the period of their greatest activity being the cold part of the year, from October to March. In the summer (August and September), baric systems with air advection from the east and north-east prevailed [22].

The forests of Jaworzyna Krynicka consist mainly of a mixed mountain forest, located mainly above an altitude of 950 m above sea level, and a mountain forest. In the study area, beech accounted for about half of the stand. Apart from beech, there are also spruce, fir, and pine trees. The role of admixture species is played here by larch, birch, and grey alder, while ash and sycamore trees also occur, but singly [23].

2.2. Sampling

Samples (ten samples per polygon) were taken from seven polygons (A, B, C, D, E, F, and G) at altitudes of 1100, 1000, 900, 800, 700, 600, and 500 m above sea level, located in an area with southern exposure (Figure 1). The samples were taken on 24 May 2021. The polygons were distributed along the hill stretching between the Szczawniczek stream in the west and the Jastrzębik stream in the east, and descending from the top of Jaworzyna towards the village of Złockie in the Muszyna-Zdrój commune.

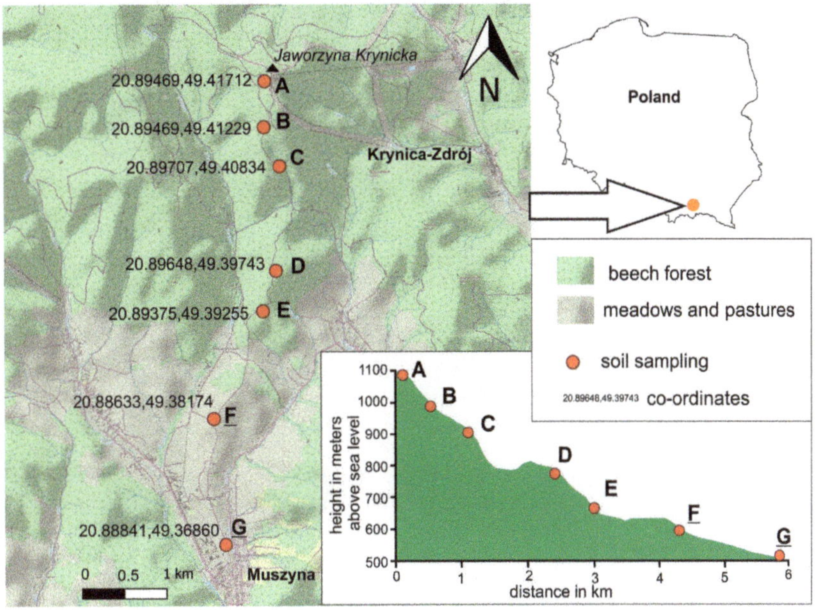

Figure 1. Location of sampling sites.

2.3. Chemical Analysis

In order to determine the content of heavy metals (Cd, Cr, Cu, Ni, Pb, and Zn) in the sampled soil material (topsoil, up to 10 cm), the following laboratory work was carried out, in accordance with the methodology used for collecting and preparing samples for chemical analyses [24,25]:

- Manually cleaning the collected samples by removing foreign material (dry leaves, twigs, grass, etc.);
- Drying the samples at 100 °C;
- Grinding soil samples in a ceramic mortar and sieving through a sieve with a mesh diameter of 2 mm;
- Mineralization, which is performed to completely break down soil samples into simple, solid compounds—1 g of the dried sample material was digested with a modified Aqua Regia solution of equal parts concentrated HCl, HNO_3, and DIH2O for one hour in a heating block or hot water bath. The resulting solution was filtered and stored in sealed polyethylene containers until sent for spectrometric analysis;
- Determination of the pseudo-total content of heavy metals using the inductively coupled plasma mass spectrometry (ICP-MS) method in the Bureau Veritas Commodities, Vancouver BC, Canada. The use of the Bureau Veritas methodology made it possible to accurately determine the metal content in the soil material, with the following detection limits (mg/kg) for Cd: 0.01, Cr: 0.5, Cu: 0.01, Ni: 0.1, Pb: 0.01, and Zn: 0.1. The STD DS11 and STD OREAS262 standards were used as reference materials.

2.4. Statistical Analysis of the Data

SAS® OnDemand for Academics software and ANOVA procedures were used for statistical analysis [26]. The software SAS® OnDemand for Academics is available on the website https://www.sas.com belonging to SAS Institute Inc. (SAS Campus Drive, Cary, NC, USA).

The ANOVA procedure was used to analyse the content of selected elements in the soil by performing the analysis of variance [27]. In the analysis of variance, the continuous response variable, considered as the dependent variable, is measured using classification variables, called independent variables. It was assumed that the variability of the obtained results follows from the adopted classification and depends on the altitude above sea level of the collected soil sample, with random error responsible for the remaining variability. The classification variable was specified in the code by the CLASS value which, unlike the GLM procedure, does not allow continuous variables on the right side of the model.

The F test indicates whether the model takes into account a sufficient amount of variability of the dependent variable [28]. The general F test is statistically significant in all examined cases of the analyzed elements, which indicates that the adopted model for a given element, as a whole, is responsible for a significant part of the variability of the dependent variable. The F test indicates that there is a difference between the mean values for individual samples according to the altitude above sea level. Interaction between the altitude of a collected soil sample and the content of selected elements in the soil is significant at a level of 95%. The significance level of this test was determined prior to the analysis. The F test does not reveal any information about the nature of the observed differences between the content of elements and altitude. Therefore, mean comparison methods were used to collect further information. The MEANS instruction requires the comparison of average levels using the Waller–Duncan K-ratio test method [29]. The Waller–Duncan K-ratio test is a multi-range test. It does not work by controlling for type I errors. Instead, it compares type I and type II error rates based on Bayesian principles [30].

In addition, the analysis used some simple statistics. The value of the R-square relationship indicates how much the model of regression between the content of metals in the soil and altitude corresponds to the variability of the variable [31]. The coefficient of variation (CV) is a measure determining the diversity of the analyzed samples together. The CV shows the degree of variation of the data in the sample relative to the mean population, even if they differ in the mean value. The mean square error (MSE) also provides information about how close the regression line is to the set of points, i.e., it proves the variability of the data in the same way as the mean of the dependent variable [32]. Root MSE (RMSE) is an estimate of the standard deviation of the dependent variable [33]. RMSE is one of the most commonly used measures to assess the quality of forecasts. It shows how far predictions deviate from the actual values measured using the Euclidean distance. In the case below, it refers to the model of regression between metal content in the soil.

3. Results

Among the analyzed elements, the best match of the model of regression according to the altitude of sampling and the content of elements in the soil was noted for Ni and Cr, where the R-Square was 0.971 and 0.963, respectively. Only in the case of Cd, the R-Square was below 0.9 and reached the value of 0.723, which is already a high value.

The results obtained for individual elements show that in most cases, the model of regression corresponds to the variability of the variable (Table 1).

Regarding the variability of the results obtained on the basis of the coefficient of variation (CV), the smallest diversity was noted among the tested elements in the case of Cr (7.6). Low values in the samples taken also occurred for Ni and Zn (around 10.8). The greatest diversity was noted in the case of Cd, which was mainly the result of two samples taken at altitudes of 1000 and 1100 m above sea level. However, in no case were the values such outliers as to be omitted from the analysis.

Considering the root mean square error (RMSE), or the square mean of errors, which is the square root of the MSE, it can be observed that the smallest difference between the obtained estimation from the model and the actual value occurs in the case of Cd (0.091). Low values of the difference between the values obtained from the model and the actual values were noted for Cr (1709), Ni (2383), and Cu (2736). In other cases, the value of the standardized difference is higher and, in the case of Zn, it is as much as 8 points.

Table 1. The results of the statistics of the content of individual metal elements in the soil according to the variable altitude above sea level.

Statistical Indicators	Cd	Cr	Cu	Ni	Pb	Zn
R-Squared	0.72	0.96	0.94	0.97	0.90	0.92
Coefficient of variation (CV)	20.61	7.59	15.70	10.78	14.32	10.82
Root mean square error (RMSE)	0.091	1.709	2.736	2.383	5.419	8.020
Mean square (MS)	0.22	795.52	1262.76	1962.12	2888.44	7998.22
F value (<0.0001)	27.35	272.33	168.74	345.61	98.37	124.36
Mean square error (MSE) of grouping model	0.008	2.921	7.483	5.677	29.362	64.314

Additionally, in the case of the mean square error (MSE) for the grouping model, the smallest difference between the model and the observed values was noted for Cd content (0.008) and, in ascending order, for Cr (2.921), Ni (5.677), and Cu (7.483). The greatest differences were noted, as before, for Zn (64.314).

Therefore, referring to the distribution of elements in the soil samples taken at specific altitudes, in the case of Cr, Cu, Ni, and Zn, the metal content increases from 500 m above sea level to 700 m above sea level, and then it usually drops below the value from 500 m above sea level (Figure 2). The change in metal content may, on the one hand, result from the altitude above sea level, but also from the type of land cover (the two lowest polygons were without any forest).

In the case of Cd, the concentration of metals increases from an altitude of 500 to 800 m above sea level, and then at 900 m, for example, it decreases to the level from 500 m above sea level, to reach at an altitude of 1100 m above sea level the same value as at 800 m above sea level.

In the case of Pb content, the values show a slow increase from an altitude of 500 to 900 m above sea level. Samples from altitudes of 700 to 800 m above sea level have the same average at a level of 0.05. On the other hand, samples from the highest polygons have the highest values of Pb content.

In the analysis of grouping according to average levels using the Waller–Duncan K-ratio coefficient method, polygons that are graphically connected by the same line have the same average value at a confidence level of 0.05. All samples have a significantly different average across all polygons only in the case of Cr. In the case of Cd and Cu, there are the smallest differences in the average values according to the altitude of sampling.

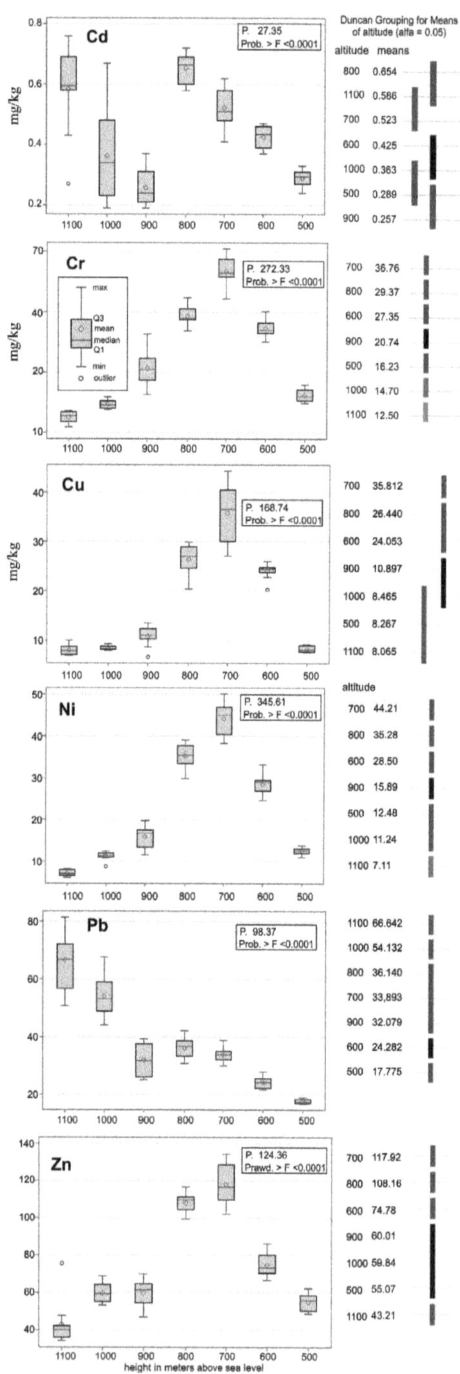

Figure 2. The content of metals in the soil according to altitude and the grouping of polygons according to the average content of elements in the soil.

4. Discussion

The first polygon (A) at an altitude of 1100 m above sea level was located below the tourist-developed peak of Jaworzyna Krynicka, overgrown mainly with beech, with blueberry with an admixture of spruce included in the meso-climatic floor of the peak area. As in the case of the polygons from A to E, there were cryptopodzolic soils or brown acidic soils. Additionally, in the case of samples from polygons B and C, these were beech forest habitats. The reaction of these soils from polygons A, B, and C was very acidic (pH in H_2O) and varies in a range of 4.0–4.3. In polygons D and E, mixed beech–spruce–pine stands in a mountain forest habitat have a greater share. The reaction of these soils was already acidic and slightly acidic, in a range of 5.5–6.3. Polygon F was grassland. It represents a grassy habitat with the predominance of fescue, bent grass, and dog's-tail grass. The grassland was characterized by a reaction close to neutral or slightly acidic (pH ranging from 6.5 to 6.9). Therefore, these were proper brown soils and acidic soils. Polygon G, located at the lowest level and surrounded by buildings located in adjacent valleys, was an agricultural area used as pastures and farmland. These were mainly leached brown soils. These soils, similar to the soils in polygon F, had a neutral and slightly acid reaction (6.2–7.2). The organic carbon content in the soils increased with altitude, starting from 2.38% (500 m above sea level), through 2.91% (600 m above sea level), then 6.53% (700 m above sea level), 7.45% (800 m above sea level), and then 9.12% (900 m above sea level), reaching 10.42% at 1000 m, and 12.34% at 1100 m above sea level. [21]. The average content of the tested metals in the soils of Jaworzyna Krynicka within each polygon was compared with the globally defined geochemical background (average content of the element in the Earth's crust according to Turkian and Wedepohl [34]) and with the value of the locally defined geochemical background for Polish soils proposed by Czarnowska [35]. The average content of the element in the Earth's crust [34] and the geochemical background for Polish soils [35] amount to, respectively: for Cd, 0.18 and 0.13 mg/kg; for Cr, 27.0 and 13.0 mg/kg; for Cu, 7.1 and 14.0 mg/kg; for Ni, 10.2 and 15.0 mg/kg; for Pb, 9.8 and 17.5 mg/kg; for Zn, 30.0 and 50 mg/kg.

The contents of Cu, Ni, Pb, and Zn are almost twice as high in Polish soils as in world soils. However, the contents of Cd and Cr are lower in Polish soils.

The average contents of Cd and Pb (for all absolute altitudes) in the soils of the southern slope of Jaworzyna Krynicka are higher than both the globally defined geochemical background (average content of the element in the Earth's crust) and the geochemical background for Polish soils. The average contents of Zn and Cu in the tested soils are higher than the globally defined geochemical background for all tested altitudes, while for altitudes of 500–1000 m above sea level, the Zn content is higher than the geochemical background for Polish soils. The average Cu content is higher than the geochemical background for Polish soils only for an altitude of 600–800 m above sea level. The average Ni content in the soils of Jaworzyna Krynicka is higher, for all altitudes except 1100 m above sea level, than the globally defined geochemical background, and for altitudes of 600–900 m above sea level, higher than the geochemical background value for Polish soils. The average Cr content is higher than the globally defined geochemical background only for altitudes of 700 and 800 m above sea level, while the tested Cr content is higher, for all altitudes except 1100 m above sea level, than the geochemical background for Polish soils.

One should consider why the concentrations of all metals tested, except for Pb, increase up to an altitude of 800 m above sea level, and then decrease. Why do metal concentrations reach the highest values for altitudes of 700 and 800 m above sea level? Why do soil Pb concentrations increase rather than decrease with increasing altitude? The logical explanation seems to be the land cover and soil reaction. In the polygons located at altitudes of 1100, 1000, and 900 m above sea level, we are dealing with beech forest habitats, while in the polygons located at altitudes of 800 and 700 m above sea level, mixed stands of beech, spruce, and pine predominate in the mountain forest habitat. Here we can see the influence of conifer needles on the pH of the soil, which is definitely more acidic compared to higher altitudes. The polygons at altitudes of 600 and 500 m above sea level are grasslands, pastures, and farmland. Neutral to slightly acidic reaction prevails here. Moving on to the increase in Pb content in the soil of the slope of Jaworzyna Krynicka with increasing absolute altitude, this result of the study can be explained by the strong affinity of Pb with the content of organic matter in the soil. Organic matter (content of organic carbon) in the soil in the analyzed area increased, similarly to the concentration of Pb, with the altitude above sea level.

Comparing the results obtained with the metal content in the soils of environmentally important and slightly polluted areas, a similar content of Cr and Pb in the soils reported by Słowik et al. [36] for the Roztocze National Park was found. However, the content of Zn was almost twice as high as that measured for Jaworzyna Krynicka. Most likely, such a high content of Zn was affected by the type of soil found in the park. Similar concentrations of Cd, Cr, Cu, and Ni were also found in the soils of Jaworzyna Krynicka, and in the soils of the Stołowe Mountains National Park [37], and in the soils of the Tatra National Park [38]. The content of Pb and Zn in the Jaworzyna Krynicka soils was almost twice as high as the content of this metal in the soils of the above-mentioned parks.

The results of the biomonitoring of atmospheric aerosol pollution are often interpreted using the enrichment factor (EF) [39–41]. However, the best results are obtained when comparing the content of metals in soil and mosses. Therefore, it was not used in this case. However, further research is planned to compare the content of the tested metals in soil and mosses.

5. Conclusions

The mean contents of all the tested metals in the soils of the southern slope of Jaworzyna Krynicka, for altitudes of 500 and 900 m above sea level were within the range of values natural in Polish soils. Cadmium was the metal which did not exceed natural levels in soils for all absolute altitudes. The metal which, for all the tested altitudes, exceeded the values natural in Polish soils was zinc.

The mean concentrations of metals for altitudes of 700 and 800 m above sea level were highest for Cd, Cr, Cu, Ni, and Zn. An increase in the content of these metals was observed from an altitude of 500 to 800 m above sea level, and then a decrease in concentration began from 900 to 1100 m above sea level. The Pb content increased with the increase in absolute altitude, reaching maximum values for an altitude of 1100 m above sea level.

Author Contributions: Conceptualization, S.D. and J.K.; methodology, J.K.; software, S.D.; validation, S.D. and J.K.; formal analysis, S.D. and J.K.; investigation, S.D. and J.K.; resources, S.D. and J.K.; data curation, S.D. and J.K.; writing—original draft preparation, S.D. and J.K.; writing—review and editing, S.D. and J.K.; visualization, S.D. and J.K.; supervision, S.D. and J.K.; project administration, S.D. and J.K.; funding acquisition, S.D. and J.K. All authors have read and agreed to the published version of the manuscript.

Funding: Co-financing of research from the Pool of the Vice-Rector for Science of the Pedagogical University of Krakow—dr hab. Michał Rogoż, prof. UP. Application No. BN.302.283.2021.RN.

Institutional Review Board Statement: Not applicable.

Informed Consent Statement: Not applicable.

Data Availability Statement: The data presented in this study are available on request from the corresponding author. The data are not publicly available due to its further use for comparative research.

Conflicts of Interest: The authors declare no conflict of interest.

References

1. Nguyen, H.T.H.; Nguyen, B.Q.; Duong, T.T.; Bui, A.T.K.; Nguyen, H.T.A.; Cao, H.T.; Mai, N.T.; Nguyen, K.M.; Pham, T.T.; Kim, K.-W. Pilot-Scale Removal of Arsenic and Heavy Metals from Mining Wastewater Using Adsorption Combined with Constructed Wetland. *Minerals* **2019**, *9*, 379. [CrossRef]
2. Garcia-Lawrence, M.L.; Crespo-Ugly, E.; Esbry, J.M.; Figs, P.; Grau, P.; Crespo, I.; Sánchez-Donoso, R. Assessment of Potentially Toxic Elements in Technosols by Tailings Derived from Pb–Zn–Ag Mining Activities at San Quintín (Ciudad Real, Spain): Some Insights into the Importance of Integral Studies to Evaluate Metal Contamination Pollution Hazards. *Minerals* **2019**, *9*, 346. [CrossRef]
3. Kabata-Pendias, A.; Pendias, H. *Trace Elements in Soils and Plants*; CRC Press: Boca Raton, FL, USA; London, UK, 2001; pp. 15–204.
4. Ciepał, R. *Kumulacja Metali Ciężkich i Siarki w Roślinach Wybranych Gatunków Oraz Glebie Jako Wskaźnik Stanu Skażenia Środowiska Terenów Chronionych Województw Śląskiego i Małopolskiego [The Accumulation of Heavy Metals and Sulphur in Plants of Selected Species and Soil as an Indicator of the State of Environmental Contamination of Protected Areas in Silesia and Małopolska Provinces]*; Wydawnictwo Uniwersytetu Śląskiego: Katowice, Poland, 1999; pp. 9–35.
5. Macías, R.; Ramos, M.S.; Guerrero, A.L.; Farfán, M.G.; Mitchell, K.; Avelar, F.J. Contamination Assessment and Chemical Speciation of Lead in Soils and Sediments: A Case Study in Aguascalientes, Mexico. *Appl. Sci.* **2022**, *12*, 8592. [CrossRef]
6. Jaishankar, M.; Tseten, T.; Anbalagan, N.; Mathew, B.B.; Beeregowda, K.N. Toxicity, mechanism and health effects of some heavy metals. *Interdiscip. Toxicol.* **2014**, *7*, 60. [CrossRef]
7. Migaszewski, Z.; Gałuszka, A. *Geochemia Środowiska [Geochemistry of the Environment]*; Wydawnictwo Naukowe PWN: Warszawa, Poland, 2016; pp. 84–289.
8. Alloway, B. *Cadmium. Heavy Metals in Soils*; Blackie Academic: London, UK, 1990; pp. 122–151.
9. Fergusson, J. *The Heavy Elements: Chemistry, Environmental Impact and Health Effects*; Pergamon Press: Oxford, UK, 1990.
10. Fakayode, S.; Onianwa, P. Heavy metal contamination of soil, and bioaccumulation in Guinea grass (*Panicum maximum*) around Ikea Industrial Estate, Lagos, Nigeria. *Environ. Geol.* **2002**, *43*, 145–150.
11. Bergbäck, B.; Johansson, K.; Mohlander, U. Urban metal flows—A case study of Stockholm. *Water Air Soil Pollut.* **2001**, *1*, 3–24. [CrossRef]
12. Sörme, L.; Bergbäck, B.; Lohm, U. Goods in the antroposphere as a metal emission source. *Water Air Soil Pollut.* **2001**, *1*, 213–227. [CrossRef]
13. Lehndorf, E.; Schwark, L. Accumulation histories of major and trace elements on pine needles in the Cologne Conurbation as function of air Quality. *Atmos. Environ.* **2008**, *42*, 833–845. [CrossRef]
14. Steinnes, E.; Friedland, A.J. Metal contamination of natural surface soils from long-range atmospheric transport: Existing and missing knowledge. *Environ. Rev.* **2006**, *14*, 169–186. [CrossRef]
15. Miśkowiec, P. The impact of the mountain barrier on the spread of heavy metal pollution on the example of Gorce Mountains, Southern Poland. *Environ. Monit Assess.* **2022**, *194*, 663. [CrossRef]
16. Dorocki, S.; Brzegowy, P. Impact of natural resources on the development of spa industry in Krynica-Zdroj. *Int. Multidiscip. Sci. GeoConf. SCEM* **2013**, *2*, 309.
17. Wójcikowski, W.K. Muszyna Zdrój-10 years of spatial changes in the town and in the spa. *SWS J. Soc. Sci. Art* **2020**, *2*, 1–13. [CrossRef]
18. Dorocki, S.; Brzegowy, P. Ski and spa tourism as local development strategy–the case of Krynica Zdrój (Poland). *Ann. Univ. Paedagog. Crac. Stud. Geogr.* **2014**, *5*, 88–116.
19. Kwiek, M.; Korzeniowska, J. Rozwój infrastruktury sportowo-rekreacyjnej w gminie Krynica-Zdrój w ostatnich 10-ciu latach. [The development of sports and recreation infrastructure in the Krynica-Zdrój commune in the last 10 years]. In *Tradycje i Perspektywy Rozwoju Kultury Uzdrowiskowej w Muszynie w Kontekście Europejskim, pod Redakcją Bożeny Płonki-Syroki, Pawła Brzego Wego, Andrzeja Syroki i Sławomira Dorockiego. [Traditions and Prospects for the Development of Health Resort Culture in Muszyna in the European context*; Płonka-Syroka, B., Brzegowy, P., Syroka, A., Dorocki, S., Eds.; Oficyna Wydawnicza Arboretum, Medical University Named after Silesian Piasts in Wrocław: Wrocław, Poland, 2020; Volume 12, pp. 289–304.
20. Margielewski, W. Formy osuwiskowe pasma Jaworzyny Krynickiej w Popradzkim Parku Krajobrazowym. [Landslide forms of the Jaworzyna Krynicka range in the Poprad Landscape Park]. *Chrońmy Przyr. Ojczystą Let's Prot. Our Mother Nat.* **1992**, *48*, 5–16.
21. Firek, A.; Zasonski, S. Wstępne badania niektórych własności różnie użytkowanych gleb w południowych partiach pasma Jaworzyny Krynickiej. [Preliminary studies of some properties of variously used soils in the southern parts of the Jaworzyna Krynicka range.]. *Rocz. Glebozn.* **1969**, *20*, 99–118.
22. Durło, G. Typologia mikroklimatyczna Jaworzyny Krynickiej i Doliny Czarnego Potoku. [The microclimatic typology of Jaworzyna Krynicka and the Czarny Potok Valley.]. *Sylwan* **2003**, *2*, 58–66.
23. Malek, S.; Wacławek, M.; Kroczek, M.; Wieczorek, T. Wpływ nartostrad stacji narciarskiej Jaworzyna Krynicka SA na otaczający drzewostan. [The impact of the ski slopes of the Jaworzyna Krynicka SA ski station on the surrounding forest stand]. *Agrar. Silvestria* **2013**, *51*, 91–101.
24. Gajec, M.; King, A.; Kukulska-Zając, E.; Mostowska-Stąsiek, J. Pobieranie próbek gleby w kontekście prowadzenia oceny zanieczyszczenia powierzchni ziemi [Soil sampling in the context of soil pollution assessment]. *Naft. Gas* **2018**, *3*, 215–225. [CrossRef]

25. Namieśnik, J.; Lukasiak, J.; Jamrógiewicz, Z. *Pobieranie Próbek Środowiskowych Do Analizy [Collecting Environmental Samples for Analysis]*; Wydawnictwo Naukowe PWN: Warsaw, Poland, 1995; pp. 67–96.
26. Pham, T.V.; Kromrey, J.D.; Chen, Y.H.; Kim, E.S.; Nguyen, D.T.; Wang, Y. ANOVA_robust: A SAS Macro for Parametric Tests of Mean Differences in One-Factor ANOVA Models. *J. Stat. Softw.* **2020**, *95*, 1–16. [CrossRef]
27. Sawyer, S.F. Analysis of variance: The fundamental concepts. *J. Man. Manip. Ther.* **2009**, *17*, 27E–38E. [CrossRef]
28. Goat, L.S.; Green, C.E. Analysis of variance: Is there a difference in means and what does it mean? *J. Surg. Res.* **2008**, *144*, 158–170.
29. Dixon, D.O.; Divine, G.W. Multiple comparisons for relative risk regression: Extension of the k-ratio method. *Stat. Med.* **1987**, *6*, 591–597. [CrossRef]
30. Komfortowski, W.; Trewhella, J.; Andre, I. Bayesian inference of protein conformational ensembles from limited structural data. *PLoS Comput. Biol.* **2018**, *14*, e1006641.
31. Bramante, R.; Zappa, D.; Petrella, G. On the interpretation and estimation of the market model R-square. *Electron. J. Appl. Stat. Anal.* **2013**, *6*, 57–66.
32. Chai, T.; Draxler, R.R. Root mean square error (RMSE) or mean absolute error (MAE). *Geosci. Model Dev. Discuss.* **2014**, *7*, 1525–1534.
33. Willmott, C.J.; Matsuura, K. Advantages of the mean absolute error (MAE) over the root mean square error (RMSE) in assessing average model performance. *Clim. Res.* **2005**, *30*, 79–82. [CrossRef]
34. Turekian, K.K.; Wedepohl, D.H. Distribution of the elements in some major units of the earth_s crust. *Bull. Geol. Soc. Am.* **1961**, *72*, 175–192. [CrossRef]
35. Czarnowska, K. Ogólna zawartość metali ciężkich w skałach macierzystych jako tło geochemiczne gleb. *Rocz. Glebozn.* **1996**, *47*, 43–50.
36. Słowik, T.; Piekarski, W.; Tarasińska, J. Analiza statystyczna wpływu odległości i głębokości poboru próbek od drogi na zawartość niektórych jonów metali ciężkich w glebie. [The statistical analysis of the influence of the distance and depth of sampling from the road on the content of some heavy metal ions in the soil.]. *Inżynieria Rol.* **2006**, *6*, 275–283.
37. Karczewska, A.; Kabała, C. Trace elements in soils in the Stołowe Mountains National Park. *Szczeliniec* **2002**, *6*, 133–160.
38. Korzeniowska, J.; Krąż, P. Heavy Metals Content in the Soils of the Tatra National Park Near Lake Morskie Oko and Kasprowy Wierch—A Case Study (Tatra Mts, Central Europe). *Minerals* **2020**, *10*, 1120. [CrossRef]
39. Barbieri, M.J. The importance of enrichment factor (EF) and geoaccumulation index (Igeo) to evaluate the soil contamination. *J. Geol. Geophys.* **2016**, *5*, 1–4. [CrossRef]
40. Li, Y.; Zhou, H.; Gao, B.; Xu, D. Improved enrichment factor model for correcting and predicting the evaluation of heavy metals in sediments. *Sci. Total Environ.* **2021**, *755*, 142437. [CrossRef]
41. Looi, L.J.; Aris, A.Z.; Yusoff, F.M.; Isa, N.M.; Haris, H. Application of enrichment factor, geoaccumulation index, and ecological risk index in assessing the elemental pollution status of surface sediments. *Environ. Geochem. Health* **2019**, *41*, 27–42. [CrossRef]

International Journal of
Environmental Research and Public Health

Article

Apportionment and Spatial Pattern Analysis of Soil Heavy Metal Pollution Sources Related to Industries of Concern in a County in Southwestern China

Xiaohui Chen [1,2], Mei Lei [1,2,*], Shiwen Zhang [3], Degang Zhang [1,2], Guanghui Guo [1] and Xiaofeng Zhao [1,2]

[1] Center for Environmental Remediation, Institute of Geographic Sciences and Natural Resources Research, Chinese Academy of Sciences, Beijing 100101, China; luokeangcxh@163.com (X.C.); zdg_biology2@126.com (D.Z.); guogh@igsnrr.ac.cn (G.G.); zhaoxf2011@163.com (X.Z.)
[2] University of Chinese Academy of Sciences, Beijing 100049, China
[3] School of Earth and Environment, Anhui University of Science and Technology, Huainan 232001, China; shwzhang@aust.edu.cn
*   Correspondence: leim@igsnrr.ac.cn

Abstract: Soil heavy metal pollution is frequent around areas with a high concentration of heavy industry enterprises. The integration of geostatistical and chemometric methods has been used to identify sources and the spatial patterns of soil heavy metals. Taking a county in southwestern China as an example, two subregions were analyzed. Subregion R1 mainly contained nonferrous mining, and subregion R2 was affected by smelting. Two factors (R1F1 and R1F2) associated with industry in R1 were extracted through positive matrix factorization (PMF) to obtain contributions to the soil As (64.62%), Cd (77.77%), Cu (53.10%), Pb (75.76%), Zn (59.59%), and Sb (32.66%); two factors (R2F1 and R2F2) also related to industry in R2 were extracted to obtain contributions to the As (53.35%), Cd (32.99%), Cu (53.10%), Pb (56.08%), Zn (67.61%), and Sb (42.79%). Combined with PMF results, cokriging (CK) was applied, and the z-score and root-mean square error were reduced by 11.04% on average due to the homology of heavy metals. Furthermore, a prevention distance of approximately 1800 m for the industries of concern was proposed based on locally weighted regression (LWR). It is concluded that it is necessary to define subregions for apportionment in area with different industries, and CK and LWR analyses could be used to analyze prevention distance.

Keywords: heavy metals; industries of concern; source apportionment; spatial patterns; prevention distance

Citation: Chen, X.; Lei, M.; Zhang, S.; Zhang, D.; Guo, G.; Zhao, X. Apportionment and Spatial Pattern Analysis of Soil Heavy Metal Pollution Sources Related to Industries of Concern in a County in Southwestern China. *Int. J. Environ. Res. Public Health* **2022**, *19*, 7421. https://doi.org/10.3390/ijerph19127421

Academic Editors: Bo Sun, Ming Liu and Yan Chen

Received: 21 March 2022
Accepted: 9 June 2022
Published: 16 June 2022

Publisher's Note: MDPI stays neutral with regard to jurisdictional claims in published maps and institutional affiliations.

1. Introduction

With the rapid development of the global economy, industrialization is intensifying, and cities are expanding. Heavy metals in industrial waste have affected the surrounding soil, making it easier for heavy metal concentrations to reach high levels of toxicity [1]. Heavy metal pollutants do not degrade easily, have poor mobility, and easily accumulate. Additionally, mining and smelting operations play a critical role in the accumulation of heavy metals in local soils, especially in areas with resource-dependent economies. For example, the mining of polymetallic ores provides not only the targeted metals but also vast amounts of by-products, such as tin and copper associated with sulfide ore [2]. The areas affected by mining are always seriously polluted by heavy metals, which are emitted from the mining waste and tailings and are transported by runoff and human activities [1,3]. The concern is that the mining sites, tailings, and associated equipment could be pollution sources for a long time [4]. In addition, emissions from smelting may be distributed widely via atmospheric deposition, and this is especially true for metals such as As, Cd, and Pb, which can form into oxides and condense into particulates at high temperature [5].

Identification and quantitation of the sources of pollutants in soils are critical for risk assessments and the development of reclamation strategies. The apportionment of pollution sources began with atmospheric particulates, and the technical system consisted of an emission inventory, mass diffusion model, and receptor model [6–9]. The main objects of the apportionment for soil pollution sources are heavy metals and some organic matter (e.g., polycyclic aromatic hydrocarbons, PAHs), which are always widespread and do not easily degrade [10,11]. Receptor models have been the primary method because of the uncertainty of emissions inventory information and the complexity of mass diffusion models [9,12–14]. Currently, many studies have estimated the spatial patterns of source factors extracted from receptor models to increase the credibility of pollution source identification [13,15]. When using receptor models, few studies have paid attention to spatial variations in the pollution sources, e.g., the areas that have a higher density of key industries that emit pollution.

In China, investigations of the soil environment around enterprises in key industries have been conducted [16]. Due to a literal misunderstanding, the key industries were named as the industries concerned in these studies. At the beginning of the survey, risk screening was carried out through attribute scores of large-scale areas [17,18], but the soil environment around the industries of concern should be investigated before health risk assessments of contaminated sites are conducted to provide suggestions about the spatial patterns of the environmental protection and the priority pollutants on the county level, which is also the target scale in this study. The study area is a remote region with predominantly silver and copper mining in the early stage. However, after the 1840s, it became a strategic area for tin, lead, and zinc mining and smelting. These mineral resources have been gradually exhausted since the 1990s, but historic mining has been inseparable from the local businesses and the livelihoods of the residents, so determination of a sustainable development scheme for this region has become a critical issue. Most environmental surveys and studies of this area have paid attention to the issues of soil heavy metal pollution, which mainly originates from mining and smelting activities [19–21]. In addition to tin, multiple minerals occur in the study area, including copper, lead, and zinc ores. Some of the mineral deposits are distributed within the tin mineralization zone, while others are distributed outside of this zone. The characteristics of soil heavy metal pollution in polymetallic mining area can help to distinguish the contribution of major industrial pollution sources and give the protection distance of soil heavy metal pollution of main industrial land in this area through its spatial distribution with enterprise land.

Specifically, the main objectives of this study are as follows: (i) to explore the relationship between the soil heavy metals and the spatial distribution of the pollution sources from industries of concern; (ii) to extract the factors of the soil heavy metals through positive matrix factorization (PMF) to estimate the contributions of the industries of concern; and (iii) to determine the spatial patterns of the heavy metals originating from certain factors through cokriging (CK), which makes use of the homology of heavy metals to improve the accuracy of the assessment, so as to determine the variations in the heavy metals with distance from the nearest source through locally weighted regression (LWR).

2. Materials and Methods

2.1. Study Area and Investigation of Pollution Sources

The study area (102°54′–103°25′ E; 23°01′–23°36′ N) contained more than 400 key enterprises and consists of five towns, with a total area of 992.05 km², which are in a supergiant tin polymetallic district located along the suture zone of the Indian and Eurasian plates on the southwestern edge of the China sub-plate. This area has a subtropical mountain monsoon climate, with an annual average temperature of 15.9 °C and an annual rainfall of 1292.8 mm. The parent materials of the area are very complex, which is mainly composed of limestone and minor dolomite, and the main soil types are yellow-brown and red soils.

Using a combination of remote sensing and historical business information, a field survey of the pollution sources related to the nonferrous metal industry was conducted, and five types of sources were confirmed and vectorized, which provided their spatial positions and areas (Table S1). The results of the field survey revealed that 3000 ha of land may have been the source of the soil pollution in subregion 1 (R1). Mining had led to destruction of the landscape, and tailings (i.e., large particles that settle at the bottom of the flotation tank), which are of lower economic value, were directly discarded in a natural depression during a time when environmental protection was not of great concern. Additionally, in subregion 2 (R2), 935 ha of land could be a potential source of soil pollution, and this land was mainly covered by slag produced by lead and zinc smelting.

2.2. Sample Collection and Chemical Analysis

The collection and analysis of 230 samples were completed in 2018, and the database used in this study also incorporated the results of geochemical surveys conducted in 2013 and 2015 (390 samples). In R1 and R2, 389 and 231 samples were collected and analyzed, respectively. All surveys were conducted using the same sampling and geochemical analysis methods, and the soil samples were collected according to the distribution of the different land use (dry land, $n = 412$; paddy land, $n = 123$; garden plot, $n = 85$). Each composite sample (1 kg) was composed of five subsamples collected at the central point and four additional points within an area of 5 m^2. The 3S technique was used for the sampling conducted around the mining area, industrial plants, mining waste heaps, smelting slag heaps, and tailings ponds. In addition, several samples were collected away from the pollution sources to study the pollution from atmospheric deposition and surface runoff. Moreover, the sampling density around the sources was four samples per square kilometer (Figure 1).

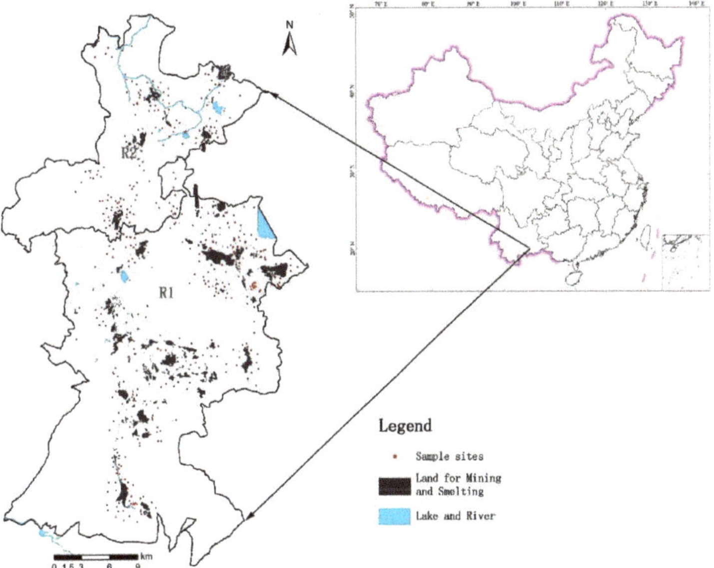

Figure 1. Land used for mining and smelting and the sampling sites in the study area.

Upon receipt, the samples were dried in a lyophilizer and sieved (2 mm mesh), and then, the stones, litter, and roots were removed. The total major element contents (K, Ca, Mn, and Fe) of the samples were analyzed using an X-ray fluorescence spectrometer (Niton FXL analyzer, Thermo-Fisher Scientific, Waltham, MA, USA) [22]. Then, the samples were digested in HNO_3 and H_2O_2 using method 3050B (USEPA, 1996). The total As concentration was analyzed using atomic fluorescence spectroscopy (AFS-9800, Haiguang Instrument Co., Beijing, China), and the Cd concentration was analyzed using graphite furnace atomic absorption spectrometry (contrAA700, Analytikjena, Jena, Germany). The other minor elements were measured using inductively coupled plasma optical emission spectrometry (Optima 5300DV, PerkinElmer, Boston, MA, USA). The detection limits of As, Cd, Cu, Cr, Ni, Pb, Sb, and Zn were 0.10, 0.05, 0.10, 0.10, 0.05, 0.10, 0.05, and 0.50 mg/kg, respectively. For quality control, blanks control, sample replicates (20%), and standard reference materials (GSS-5/GBW07405) were included in each batch of sample digestion and chemical analysis. And the relative standard deviations were less than 5%.

2.3. Methodology

2.3.1. Exploratory Analysis

Identification of soil heavy metal sources on the regional scale (i.e., up to 1000 km^2, almost the area of a county) using receptor models is difficult because of the heterogeneity of the parent materials of the soil and the high variability of anthropogenic activities. Thus, in this study, the entire region was partitioned using the spatial distribution of the pollution sources based on the enterprise land survey and the collection of data on the environmental factors. Moreover, this idea could help deal with the rotation ambiguity (i.e., different results obtained via PMF may generate similar model fitting) by decreasing the number of columns in the factor contribution matrix and the number of rows in the factor profile matrix.

After this, correlation analysis was conducted on the concentrations of the elements in the samples and the environmental factors. Because of the heavy rainfall in the study area, surface runoff and soil water flow must affect the spatial distribution of soil heavy metals, and water flow is closely related to topographic characteristics. Elevation (EL) directly reflects topographic features, while humidity index (HI) quantifies topographic control over basic hydrological processes. EL and HI were obtained via digital elevation model (DEM) data processing. DEM data came from a geospatial data cloud platform (https://www.gscloud.cn/, accessed on 8 July 2021), with a resolution of 30 m. The ELs of the samples were obtained directly using a spatial overlay, while the HI was determined from the topographic HI [23]:

$$HI_i = \ln(a_i/\tan\beta_i), \tag{1}$$

where HI_i represents the humidity index at surface point i; a_i represents the specific catchment area, i.e., the contributing upslope area per unit width of the contour; and β_i is the gradient at point i.

The other environmental factors considered were distance (Dist) and direction (Dir) from the nearest pollution source, which were estimated using the Near tool in ArcGIS version 10.4.1 (Esri, Redlands, CA, USA).

2.3.2. Source Apportionment via PMF Model

Similar to principal component analysis, PMF is a typical analytical factoring technique, which is based on the study of Paatero and Tapper [24,25]. In this study, the PMF 5.0 program was used to conduct the source apportionment of the soil heavy metals. The receptor sites were defined as the matrix relationship of a two-dimensional factor analytic with a residue matrix,

$$X = GF + E, \tag{2}$$

or in the component form,

$$x_{ij} = \sum_{k=1}^{K} g_{ik}f_{kj} + e_{ij}, \tag{3}$$

where x_{ij} represents the measured sample concentration, g_{ik} represents the contribution of the kth factor for the ith sample, f_{kj} represents the composition of the jth element within the kth factor, and e_{ij} is the residual error.

Matrix G and matrix F were approximated using the PMF model to minimize the objective function Q under the constraint of non-negative contributions, which relies on more physically significant assumptions than other factor analysis methods [25].

$$Q = \sum_{i=1}^{n} \sum_{j=1}^{m} \left(\frac{e_{ij}}{u_{ij}}\right)^2, \tag{4}$$

where u_{ij} represents the uncertainty of the jth chemical element for sample i.

2.3.3. Spatial Pattern Analysis

Once contents of the heavy metals, which originated from industries of concern, were confirmed approximately through the above-described processes, these data were used to estimate the heavy metals of critical concern on a 200 m interval grid across the study area. Multivariable CK was chosen because homologous heavy metals could help each other to improve the accuracy through this method [26,27], and the estimation function was as follows:

$$Z^*(x) = \sum_{i=1}^{n} Z(x_i)\Gamma_i, \tag{5}$$

where $x_1, \cdots, x_n$ represent the locations of the samples, and $Z_1(x), \cdots, Z_m(x)$ represent the values of the multivariates at location x. Γ_i represents the weighted vector. To determine Γ_i, an unbiased estimation was made with the smallest variance of error. Furthermore, by introducing Lagrangian multipliers and determining the derivation of Γ_i, linear equations were derived with semi-variogram and cross-variogram functions [28]. The semi-variogram was obtained by fitting the models, including Matérn, spherical, exponential, and power function models, while the cross-variogram functions were calculated as follows:

$$\gamma_{ij}(h) = \frac{1}{2}\left[\gamma_{ij}^+(h) - \gamma_{ii}(h) - \gamma_{jj}(h)\right], \tag{6}$$

where $\gamma_{ii}(h)$ represents semi-variogram of the ith variate $Z_i(x)$, $\gamma_{ij}(h)$ represents the cross-variogram between the ith and jth variates, and $\gamma_{ij}^+(h)$ represents the semi-variogram of $Z_{ij}^+(x)$, which is equivalent to $Z_i(x) + Z_j(x)$.

It must be emphasized that the auxiliary variables used in this study not only have a significant correlation, but they also have the same origin as the heavy metals from the mining and smelting sources, which increases the accuracy of the spatial prediction and provides a good foundation for the subsequent analysis using the environmental factors. In order to measure the error, the method of Zhang and Wang (2009) [29] and cross-validation were used depending on the predictions and variances of the predictions derived from the remaining observations. The predictive accuracy scores were calculated as follows:

$$z_i = \frac{Z(x_i) - \hat{Z}_{-i}(x_i)}{\sigma_{-i}}, \tag{7}$$

where $Z(x_i)$ represents the observation made at location x_i for $i = 1, \cdots, n$; and $\hat{Z}_{-i}(x_i)$ is the drop-one prediction based on all of the data $Z(x_j)$ for $j \neq i$. σ_{-i} is the corresponding standard deviation. Then, the mean of z_i becomes the z-score for the first index for evaluating the error conditions. Another predictive score is the root mean square error (RMSE):

$$\text{RMSE} = \left[\frac{1}{n}\sum_{i-1}^{n}(Z(x_i) - \hat{Z}_{-i}(x_i))^2\right]^{1/2}, \tag{8}$$

The RMSE should be as low as possible value, while the z-score should be close to 0.

Then, in this study, the heavy metals from the industries of concern were analyzed based on the environmental factor, i.e., the distances to the nearest pollution sources. The LWR method was used [30], and the target equation was minimized as follows:

$$\sum \omega^{(i)} \left(y^i - \theta^T x^{(i)} \right)^2,\qquad(9)$$

where $\omega^{(i)} = exp\left(-\frac{\left(x^{(i)}-x\right)^2}{2\tau^2}\right)$, which is the weight based on the degree of proximity to the predicted point. If the value of $\left| x^{(i)} - x \right|$ is very small, $\omega^{(i)} \approx 1$; while for a very large value, $\omega^{(i)} \approx 0$. τ indicates the rate of decrease with the degree of proximity. Finally, this regression method was used to analyze the variations in the heavy metal contents with distance from the nearest source, and two-dimensional fitting lines for the 95% confidence interval were obtained.

3. Results

3.1. Descriptive Statistic and Analysis of Variance Analysis of Samples

The descriptive statistics of the topsoil heavy metals in the two subregions are presented in Table 1. Higher heavy metal concentrations were observed in R1, including higher As (108.61%), Ni (123.47%), Cr (114.66%), Sb (164.03%), and Cu (116.80%) levels than those in R2; while the Cd, Pb, and Zn concentrations in the two subregions were similar. The coefficients of variation (CVs) of the two regions for As were significantly different than those of the other heavy metals, which may indicate that As and Cu concentrations in the two subregions have different spatial variabilities. Furthermore, the CVs of Cd, Pb, and Sb were all high (>50%), indicating that extrinsic factors strongly affected the enrichment of these heavy metals. Additionally, with the exceptions of Sb, Cr, and Ni, soil heavy metals of the different land use greatly surpassed risk screening values [31,32], indicating that As, Cd, Pb, Zn, and Cu may pose a threat to human and plants (Table S2). Specifically, 92.32%, 100%, 80.12%, 62.54%, and 82.67% of the dry land soil samples surpassed the risk screening values for As, Cd, Pb, Zn, and Cu. For paddy land, As (97.14%), Cd (100%), Pb (72.14%), Zn (45.23%), and Cu (87.55%) exceeded their corresponding risk screening values. In garden plot samples, As (82.22%), Cd (100%), Pb (74.58%), Zn (49.34%), and Cu (70.52%) exceeded risk screening values.

Table 1. Descriptive statistics of the soil heavy metals.

Subregion	Element	Minimum	Maximum	Mean	SD	CV(%)
Subregion 1 (R1) (n = 389)	As (mg kg^{-1})	94.8	633.9	233.14	92.87	39.83
	Cd (mg kg^{-1})	1.33	11.24	2.77	1.4	50.51
	Pb (mg kg^{-1})	30.4	1236.4	322.46	183.5	56.91
	Zn (mg kg^{-1})	176.3	1134.5	503.72	139.06	27.61
	Ni (mg kg^{-1})	30.76	294.63	113.66	46.7	41.09
	Cr (mg kg^{-1})	14.86	463.4	154.57	70.68	45.73
	Sb (mg kg^{-1})	2.31	48.4	9.9	6.5	65.66
	Cu (mg kg^{-1})	119.42	897.05	277.29	119.62	43.14
	pH	5.42	8.06	6.52	1.3	19.94
Subregion 1 (R1) (n = 231)	As (mg kg^{-1})	18.8	538.2	214.65	131.41	61.22
	Cd (mg kg^{-1})	0.92	9.3	3.04	1.77	58.32
	Pb (mg kg^{-1})	161.34	1353.3	333.31	198.14	59.45
	Zn (mg kg^{-1})	348.7	1018.7	528.8	153.79	29.08
	Ni (mg kg^{-1})	32.41	220.9	92.05	40.58	44.08
	Cr (mg kg^{-1})	47.1	356.9	134.81	66.12	49.05
	Sb (mg kg^{-1})	0.27	19.18	6.04	5.08	84.19
	Cu (mg kg^{-1})	28.32	715.57	237.41	178.23	75.07
	pH	5.13	7.92	6.32	1.1	17.41

SD, standard deviation; CV, coefficient of variation.

An overview of the exploratory analysis dealing with the correlations between the elements and the environmental factors in R1 and R2 are presented in Tables S3 and S4. The Pearson's correlation coefficients between the soil heavy metals, major elements (K, Ca, Mn, and Fe), and environmental factors (Dist, HI, and EL) for the samples from the two subregions were calculated (Table S3). In R1, the correlations between As, Cd, Pb, Zn, and Cu were all significant ($p < 0.01$) moderately or strong positive (Table S5), and Sb was moderately correlated with As and Cd. Dist exhibited low negative correlations with Pb, Zn, and Cu. The other correlations between the environmental factors and soil heavy metals were negligible. In R2, Pb was strongly positively correlated with Zn, while As was strongly positively correlated with Cd, Cu, and Sb. Similarly, Dist was slightly negatively correlated with Pb, Zn, and Cd, while the other correlations were negligible. In both R1 and R2, the major elements K, Mn, and Fe all exhibited at least moderate correlations, but Ca performed differently, which may be due to the different geological environments. Next, analysis of variance (ANOVA) [33] was applied to examine the differences in the soil heavy metal contents in different directions from the nearest pollution sources (Table S4). The ANOVA results revealed that the number of differences between the means ($p < 0.05$) was always larger toward the south. For example, in R1, all of the means of the heavy metal contents in the samples whose nearest sources were located to the south were larger than those whose nearest sources were located to the west, and significant differences were identified for six heavy metals. Generally, samples with pollution sources located to the south always had positive significant differences compared with the other directions.

3.2. Source Identification

The receptor data for the two subregions, including eight heavy metals and four major elements, were used as the input data for the PMF. Originally, various trials with different numbers of factors were conducted, and the results revealed that a small number of factors (i.e., three or four factors) resulted in poor fitting, while six factors were excessive because the samples were mainly located within reach of the pollution sources, which was also indicated by the loss function's Q value. Thus, the results for the two subregions using five factors both provided a reasonable interpretation with good fitting (i.e., R2 for heavy metals ≥ 0.50) and consistency with the field survey. As was previously mentioned, the model dealt with the rotational ambiguity by exploring different values of the rotational parameter Fpeak (between -1 and $+1$, step $= 0.1$), and -0.5 was adopted for R1, while no change was adopted for R2. The contributions are presented in Figure 2.

3.2.1. Factors Related to the Industries of Particular Concern

The industries of particular concern in this study were nonferrous metal mining and smelting, which were also the focus of the field survey. The factors were categorized into two groups: related to industries of concern and others (Figure 2). The main categorization rule was the pollution characteristics of the soil heavy metals, including As, Cd, Cu, Pb, Zn, and Sb, which have been reported in previous studies [19–21]. In R1, factor 1 (R1F1) mainly explained the Pb (69.55%) and Zn (44.02%), and it also made moderate contributions (approximately 20%) to Cr, Ni, As, Cd, and Cu, indicating that it may be a key contamination source for the study area. Factor 2 (R1F2) exhibited the same characteristics and made large contributions to Cd (61.92%), As (47.82%), Cu (36.77%), and Sb (27.92%). Furthermore, R1F2 was determined to be robust because it was almost unaffected by the rotations in many trials. In R2, R2F1 predominantly contributed to the Pb (41.84%) and Zn (39.46%), but it also contributed to the (24.91%) and As (22.12%). R2F2 contributed to the As, Cd, Pb, Zn, and Ni, with at least 28% contributions for each metal, which demonstrates that it was very likely a contamination source related to smelting. Therefore, the above factors contributed greatly to the As, Cd, Cu, Pb, Zn, and Sb, which were the main pollutants in the study area. The specific judgment will be discussed in the next section.

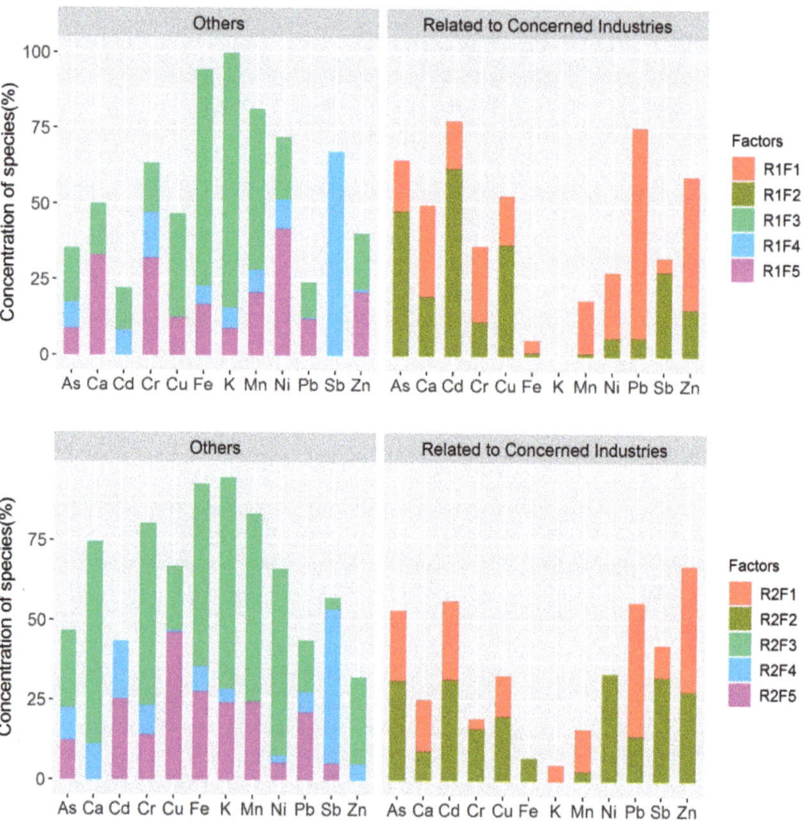

Figure 2. Contributions of the profiles for R1 (**top**) and R2 (**bottom**).

3.2.2. Other Factors

First of all, R1F3 and R2F3 deserve more attention because they are both very conspicuous, with large contributions to the major elements (K, Mn, Fe, and Ca). It is almost certain that R1F3 and R2F3 come from the weathering of local minerals. They also make moderate contributions to some of the heavy metals due to the high background values in the study area. In addition, some of the samples were collected from farmland and gardens, so R1F4 and R2F4 could be related to agricultural activity. The chemical plants, which were part of the industry chains of the local mining and smelting, may be related to R1F5 and R2F5.

3.3. Spatial Pattern of Key Heavy Metals Related to Industries of Concern

In this section, only the heavy metals originating from factors related to the industries of particular concern, i.e., nonferrous industry, are discussed. Based on their contents, the spatial distributions of the affected areas were estimated on a 200 m interval grid. The CK method was used to build models for auto-variograms and cross-variograms, which were found to have exponential forms in this study (Figure S1). Then, based on the spatial distributions of the heavy metals (Figure 3), the variation trends of the heavy metals with distance from the nearest source were analyzed, and two-dimensional fitting lines for the 95% confidence interval were obtained (Figure 4).

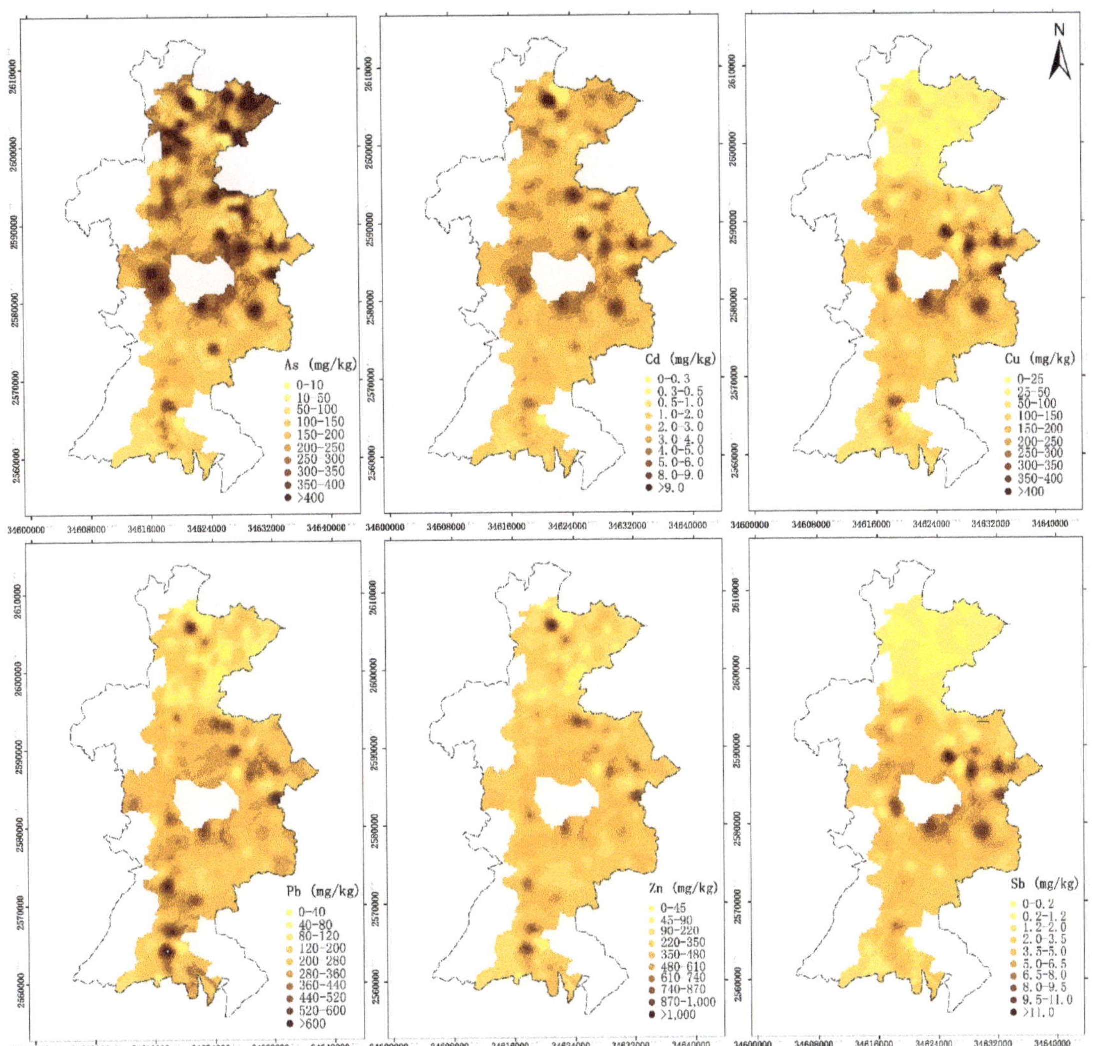

Figure 3. Spatial distributions of heavy metals obtained via PMF.

In order to determine the effective auxiliary variables for CK analysis of each heavy metal, the error conditions were tested through numerous trials (Table 2), and the variogram models were constructed to identify the homogeneous characteristics (Figure S1). As can be seen from Table 2, all RMSEs of the CK results are less than those of Universal Kriging (UK). The z-scores of the CK results are closer to 0 except for the CK for Cu. Figure 3 shows that the heavy metal concentrations around the study area all severely exceed the background levels (Table S5) due to the presence of nonferrous industry enterprises. However, the spatial distribution patterns of these elements were different. Before estimating the spatial distribution, the area with a small sampling density was excluded. The results revealed that the spatial patterns of the effects of the pollution sources were different in R1 and R2. In R1, the soil Cu and Sb contents were high, while the As, Cd, Pb, and Zn all exhibited spatial patterns consistent with the pollution sources in the two subregions. Soil As pollution was high in the entire area, and the high-value areas were consistent with the source distribution, especially in the region with clusters of smelters in the north. The excess degree of Cd was the highest, and the value in the highest value area exceeds the background value by more than 40 times. Heavy Pb and Zn pollution were observed in both R1 and R2, with values approximately 10 times higher than the background values. Heavy Cu and Sb pollution were observed in R1 but were not detected in R2.

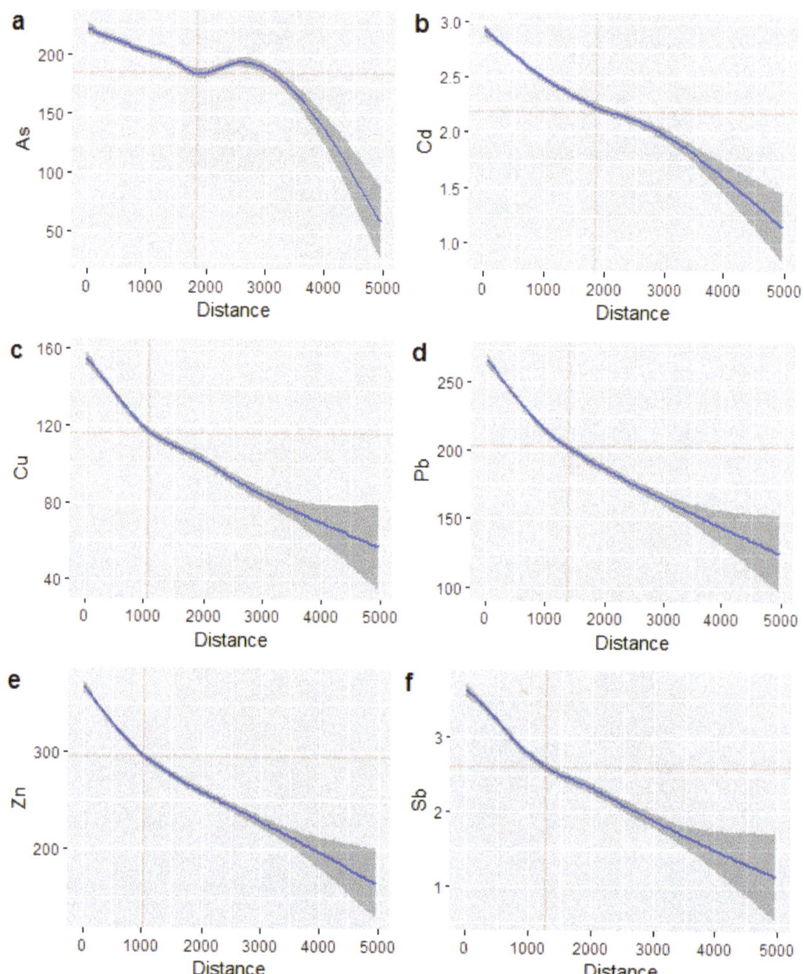

Figure 4. Variations in the contents of the heavy metals related to the industries of concern with distance from the nearest source. (**a–f**) The red crosses denote the changing nodes of the variation trends.

Figure 4 shows the variations in the heavy metals with distance from the nearest source, and the contents decrease with increasing distance. In Figure 4a, the trend flattens from 1800 m to 2500 m. However, the decreasing trends of the other heavy metals did not flatten, and the changing nodes are denoted by red crosses in Figure 4. When the distance is greater than the distance of the change node, the heavy metal contents decreased more slowly, and the uncertainty increased. The changing nodes of As and Cd were both located at approximately 1800 m, while those of Cu, Pb, Zn, and Sb were all located between 1000 m and 1500 m.

Table 2. Cross-validation results for cokriging and universal kriging.

Title 1	As		Cd		Cu	
entry 1	CK with Cd and Cu	UK	CK with As	UK	CK with Sb	UK
z-score	7.57×10^{-4}	1.74×10^{-3}	6.77×10^{-3}	7.09×10^{-3}	5.25×10^{-3}	4.54×10^{-3}
RMSE (mg/kg)	79.74	83.38	0.95	1.01	50.51	53.55
entry 2	Pb		Zn		Sb	
	CK with Zn	UK	CK with Pb	UK	UK with Cu	UK
z-score	1.05×10^{-2}	1.08×10^{-2}	1.30×10^{-2}	1.35×10^{-2}	-1.25×10^{-3}	-1.46×10^{-3}
RMSE (mg/kg)	90.42	102.19	113.18	127.22	1.39	1.48

4. Discussion

This integrated approach provides a method of apportioning the contributions of industries of concern to soil heavy metals and of estimating the spatial patterns. The information obtained can be used to develop suggestions for the prevention and control of pollution around polluted industrial land. Several studies have been performed on source apportionment and geostatistics [34–36], focusing on clarifying the sources. However, this study emphasizes the characteristics of the soil pollution around industries of concern by extracting certain factors via PMF and estimating the spatial distributions using CK with the homology of the elements.

In the field survey and soil sampling, the object of the study is the soil within a distance of 5 km around the industries of concern. With regard to the public information about the protective distance of the soil environment from polluting enterprises [37], it is suggested that the scope of influence of nonferrous industries is within 5 km, and some studies [38,39] have also collected samples within 4 km or 5 km. Through exploratory analysis, it was found that the soil pollution to the north of the pollution sources is more serious, which can be explained by the fact that the dominant wind is from the south. In addition, the Dist exhibited a low correlation with Pb, Zn, and Cu, and the correlation with the other heavy metals is negligible. Therefore, further research is needed to determine the trend based on a receptor model and regression analysis.

Apportionment of soil pollution sources, which clarifies all of the potential sources, requires uniform sampling of the entire area [40–42], but in this study, only the sources related to mining and smelting are considered. Therefore, we provide a brief discussion of the other factors, which mainly depend on the characteristics of the major elements and the land-use types of the sampling locations. As was previously stated, the main migration pathways in R1 were likely dry and wet deposition from mining and concentration and surface runoff from the tailings. Furthermore, the pollution related to the tailings left over from historical mining is very serious, and leaching tests in the study area have demonstrated that the pollution derived from the tailings contains large amounts of As, Zn, and Pb. Therefore, R1F1 may be associated with the tailings. R1F2 made considerable contributions to all of the heavy metals, so it may be related to dry and wet deposition from mining and concentration. In R2, R2F1 was dominated by Pb and Zn, so it clearly originated from the emissions from smelters based on the survey. In addition, R2F1 is also associated with the accumulation of Cd and As, which is attributed to the atmospheric deposition of emissions from smelters. This conclusion is consistent with the results of previous studies [43,44]. In addition, the receptor model can be regarded as a tool for extracting factors of concern, which helps to illustrate the variation trends of the spatial patterns described below.

While some reports provided the spatial patterns based on the factors' contributions, in this study, the contents of heavy metals related to industries of concern were extracted based on a receptor model, and the similar characteristics of some of the heavy metals (e.g., As and Cd have similar profiles related to the nonferrous industry) were taken advantage of to improve the accuracy of the spatial estimation through CK. The spatial variation trends of the heavy metals around polluting enterprises were studied [38,45,46], and the problem of contingency cannot be avoided. Based on statistical analysis, when the variation in the studied objects is large, the more samples are used, and the clearer the trend will be. The spatial patterns obtained via PMF and CK exhibit clear trends in the heavy metal contents with distance from the nearest sources. Then, the trends flattened, or change nodes of the trendlines occurred due to weakening of certain sources' influences as the homologous nature of the other sources increased. In addition, the uncertainty increased significantly, which may be due to the weakened influences or the decreased sampling density. The above data show that 1800 m may be a satisfactory pollution protective distance in the study area. The LWR method has the ability to process a large amount of data, so it is considered to be a better tool based on trials using many regression methods.

5. Conclusions

In this study, geostatistical and chemometric methods were combined to identify pollution sources and estimate the spatial patterns of the soil heavy metals around industries of concern. It was found that at the county level, subregions should be defined before the apportionment of the sources due to the large spatial variations in the polluting enterprises. The main factors related to the industries of concern were extracted using the characteristics of the key heavy metals, which have been reported. The contents of the heavy metals from the main factors can help each other improve the estimation accuracy of the CK due to their homology. The trendlines of the variations with distance from the nearest sources can be used to determine a pollution protective distance around the industries of concern, but additional research is required.

Supplementary Materials: The following supporting information can be downloaded at: https://www.mdpi.com/article/10.3390/ijerph19127421/s1, Figure S1: Auto-variograms and cross-variograms between the heavy metals extracted by PMF; Table S1: The results of field survey for the soil pollution sources in study area; Table S2: Descriptive statistics of soil heavy metals based on different land use; Table S3: Pearson coefficients for heavy metals, major elements, and the environmental factors of the two subregion; Table S4: Comparison of means between different directions for soil heavy metals; Table S5: Rule of thumb for interpreting the size of a correlation coefficient; Table S6: Background values of soil heavy metals in study area. Refs. [31,32,47] are cited in Supplementary Materials.

Author Contributions: Conceptualization, M.L. and X.C.; methodology, S.Z. and X.C.; software, X.C. and X.Z.; validation, M.L., S.Z. and G.G.; formal analysis, D.Z.; investigation, X.C. and D.Z.; resources, M.L.; data curation, X.C.; writing—original draft preparation, X.C.; writing—review and editing, M.L.; visualization, S.Z.; supervision, S.Z.; project administration, M.L. All authors have read and agreed to the published version of the manuscript.

Funding: This research was funded by the National Key R&D Program of China (Grant No. 2018YFC1800100).

Institutional Review Board Statement: Not applicable.

Informed Consent Statement: Not applicable.

Data Availability Statement: Data are contained within the article and Supplementary Material.

Acknowledgments: Gratitude is extended to Yang Haikun, Zhan Mingxu, and Wang Caiyun, who supported our study during the field survey, discussion, and paper writing.

Conflicts of Interest: The authors declare no conflict of interest.

References

1. Li, Z.; Ma, Z.; van der Kuijp, T.J.; Yuan, Z.; Huang, L. A review of soil heavy metal pollution from mines in China: Pollution and health risk assessment. *Sci. Total Env.* **2014**, *468–469*, 843–853. [CrossRef]
2. Kossoff, D.; Hudson-Edwards, K.A.; Howard, A.J.; Knight, D. Industrial mining heritage and the legacy of environmental pollution in the Derbyshire Derwent catchment: Quantifying contamination at a regional scale and developing integrated strategies for management of the wider historic environment. *J. Archaeol. Sci. Rep.* **2016**, *6*, 190–199. [CrossRef]
3. Guo, L.; Zhao, W.; Gu, X.; Zhao, X.; Chen, J.; Cheng, S. Risk Assessment and Source Identification of 17 Metals and Metalloids on Soils from the Half-Century Old Tungsten Mining Areas in Lianhuashan, Southern China. *Int. J. Environ. Res. Public Health* **2017**, *14*, 1475. [CrossRef]
4. Gamarra, J.G.P.; Brewer, P.A.; Macklin, M.G.; Martin, K. Modelling remediation scenarios in historical mining catchments. *Environ. Sci. Pollut. Res. Int.* **2014**, *21*, 6952–6963. [CrossRef]
5. Wuana, R.A.; Okieimen, F.E. Heavy Metals in Contaminated Soils: A Review of Sources, Chemistry, Risks and Best Available Strategies for Remediation. *Isrn Ecol.* **2011**, *2011*, 402647. [CrossRef]
6. Zhang, Y.; Dore, A.J.; Ma, L.; Liu, X.; Ma, W.; Cape, J.; Zhang, F. Agricultural ammonia emissions inventory and spatial distribution in the North China Plain. *Environ. Pollut.* **2009**, *158*, 490–501. [CrossRef]
7. Shi, G.L.; Liu, G.R.; Peng, X.; Wang, Y.N.; Tian, Y.Z.; Wang, W.; Feng, Y.C. A Comparison of Multiple Combined Models for Source Apportionment, Including the PCA/MLR-CMB, Unmix-CMB and PMF-CMB Models. *Aerosol Air Qual. Res.* **2014**, *14*, 2040–2050. [CrossRef]
8. Burr, M.J.; Zhang, Y. Source apportionment of fine particulate matter over the Eastern U.S. Part I: Source sensitivity simulations using CMAQ with the Brute Force method. *Atmos. Pollut. Res.* **2011**, *2*, 300–317. [CrossRef]
9. Huang, Y.; Deng, M.H.; Wu, S.F.; Jan, J.P.G.; Li, T.Q.; Yang, X.E.; He, Z.L. A modified receptor model for source apportionment of heavy metal pollution in soil. *J. Hazard. Mater.* **2018**, *354*, 161–169. [CrossRef]
10. Ghosal, D.; Ghosh, S.; Dutta, T.K.; Ahn, Y. Current State of Knowledge in Microbial Degradation of Polycyclic Aromatic Hydrocarbons (PAHs): A Review. *Front. Microbiol.* **2016**, *7*, 27. [CrossRef]
11. Toth, G.; Hermann, T.; Da Silva, M.R.; Montanarella, L. Heavy metals in agricultural soils of the European Union with implications for food safety. *Environ. Int.* **2016**, *88*, 299–309. [CrossRef] [PubMed]
12. Chen, R.; Chen, H.; Song, L.; Yao, Z.; Meng, F.; Teng, Y. Characterization and source apportionment of heavy metals in the sediments of Lake Tai (China) and its surrounding soils. *Sci. Total Environ.* **2019**, *694*, 133819. [CrossRef] [PubMed]
13. Qu, M.; Wang, Y.; Huang, B.; Zhao, Y. Source apportionment of soil heavy metals using robust absolute principal component scores-robust geographically weighted regression (RAPCS-RGWR) receptor model. *Sci. Total Environ.* **2018**, *626*, 203–210. [CrossRef] [PubMed]
14. Salim, I.; Sajjad, R.U.; Paule-Mercado, M.C.; Memon, S.A.; Lee, B.-Y.; Sukhbaatar, C.; Lee, C.-H. Comparison of two receptor models PCA-MLR and PMF for source identification and apportionment of pollution carried by runoff from catchment and sub-watershed areas with mixed land cover in South Korea. *Sci. Total Environ.* **2019**, *663*, 764–775. [CrossRef]
15. Lv, J. Multivariate receptor models and robust geostatistics to estimate source apportionment of heavy metals in soils. *Environ. Pollut.* **2018**, *244*, 72–83. [CrossRef]
16. MEP-PRC. *Ministry of Environmental Protection Detailed "Soil Ten": Resolutely Declared War on Pollution*; Ministry of Environmental Protection of the People's Republic of China: Beijing, China, 2016.
17. Ding, G.; Xin, L.; Guo, Q.; Wei, Y.; Li, M.; Liu, X. Environmental risk assessment approaches for industry park and their applications. *Resour. Conserv. Recycl.* **2020**, *159*, 104844. [CrossRef]
18. Teng, Y.; Zuo, R.; Xiong, Y.; Wu, J.; Zhai, Y.; Su, J. Risk assessment framework for nitrate contamination in groundwater for regional management. *Sci. Total Environ.* **2019**, *697*, 134102. [CrossRef]
19. Cheng, Q.; Xia, Q.; Li, W.; Zhang, S.; Chen, Z.; Zuo, R.; Wang, W. Density/area power-law models for separating multi-scale anomalies of ore and toxic elements in stream sediments in Gejiu mineral district, Yunnan Province, China. *Biogeosciences* **2010**, *7*, 3019–3025. [CrossRef]
20. Ma, J.; Lei, E.; Lei, M.; Liu, Y.; Chen, T. Remediation of Arsenic contaminated soil using malposed intercropping of Pteris vittata L. and maize. *Chemosphere* **2018**, *194*, 737–744. [CrossRef]
21. Zhan, F.; He, Y.; Zu, Y.; Zhang, N.; Yue, X.; Xia, Y.; Luo, Y. Heavy metal and sulfur concentrations and mycorrhizal colonizing status of plants from abandoned lead/zinc mine land in Gejiu, Southwest China. *Afr. J. Microbiol. Res.* **2013**, *7*, 3943–3952. [CrossRef]
22. Du, G.D.; Mei, L.; Zhou, G.D.; Chen, T.B.; Qiu, R.L. Evaluation of Field Portable X-Ray Fluorescence Performance for the Analysis of Ni in Soil. *Spectrosc. Spectr. Anal.* **2015**, *35*, 809–813. [CrossRef]
23. Woods, R.A.; Sivapalan, M.; Robinson, J.S. Modeling the spatial variability of subsurface runoff using a topographic index. *Water Resour. Res.* **1997**, *33*, 1061–1073. [CrossRef]
24. Paatero, P. Least Squares Formulation of Robust Non-Negative Factor Analysis. *Chemom. Intell. Lab. Syst.* **1997**, *37*, 23–35. [CrossRef]
25. Paatero, P. The Multilinear Engine: A Table-Driven, Least Squares Program for Solving Multilinear Problems, including the n-Way Parallel Factor Analysis Model. *J. Comput. Graph. Stat.* **1999**, *8*, 854–888. [CrossRef]

26. Emery, X. Cokriging random fields with means related by known linear combinations. *Comput. Geosci.* **2012**, *38*, 136–144. [CrossRef]
27. Wang, L.; Dai, L.; Li, L.; Liang, T. Multivariable cokriging prediction and source analysis of potentially toxic elements (Cr, Cu, Cd, Pb, and Zn) in surface sediments from Dongting Lake, China. *Ecol. Indic.* **2018**, *94*, 312–319. [CrossRef]
28. Chilès, J.P.; Delfiner, P. *Geostatistics: Modeling Spatial Uncertainty*; John Wiley & Sons: New York, NY, USA, 2012.
29. Zhang, H.; Wang, Y. Kriging and cross-validation for massive spatial data. *Environmetrics* **2009**, *21*, 290–304. [CrossRef]
30. Linton, O.B. Local Regression Models. In *The New Palgrave Dictionary of Economics*; Palgrave Macmillan: London, UK, 2008; pp. 1–4.
31. *GB15618-2018*; Environmental Quality Standards for Soil. MEP-PRC: Beijing, China, 2018.
32. Chang, A.C.; Page, A.L.; Asano, T.; Hespanhol, I. Developing human health-related chemical guidelines for reclaimed wastewater irrigation. *Water Sci. Technol.* **1996**, *33*, 463–472. [CrossRef]
33. Iorio, M.D.; Muller, P.; Rosner, G.L.; Maceachern, S.N. An ANOVA model for dependent random measures. *Publ. Am. Stat. Assoc.* **2004**, *99*, 205–215. [CrossRef]
34. Schaefer, K.; Einax, J. Source Apportionment and Geostatistics: An Outstanding Combination for Describing Metals Distribution in Soil. *CLEAN-Soil Air Water* **2016**, *44*, 877–884. [CrossRef]
35. Guo, B.; Su, Y.; Pei, L.; Wang, X.F.; Zhang, B.; Zhang, D.M.; Wang, X.X. Ecological risk evaluation and source apportionment of heavy metals in park playgrounds: A case study in Xi'an, Shaanxi Province, a northwest city of China. *Environ. Sci. Pollut. Res.* **2020**, *27*, 24400–24412. [CrossRef] [PubMed]
36. Siddiqui, A.U.; Jain, M.K.; Masto, R.E. Pollution evaluation, spatial distribution, and source apportionment of trace metals around coal mines soil: The case study of eastern India. *Environ. Sci. Pollut. Res.* **2020**, *27*, 10822–10834. [CrossRef]
37. MEP-PRC. *Reference of Identification of Type of Industry and Sphere of Influence for Soil Polluting Enterprises*; Ministry of Environmental Protection of the People's Republic of China: Beijing, China, 2017.
38. Martley, E.; Gulson, B.L.; Pfeifer, H.R. Metal concentrations in soils around the copper smelter and surrounding industrial complex of Port Kembla, NSW, Australia. *Sci. Total Environ.* **2004**, *325*, 113–127. [CrossRef]
39. Liu, D. Research on Environmental Protection Distance of Lead and Zinc Smelting Industry Based on Actural Measurement. Ph.D. Thesis, HeFei University of Technology, Hefei, China, 2015.
40. Jin, G.; Fang, W.; Shafi, M.; Wu, D.; Li, Y.; Zhong, B.; Ma, J.; Liu, D. Source apportionment of heavy metals in farmland soil with application of APCS-MLR model: A pilot study for restoration of farmland in Shaoxing City Zhejiang, China. *Ecotoxicol. Environ. Saf.* **2019**, *184*, 109495. [CrossRef] [PubMed]
41. Jiang, H.; Cai, L.; Wen, H.; Hu, G.; Chen, L.; Luo, j. An integrated approach to quantifying ecological and human health risks from different sources of soil heavy metals. *Sci. Total Environ.* **2020**, *701*, 134466. [CrossRef] [PubMed]
42. Qi, P.; Qu, C.; Albanese, S.; Lima, A.; Cicchella, D.; Hope, D.; Cerino, P.; Pizzolante, A.; Zheng, H.; Li, J.; et al. Investigation of polycyclic aromatic hydrocarbons in soils from Caserta provincial territory, southern Italy: Spatial distribution, source apportionment, and risk assessment. *J. Hazard. Mater.* **2020**, *383*, 121158. [CrossRef]
43. Pereira, P.A.D.P.; Lopes, W.A.; Carvalho, L.S.; Rocha, G.O.d.; Bahia, N.d.C.; Loyola, J.; Quiterio, S.L.; Escaleira, V.; Arbilla, G.; Andrade, J.B.d. Atmospheric concentrations and dry deposition fluxes of particulate trace metals in Salvador, Bahia, Brazil. *Atmos. Environ.* **2007**, *41*, 7837–7850. [CrossRef]
44. Yi, K.; Fan, W.; Chen, J.; Jiang, S.; Huang, S.; Peng, L.; Zeng, Q.; Luo, S. Annual input and output fluxes of heavy metals to paddy fields in four types of contaminated areas in Hunan Province, China. *Sci. Total Environ.* **2018**, *634*, 67–76. [CrossRef]
45. Keegan, T.J.; Farago, M.E.; Thornton, I.; Hong, B.; Colvile, R.N.; Pesch, B.; Jakubis, P.; Nieuwenhuijsen, M.J. Dispersion of As and selected heavy metals around a coal-burning power station in central Slovakia. *Sci. Total Environ.* **2006**, *358*, 61–71. [CrossRef]
46. Rogival, D.; Scheirs, J.; Blust, R. Transfer and accumulation of metals in a soil–diet–wood mouse food chain along a metal pollution gradient. *Environ. Pollut.* **2007**, *145*, 516–528. [CrossRef]
47. Mukaka, M.M. Statistics corner: A guide to appropriate use of correlation coefficient in medical research. *Malawi Med. J.* **2012**, *24*, 69–71. [CrossRef] [PubMed]

International Journal of
*Environmental Research
and Public Health*

Article

The Role of Grass Compost and *Zea Mays* in Alleviating Toxic Effects of Tetracycline on the Soil Bacteria Community

Jadwiga Wyszkowska *, Agata Borowik and Jan Kucharski

Department of Soil Science and Microbiology, University of Warmia and Mazury in Olsztyn,
10-727 Olsztyn, Poland; agata.borowik@uwm.edu.pl (A.B.); jan.kucharski@uwm.edu.pl (J.K.)
* Correspondence: jadwiga.wyszkowska@uwm.edu.pl

Abstract: Given their common use for disease treatment in humans, and particularly in animals, antibiotics pose an exceptionally serious threat to the soil environment. This study aimed to determine the response of soil bacteria and oxidoreductases to a tetracycline (Tc) contamination, and to establish the usability of grass compost (G) and *Zea mays* (Zm) in mitigating adverse Tc effects on selected microbial properties of the soil. The scope of microbiological analyses included determinations of bacteria with the conventional culture method and new-generation sequencing method (NGS). Activities of soil dehydrogenases and catalase were determined as well. Tc was found to reduce counts of organotrophic bacteria and actinobacteria in the soils as well as the activity of soil oxidoreductases. Soil fertilization with grass compost (G) and *Zea mays* (Zm) cultivation was found to alleviate the adverse effects of tetracycline on the mentioned group of bacteria and activity of oxidoreductases. The metagenomic analysis demonstrated that the bacteria belonging to *Acidiobacteria* and *Proteobacteria* phyla were found to prevail in the soil samples. The study results recommend soil fertilization with G and Zm cultivation as successful measures in the bioremediation of tetracycline-contaminated soils and indicate the usability of the so-called core bacteria in the bioaugmentation of such soils.

Keywords: antibiotics in soils; microbial community; soil degradation; soil oxidoreductases; compost fertility

Citation: Wyszkowska, J.; Borowik, A.; Kucharski, J. The Role of Grass Compost and *Zea Mays* in Alleviating Toxic Effects of Tetracycline on the Soil Bacteria Community. *Int. J. Environ. Res. Public Health* **2022**, *19*, 7357. https://doi.org/10.3390/ijerph19127357

Academic Editors: Bo Sun, Ming Liu and Yan Chen

Received: 2 May 2022
Accepted: 14 June 2022
Published: 15 June 2022

Publisher's Note: MDPI stays neutral with regard to jurisdictional claims in published maps and institutional affiliations.

1. Introduction

Tetracyclines (Tc) represent a group of antibiotics with a broad spectrum of activities [1,2]. The first antibiotics from the class of tetracyclines (chlorotetracycline and oxytetracycline) were discovered at the end of 1940 s, with ultimately over 20 tetracyclines in use today [3]. They can be divided into three categories: I generation tetracyclines (chlorotetracycline, oxytetracycline, tetracycline, demeclocycline), II generation tetracyclines (doxycycline, minocycline, lymecycline, meclocycline, metacycline, rolitetracycline), and III generation ones (tigecycline, omadacycline, sarecycline) [2,4]. The I generation antibiotics are naturally synthesized, whereas the II and III generation ones are semi-synthetic.

Considering data derived from 103 countries, the World Organization for Animal Health [5] has reported tetracyclines to be the most frequently used antimicrobial agents for animals worldwide. In 2017, they accounted for 33.9% of all administered drugs, on average, followed by polypeptides (11.1%), penicillins (10.9%), and macrolides (10.4%). A slightly different situation is observed in the European countries [6]. Based on data derived from 31 European countries (30 UE/UOG Member States and Switzerland), the European Medicines Agency has demonstrated tetracyclines (30.4%), followed by penicillins (26.9%), and sulfonamides (9.2%) to be the greatest contributors to the total sales of antimicrobial agents in 2017. In 2020, the demand for tetracyclines changed, making them second (26.7%) after penicillins (31.1%) and followed by sulfonamides (9.9%) in the total sales of these agents. This was the first year when the sales of penicillins surpassed those of tetracyclines. Noteworthy is that the World Health Organization classifies tetracyclines as antimicrobial agents of great importance to human medicine [7].

Their beneficial antimicrobial properties and low production costs compared to other antibiotics have made them widely applicable not only in human [8,9] and animal therapies [10] but also in agriculture [3,11]. The antibiotics from the class of tetracyclines deployed most often for therapeutic purposes in animal production include chlorotetracycline, oxytetracycline, and tetracycline [3]. In many countries, tetracyclines are also often added to feedstuffs for livestock and poultry, and are used in aquaculture as growth stimulants [10,12–14]. Antibiotics are poorly metabolized under in vivo conditions [1,15]. According to Daghrir and Drogui [16], 70% of the tetracycline group antibiotics are excreted with human and animal urine and excreta and released in the active form to the natural environment. The continuous release of Tc residues to the environment [17,18] and the fact that low temperature contributes to the retardation and even complete inhibition of their degradation in the soil [19] make them ubiquitous in the soil environment (Table 1).

Table 1. The concentrations of tetracycline in soil reported in other countries and regions.

Location	Soil Sample Description	Maximum Antibiotics, Concentration, $\mu g\ kg^{-1}$	References
city of Shenyan, northeast China	depth of 0–15 cm	976.17	[20]
China, Beijing,	soils from vegetable greenhouses	74.4	[21]
Spain, South surrounding of Valencia	agricultural soils with 20% of clay	64.4	[22]
Korea, Hongcheon, Gangwon province,	paddy and upland soils	177.6	[19]
Spain, NW, Galicia	sand 40%, silt—38%, clay—22%	600	[23]
Germany	0–10, 10–20 and 20–30 cm soils sand 91.6%, silt—6.0%, clay—2.4%	86.2; 198.7; 171.7	[24]
Poland, areas of Pomeranian Voivodeship	soils from agricultural areas of Pomeranian Voivodeship	14.5	[25]

Tetracyclines usually pervade the soil environment after the use of fertilizers contaminated with these antibiotics [17,26]. Types of veterinary drug residues in soil are associated with animal species being manure providers. The greatest amounts of antibiotics are found in the manure derived from poultry and livestock production, while lesser ones in the cow's manure [23–29]. Antibiotics pervade the arable soils also as a result of field irrigation with wastewater [30–32]. The highest detection rate of antibiotics in the world, reaching almost 100% in soils and 98% in surface water, has been reported in China [27]. A far better situation is observed in the European countries [33]. According to Conde-Cid et al. [23], antibiotics can be detected in approximately 40% of manures and 17% of soils therein. In addition, the highly hydrophilic character and low volatility of antibiotics make them highly stable in surface and ground waters [16,34,35], and even in drinking water [36].

Given the above, they pose severe environmental problems, including ecological hazards and adverse effects on human health. These problems additionally lead to the emergence of antibiotic resistance genes in the environment [34,37]. Considering high amounts of tetracyclines used worldwide, their monitoring, including identifying responses of soil microorganisms and determining enzymatic activity, seems to be of key importance. Today, little is known about the basic changes in the community of microorganisms involved in the accelerated degradation of tetracyclines in the soil. Therefore, the aims of this research were to: (1) determine the response of soil bacteria and oxidoreductases to tetracycline (Tc), (2) establish the effect of *Zea mays* (Zm) cultivation on bacteria and oxidoreductases in the soil contaminated with Tc, and (3) identify the usability of compost from various grass species (G) in restoring the microbiological homeostasis of tetracycline contaminated soil. The gathered information may facilitate the choice of bacteria most active in tetracycline degradation and an effective bioremediation method.

2. Materials and Methods

2.1. Study Objective

The soil (Eutric Cambisol) to be analyzed was sampled from the topsoil layer (0–20 cm) of an arable field located in North-Eastern Poland (53.7167° N, 20.4167° E), where organic fertilizers have not been administered five years prior to soil sampling. According to fraction size classification by the International Union of Soil Sciences and the United States Department of Agriculture [38], the soil was classified as sandy loam (sand—0.0–2.0 mm—63.61%, dust 0.02–0.05 mm—32.68%, and loam <0.002 mm—3.71%). Compost used in the study derived from aerobic composting of cut and pre-dried grass. Table 2 presents the basic properties of the soil and compost used in the study.

Table 2. Some properties of the soil and compost used in the experiment, the test units and literature according to which the analyses were made.

Abbreviation	Properties	Unit	Soil	Compost	References
		Chemical and physicochemical properties			
N_{tot}	total nitrogen		0.83 ± 0.05	20.18 ± 0.86	[39]
C_{org}	organic carbon	g kg^{-1} DM	10.00 ± 0.60	146.61 ± 6.19	[40]
SOM	soil organic matter		17.24 ± 1.03	252.76 ± 10.67	[40]
P	phosphorus		81.10 ± 4.13	3.41 ± 0.18	[41]
K	potassium	mg kg^{-1} DM	145.25 ± 7.18	9.25 ± 0.24	[41]
Mg	magnesium		71.00 ± 3.00	5.69 ± 0.17	[42]
pH	pH_{KCl}—soil reaction	1 mol KCl dm^{-3}	4.40 ± 0.15	6.1 ± 0.15	[43]
EBC	sum of exchangeable base cations		63.60 ± 4.50	659.76 ± 32.01	[44]
HAC	hydrolytic acidity	mmol (+) kg^{-1} DM	26.10 ± 1.92	82.04 ± 3.98	[44]
CEC	cation exchange capacity		89.70	741.80	[44]
ACS	alkaline cation saturation	%	70.90	88.94	[44]
		Exchangeable cations			
K$^+$	potassium		168.00 ± 8.30	NA	[45]
Ca^{++}	calcium	mg kg^{-1} DM	1190.5 ± 52.70	NA	[45]
Na$^+$	sodium		10.00 ± 0.43	NA	[45]
Mg^{++}	magnesium		82.10 ± 4.70	NA	[45]
		Microorganisms number per 1 kg DM			
Org	organotrophic bacteria		$36.728 \pm 1.986 \times 10^9$	NA	[46]
Olig	oligotrophic bacteria	cfu	$7.740 \pm 0.581 \times 10^9$	NA	[47]
Cop	copiotrophic bacteria		$11.240 \pm 0.127 \times 10^9$	NA	[47]
Act	actinomyces		$17.008 \pm 0.537 \times 10^9$	NA	[48]
		Enzymatic activity per 1 kg DM h^{-1}			
Deh	dehydrogenases	μmol TFF	4.042 ± 0.136	NA	[49]
Cat	catalase	mol O$_2$	0.212 ± 0.001	NA	[50]

NA—not analized.

Tetracycline ($C_{22}H_{24}N_2O_8$) was purchased at Sigma-Aldrich (Saint Louis, MO, USA). Its molecular weight is 444.4 g moL^{-1}. It is highly soluble in water (231 mg dm^{-3}) and features a low octanol-water partition coefficient (Log K_{OW}), i.e., -1.30, pointing to its hydrophilic nature. A tetracycline (Tc) molecule has acidic functional groups, namely: tricarbonyl methane (pKa~3.3) and phenolic diketone (pKa~7.8) [23,51–54].

2.2. Experiment Design

A pot experiment was carried out in a greenhouse of the University of Warmia and Mazury in Olsztyn (Poland) in 2021, in polyethylene pots having the following sizes: 15 cm (height), 14 cm (pot bottom diameter), and 16 cm (pot upper diameter). Each pot was filled with 3.5 kg of sandy loam sieved through a screen with a mesh diameter of 5 mm. The experiment was established in a randomized block design and was conducted in five replications for the control soil and the soil with tetracycline addition. The following experimental variants were tested: (1) non-sown soil (C); (2) non-sown soil with the addition of tetracycline (CTc); (3) non-sown soil fertilized with grass compost (CG); (4) non-sown soil with the addition of tetracycline fertilized with grass compost (CTcG); (5) soil sown with maize (Zm); (6) soil with tetracycline and sown with maize (ZmTc); (7) soil sown with maize and fertilized with grass compost (ZmG); and (8) soil sown with maize, contaminated with tetracycline, and fertilized with grass compost (ZmTcG). The same mineral fertilization was applied in all soil variants, including both the non-sown and sown ones. Fertilizer doses were established considering nutritional demands of maize (*Zea mays* L.) of LG 32.58 variety (variety registered in the European Union), which served as the experimental crop. Maize was selected for this study because it is one of the two crops most often cultivated on all continents [55], and also because its acreage and yield continuously increase, expecting to grow by 4% and 10% until 2030, respectively, compared to 2020 [56]. The following fertilization was applied in the experiment, in mg kg^{-1} DM soil: N—140, P—50, K—140, and Mg—20. Nitrogen was used in the form of urea, phosphorus as potassium dihydrophosphate, potassium as potassium dihydrophosphate and potassium chloride, and magnesium as magnesium sulfate heptahydrate. Tetracycline (Tc) was applied in a dose of 100 mg kg^{-1} DM soil in respective experimental variants (CTc; TcG; ZmTc; ZmTcG). This Tc dose reflects adverse scenarios involving high doses of antibiotics pervading the soil as a result of their uncontrolled disposal to municipal sewage or their storage in landfills [57,58]. Compost from grasses (G) was used in the amount of 4 g C kg^{-1} DM soil to ensure natural biostimulation of the soil microbiota in respective experimental variants (CG; CTcG; ZmG; ZmTcG).

Mineral fertilizers, tetracycline, and compost were applied to the soil once before it had been placed in pots. Afterward, the soil was hydrated using distilled water to 60% of the water capillary capacity, and soil from selected experimental variants (Zm; ZmTc; ZmG; ZmTcG) was sown with *Zea mays* L. (Zm) in the amount of 6 seeds. After the emergence (at the BBCH 10 stage), maize plants were thinned and only 4 were left in each pot. The experiment lasted 50 days, and soil moisture content was kept stable throughout this period at 60% of the water capillary capacity. Daytime ranged from 15 h 5 min to 17 h 5 min, the average temperature was at 14.9 °C, and the average air humidity was at 67.5%. The plants were grown under natural light. The leaf greenness index (SPAD, Soil and Plant Analysis Development) was determined twice (on day 25 and day 50 just before the plants had been cut). Determinations were carried out in 8 replications, using a SPAD 502 Chlorophyll Meter 2900P. At the BBCH 51 stage (beginning of panicle emergence), analyses were conducted to determine the yield of aerial parts and roots of maize. The plants were dried at a temperature of 60 °C for 4 days.

2.3. Methods of Laboratory Analyses

The count of microorganisms and the activity of enzymes were determined in fresh soil, moist soil, and soil sieved through a screen with 2 mm mesh diameter, whereas the physicochemical properties were tested in dry soil.

2.3.1. Microbiological Analyses

- Count of microorganisms

On days 25 and 50 of the experiment, the numbers of soil microorganism colonies (cfu) were determined with the serial dilution method. In order to extract a microbial community, a soil suspension was prepared (10 g of dry soil bulk in 90 cm^3 of a sterile 0.85% NaCl solution) and shaken for 30 min. Afterward, a series of dilutions were prepared and used to isolate organotrophic bacteria [46], oligotrophic and copiotrophic bacteria [47], and actinobacteria [48]. Culture conditions and detailed procedure of microorganism isolation were described in our previous work [59]. The colony numbers (cfu) of all groups of microorganisms were determined in 4 replications.

- DNA isolation

DNA was isolated from the soil using a Genomic Mini AX Bacteria+" kit (A&A Biotechnology, Gdansk, Poland). DNA isolation was followed by enzymatic lysis conducted using Lyticase (cat. 1018-10, A&A Biotechnology, Gdansk, Poland) and mechanical lysis performed on a FastPrep—24 type apparatus using zirconia beads. The resulting bacterial DNA was additionally purified by means of an Anti-Inhibitor Kit (A&A Biotechnology, Gdansk, Poland). The presence of bacterial DNA in the tested samples was confirmed in the Real-Time PCR performed in a CFX Connect thermocycler (Biorad), using a SYBR Green dye as fluorochrome and universal primers 1055F (5′-ATGGCTGTCGTCAGCT-3′) and 1392R (5′-ACGGGCGGTGTGTAC-3′) [60].

- Amplicon sequencing

Amplicons of the taxonomic groups Bacteria and Archea were sequenced based on the hypervariable V3-V4 region of the 16S rRNA gene. In the case of bacteria, amplification was performed using specific primer sequences 341F (5′-CCTACGGGNGGCWGCAG-3′) and 785R (5′-GACTACHVGGGTATCTAATCC-3′). PCR conditions were provided in detail in our previous work [61]. The sequencing was conducted at the Genomed SA company (Warsaw, Poland). DNA was sequenced on a MiSeq sequencer (Illumina, San Diego, CA, USA), in the paired-end (PE) technology using an Illumina v2 kit (San Diego, CA, USA).

- Bioinformatic analysis

The sequences obtained were subjected to quality control. Incomplete and chimeric sequences were discarded. The bioinformatic analysis ensuring the classification of bacteria to particular taxonomic levels was performed using the QIIME software package based on reference sequence database GreenGenes ver. 13_8. The results were presented as the percentage of relative abundance of sequences identified at selected taxonomic levels (phylum and genus). This manuscript presents results with OTU (operational taxonomic units) exceeding 1%. The sequencing data were deposited in the GenBank NCBI (https://www.ncbi.nlm.nih.gov/) (accessed on 27 March 2022) under accession numbers of ON042235-ON042332.

2.3.2. Biochemical Analyses

As in the case of microbial count determination, i.e., on days 25 and 50 of the experiment, soil samples from each replication were determined in another 3 replications for the activity of selected enzymes from the class of oxidoreductases: Deh—dehydrogenases (EC 1.1), and Cat—catalase (EC 1.11.1.6). All determinations were carried out with standard methods presented in Table 2. The specific procedures of the enzymatic test (buffers, incubation temperature and duration, reaction arrest time) were provided in detail in our previous works [59,62]. The activity of dehydrogenases was determined using a Perkin-Elmer Lambda 25 spectrophotometer (Woburn, MA, USA) at a wavelength (λ) of 485 nm. The enzymatic activity was defined as the amount of product released by 1 kg DM soil per 1 h, And hence, Deh activity was expressed in μMol triphenylformazan, whereas Cat activity—in Mol O_2.

2.3.3. Chemical and Physicochemical Analyses

Before the experiment, the collected soil was determined for fraction size with the aerometric method and for contents of: total nitrogen (N_{total}); organic carbon (C_{org}); soil organic matter (SOM); available P, K, and Mg; and exchangeable cations Ca^{2+}, Mg^{2+}, K^+, and Na^+. In addition, before the experiment had been established and after plant harvest, soil pH was measured in 1 mol KCl dm^{-3} and soil samples were determined for hydrolytic acidity (HAC) and sum of exchangeable base cations (EBC). The HAC and EBC values obtained were used to compute the cation exchange capacity (CEC) and alkaline cation saturation (ACS). All determinations were made in three replications. The above-mentioned determinations were carried out with standard methods presented in Table 2 and described in our previous work [62].

2.4. Data and Statistical Analysis

The counts of organotrophic bacteria and actinobacteria (cfu) were used to determine the colony development index (CD) and the ecophysiological diversity index (EP) [63]. The CD and EP values were computed from Equations (1) and (2), respectively:

$$CD = [N1/1 + N2/2 + 3/3 \ldots .. N10/10] \times 100; \tag{1}$$

where: N1, N2, N3,...N10—sum of the quotients of colony numbers of microorganisms identified in particular days of the study (1, 2, 3, ... 10) and the sum of all colonies in the entire study period;

$$EP = -\Sigma(pi \times log10\ pi) \tag{2}$$

where: pi—the quotient of the number of colonies of microorganisms from particular days of the study and the sum of all colonies from the entire study period.

The effect of tetracycline on soil microorganisms and enzymes was determined based on the index of tetracycline effect (IF_{Tc}) on soil microorganisms and enzymes (3):

$$IF_{Tc} = \frac{A_{Tc}}{A} - 1 \tag{3}$$

where: A_{Tc}—the number of microorganisms or the activity of enzymes in the soil with the addition of tetracycline, A—the number of microorganisms or the activity of enzymes in the control soil.

Additional analyses were conducted to evaluate the effect of maize cultivation (IF_{Zm}) and soil fertilization with grass compost (IF_G) on the colony numbers of microorganisms and activities of soil enzymes. The IF_{Zm} and IF_G values were computed from Equations (4) and (5), respectively:

$$IF_{Zm} = \frac{A_{Zm}}{A} - 1 \tag{4}$$

where: A_{Zm}—the number of microorganisms or the activity of enzymes in the soil sown with maize, A—the number of microorganisms or the activity of enzymes in the non-sown soil.

$$IF_G = \frac{A_G}{A} - 1 \tag{5}$$

where: A_G—the number of microorganisms or the activity of enzymes in the soil fertilized with grass compost, A—the number of microorganisms or the activity of enzymes in the non-fertilized soil.

The operational taxonomic units (OTU) of bacteria established at all taxonomic levels allowed computing the Shannon-Wiener index (H′) of bacterial diversity (6):

$$H' = -\Sigma p_i \times (ln p_i) \tag{6}$$

where: p is the ratio of OTU number of one representative of the tested taxon to the total OTU number of the entire taxon.

All results were subjected to a statistical analysis using Statistica 13.3 package [64]. Firstly, Kruskal-Wallis and Shapiro-Wilk tests were conducted to establish the normality of the data distribution. Next, two-way analysis of variance (ANOVA) was carried out to determine the effect of tetracycline and compost on the microbiological, enzymatic, and physicochemical properties of the soil. All significant differences between mean values were determined using the post-hoc Duncan test or post-hoc Dunn test with Bonferroni correction at a 95% confidence interval.

The data from the metagenomic analysis were developed using bioinformatic software for statistical calculations and graphical visualization. The phylum of bacteria was compared statistically by means of the G-test (w/Yates') + Fisher test, using STAMP 2.1.3 software [65]. The analyzed data was presented with the confidence interval of 95%. The collected metagenomic data related to the phylum and genus of bacteria was analyzed using the R v1.2.5033 software [66] with R v3.6.2 [67] and a gplots library [68]. The OTU $\geq 1\%$ data was presented on heat maps with a dendrogram of their similarities, whereas unique and common genera of bacteria identified in the soils non-sown and sown with *Zea mays*, OTU $\geq 1\%$, were presented on a Venn diagram plotted using InteractiVenn [69].

3. Results

3.1. Response of Zea Mays to Soil Contamination with Tetracycline

In the soil not fertilized with grass compost (G), tetracycline (Tc) did not inhibit the growth and development of *Zea mays* (Zm) (Table 3), while it increased the leaf greenness intensity at the 8th leaf stage of Zm development, increasing the SPAD value by as much as 23.8%. The aerial parts of Zm produced significantly greater biomass in the soil fertilized with G than in the non-fertilized soil. In turn, Tc effect in the fertilized soil was insignificant, just as in the non-fertilized one. In this experimental variant, Tc did not affect the SPAD value. The Zm yield was positively affected by soil fertilization with G, which increased the biomass of aerial parts by 9.3% in the soil not contaminated with Tc to 10.6% in the Tc-contaminated soil.

Table 3. The yield of *Zea mays* and the value of leaf greenness index (SPAD).

Tc Content (mg kg^{-1} DM of soil)	Yield, (DM g pot^{-1})			*Zea Mays* Development Phase SPAD	
	Shoots	Roots	Together	4 Leaves	8 Leaves
			−G		
0	55.33 ^{b} ± 0.92	13.51 ^{bc} ± 0.44	68.84 ^{c} ± 0.44	38.10 ^{a} ± 2.74	22.62 ^{b} ± 4.89
100	56.49 ^{b} ± 1.55	14.69 ^{ab} ± 0.42	71.18 ^{bc} ± 0.64	38.69 ^{a} ± 3.62	28.00 ^{a} ± 2.36
			+G		
0	60.46 ^{a} ± 0.93	12.07 ^{c} ± 1.44	72.53 ^{b} ± 1.12	34.36 ^{c} ± 2.36	29.01 ^{a} ± 1.54
100	62.48 ^{a} ± 1.03	15.83 ^{a} ± 0.72	78.31 ^{a} ± 1.37	36.33 ^{b} ± 2.61	27.48 ^{a} ± 1.58

Tc—tetracycline; −G—soil without grass compost; +G—soil with grass compost. Homogeneous groups denoted with letters (*a–c*) were calculated separately for each of the columns.

3.2. Response of Bacteria to Soil Contamination with Tetracycline

The negative values of the IF_{Tc} index proved the adverse effect of Tc on the communities of organotrophic bacteria and actinobacteria (Table 4), whereas its positive value indicated the positive impact of Tc on copiotrophic bacteria (Table S1).

Table 4. The index of tetracycline effect on the number of microorganisms (IF_{Tc}).

The Dose of Grass Compost in g C kg⁻¹ DM Soil	−Zm		+Zm	
	Analysis Day			
	25	50	25	50
Organotrophic bacteria (Org)				
0	−0.352 [d] ± 0.011	−0.514 [e] ± 0.046	−0.122 [ab] ± 0.007	−0.123 [ab] ± 0.014
4	−0.093 [ab] ± 0.005	−0.161 [bc] ± 0.012	−0.236 [c] ± 0.025	−0.061 [a] ± 0.002
Oligotrophic bacteria (Olig)				
0	−0.262 [f] ± 0.022	−0.009 [d] ± 0.001	−0.469 [g] ± 0.027	−0.052 [de] ± 0.005
4	−0.102 [e] ± 0.008	0.223 [b] ± 0.017	0.687 [a] ± 0.029	0.134 [c] ± 0.008
Copiotrophic bacteria (Cop)				
0	0.214 [c–e] ± 0.015	0.375 [bc] ± 0.053	0.300 [cd] ± 0.064	0.591 [a] ± 0.007
4	0.070 [e] ± 0.006	0.388 [bc] ± 0.051	0.139 [de] ± 0.005	0.529 [ab] ± 0.051
Actinomycetes (Act)				
0	−0.633 [c] ± 0.060	−0.265 [b] ± 0.067	−0.063 [b] ± 0.006	−0.259 [b] ± 0.016
4	−0.222 [ab] ± 0.021	−0.217 [ab] ± 0.037	−0.054 [a] ± 0.005	−0.118 [ab] ± 0.53

−Zm—unsown soil; +Zm—soil sown with *Zea mays*. Homogeneous groups denoted with letters (*a–g*) were calculated separately for each group of microorganisms.

Definitively lower negative IF_{Tc} values were obtained in the soil fertilized using G than in the non-fertilized soil, proving that organic matter provided to the soil with compost mitigates the adverse effects of Tc on the mentioned bacterial groups. In the case of copiotrophic bacteria, higher IF_{Tc} values were usually determined in the soil not fertilized with G (0.214–0.591) than in the fertilized soil (0.070–0.529). On day 25 of the experiment, Tc elicited an adverse effect on oligotrophic bacteria, as the IF_{Tc} values reported in this analytical period ranged from −0.262 to −0.469. The adverse Tc effect was neutralized on day 50, as indicated by IF_{Tc} values ranging from −0.009 to −0.052. An explicitly positive effect on this bacterial group was observed upon the fertilization of soil sown with Zm (IF_{Tc} = 0.134—0.687).

In the non-sown soil, the fertilization with G contributed to the greatest increase in the population numbers of actinobacteria and organotrophs (Table 5), with the increase being significantly greater in the soil contaminated with Tc than in the non-contaminated soil. The population numbers of copiotrophic bacteria remained relatively stable, whereas negative IF_G values were noted in most variants for oligotrophic bacteria. In the case of soil sown with Zm, fertilization with compost positively affected actinobacteria, whereas it adversely influenced oligotrophic bacteria in the soil not contaminated with tetracycline. Fertilization with compost triggered greater positive changes in the actinobacteria count in the soil contaminated with Tc than non-contaminated with Tc. In this experimental variant, compost elicited a minor effect on the other groups of the analyzed bacteria. On day 50 of the experiment, the IF_G values ranged from −0.060 to 0.005 for organotrophic bacteria, from −0.189 to −0.031 for oligotrophic bacteria, and from 0.032 to 0.074 for copiotrophic bacteria.

Table 5. The index of compost effect on the number of microorganisms (IF_G).

Tc Content (mg kg^{-1} DM of soil)	−Zm		+Zm	
	Analysis Day			
	25	50	25	50
	Organotrophic bacteria (Org)			
0	0.289 [c] ± 0.060	0.150 [c–e] ± 0.015	0.203 [cd] ± 0.021	−0.060 [f] ± 0.014
100	0.804 [b] ± 0.083	0.988 [a] ± 0.047	0.046 [d–f] ± 0.004	0.005 [ef] ± 0.002
	Oligotrophic bacteria (Olig)			
0	−0.333 [e] ± 0.029	−0.118 [cd] ± 0.019	−0.353 [e] ± 0.045	−0.189 [d] ± 0.017
100	−0.189 [d] ± 0.007	0.089 [b] ± 0.023	1.057 [a] ± 0.036	−0.031 [c] ± 0.010
	Copiotrophic bacteria (Cop)			
0	0.190 [ab] ± 0.019	0.058 [c] ± 0.020	0.278 [a] ± 0.033	0.074 [c] ± 0.036
100	0.049 [c] ± 0.005	0.068 [c] ± 0.015	0.120 [ab] ± 0.053	0.032 [c] ± 0.021
	Actinomycetes (Act)			
0	1.020 [b] ± 0.075	0.878 [bc] ± 0.034	0.311 [d] ± 0.029	0.528 [cd] ± 0.078
100	3.278 [a] ± 0.374	1.000 [b] ± 0.146	0.323 [d] ± 0.043	0.819 [bc] ± 0.089

Tc—tetracycline; −Zm—unsown soil; +Zm—soil sown with *Zea mays*. Homogeneous groups denoted with letters (*a–f*) were calculated separately for each group of microorganisms.

Among all independent variables tested, bacterial communities were most strongly affected by Zm cultivation. The mean results achieved in the assumed analytical periods indicate that the IF_{Zm} values determined for organotrophic and copiotrophic bacteria and actinobacteria in the soil not fertilized with G and contaminated with tetracycline were generally high, and higher than in the non-contaminated soil (Table 6), reaching 2.233, 0.168, and 3.208, respectively. In the soil fertilized with G, the IF_{Zm} values were higher for copiotrophic bacteria, actinobacteria and additionally for oligotrophic bacteria under conditions of soil exposure to Tc compared to the non-exposed soil. The mean IF_{Zm} values reached 0.186, 0.969, and 0.889, respectively, for the mentioned groups of bacteria.

Table 6. The index of *Zea mays* effect on the number of microorganisms (IF_{Zm}).

Tc Content (mg kg^{-1} DM of Soil)	Microorganisms							
	Org		Olig		Cop		Act	
	Analysis Day							
	25	50	25	50	25	50	25	50
	−G							
0	0.563 [e] ± 0.049	1.405 [b] ± 0.049	0.417 [b] ± 0.008	0.402 [b] ± 0.019	0.071 [bc] ± 0.007	0.027 [c] ± 0.009	1.429 [b] ± 0.114	1.204 [ab] ± 0.066
100	1.120 [c] ± 0.016	3.345 [a] ± 0.069	0.018 [d] ± 0.005	0.342 [ab] ± 0.027	0.147 [a–c] ± 0.057	0.188 [ab] ± 0.022	5.194 [a] ± 0.547	1.222 [ab] ± 0.049
	+G							
0	0.459 [f] ± 0.025	0.965 [d] ± 0.029	0.375 [ab] ± 0.038	0.289 [ab] ± 0.039	0.150 [a–c] ± 0.048	0.042 [c] ± 0.012	0.576 [c] ± 0.095	0.793 [ab] ± 0.093
100	0.229 [g] ± 0.039	1.198 [c] ± 0.015	1.583 [a] ± 0.045	0.194 [cd] ± 0.018	0.224 [a] ± 0.028	0.148 [a–c] ± 0.038	0.916 [ab] ± 0.098	1.021 [ab] ± 0.101

Tc—tetracycline; −G—soil without grass compost; +G—soil with grass compost. Org—organotrophic bacteria, Olig—oligotrophic bacteria, Cop—copiotrophic bacteria, Act—actinomycetes. Homogeneous groups denoted with letters (*a–g*) were calculated separately for each group of microorganisms.

The value of the colony development index (CD) determined for organotrophic bacteria was higher than that computed for actinobacteria (Figure 1). Regardless of the analytical period, the mean value of soil fertilization with G and soil contamination with Tc reached 29.720 for Org and 27.326 for Act. Soil contamination with tetracycline caused no changes in CD values determined for Org and Act. Mean differences in CD values between the soil samples non-contaminated and contaminated with tetracycline were not statistically significant despite fluctuations in certain analytical terms. Soil fertilization with G evoked an inexplicit effect on CD values calculated for the analyzed microorganisms as it caused no significant changes in CD value of Org and increased CD in Act from 25.962 to 28.689. The analysis of data collated in Figure 1 shows that CD values of Org and Act did not change significantly also upon Zm cultivation.

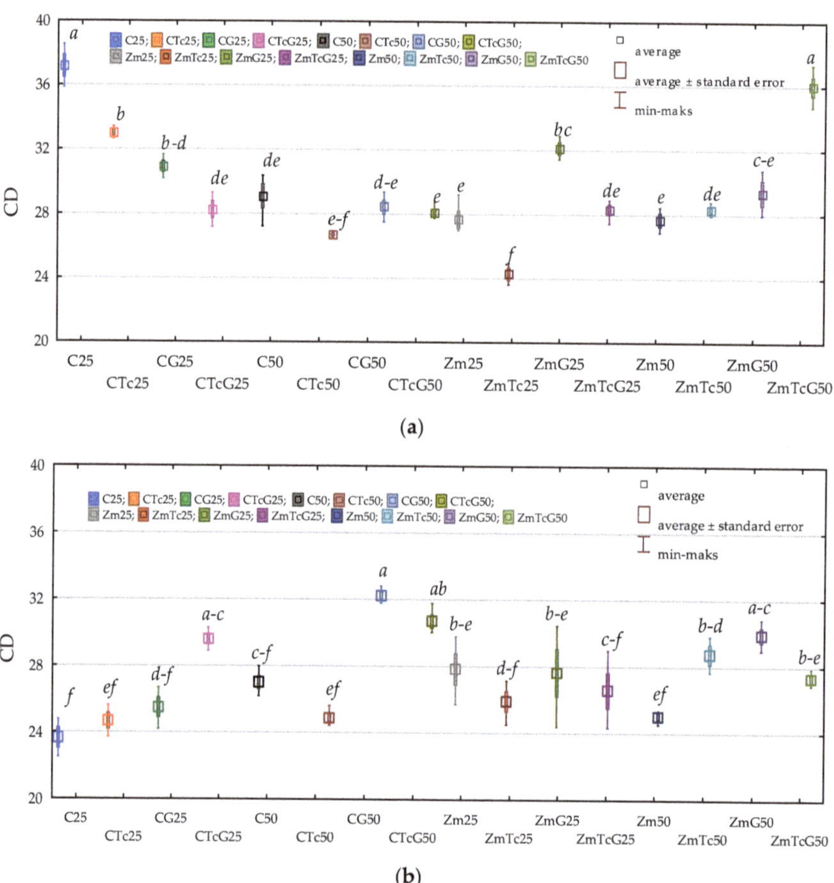

(a)

(b)

Figure 1. Colony development index (CD) (**a**) organotrophic bacteria (**b**) actinomycetes in soil. C—non-sown soil (control), CTc—non-sown soil contaminated with tetracycline, CG—non-sown soil fertilized with compost, CTcG—non-sown soil contaminated with tetracycline and fertilized with compost, Zm—soil sown with maize (*Zea mays*), ZmTc—soil sown with maize and contaminated with tetracycline, ZmG—soil sown with maize and fertilized with compost, and ZmTcG—soil sown with maize, contaminated with tetracycline, and fertilized with compost. 25—the date of the analysis, 25 days, 50—the date of the analysis, 50 days. Homogeneous groups denoted with letters (*a–f*) were calculated separately for each group of microorganisms.

In turn, data presented in Figure 2 indicates that the mean values of the ecophysiological diversity index (EP) computed for both Org and Act were high and reached 0.828 and 0.855, respectively. Regardless of analytical term, fertilization with G compost, and *Zea mays* cultivation, the soil contamination with Tc did not disturb the ecophysiological diversity of Org and Act. The mean EP value computed for Org was 0.824 in the non-contaminated soil and 0.832 in the Tc-contaminated soil, whereas the respective values calculated for Act were 0.851 and 0.858. Soil fertilization with G also did not disturb the ecophysiological diversity of the microorganisms tested, whereas Zm cultivation increased EP values from 0.813 to 0.844 for Org and from 0.838 to 0.872 for Act.

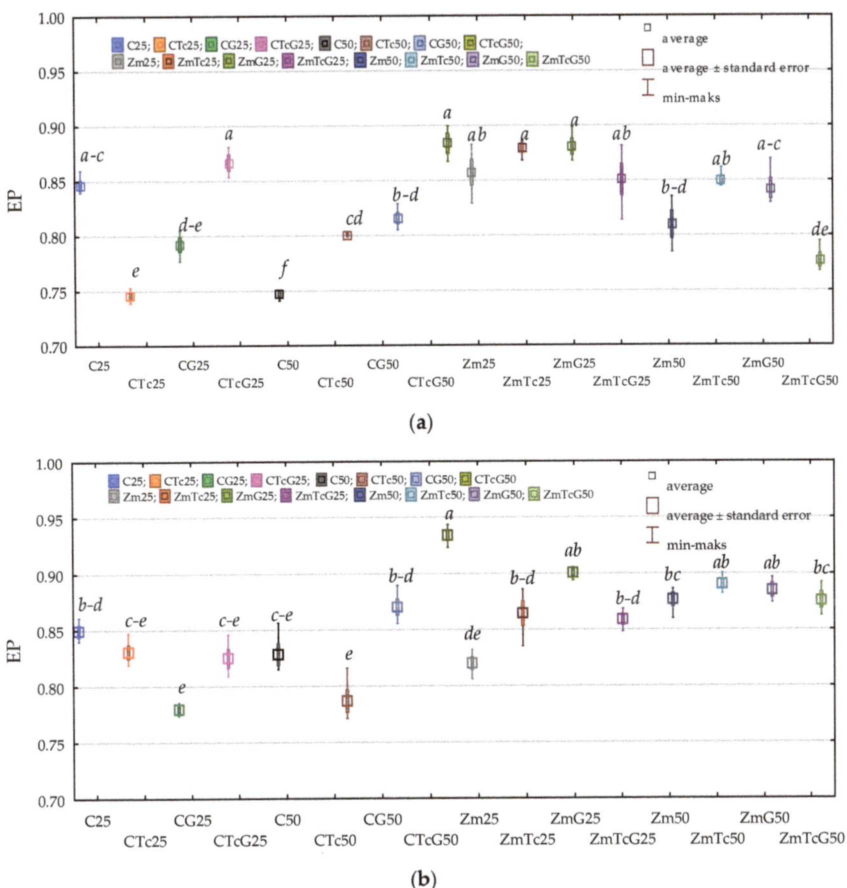

Figure 2. Ecophysiological diversity index (EP) (**a**) organotrophic bacteria (**b**) actinomycetes in soil. C—non-sown soil (control), CTc—non-sown soil contaminated with tetracycline, CG—non-sown soil fertilized with compost, CTcG—non-sown soil contaminated with tetracycline and fertilized with compost, Zm—soil sown with maize (*Zea mays*), ZmTc—soil sown with maize and contaminated with tetracycline, ZmG—soil sown with maize and fertilized with compost, and ZmTcG—soil sown with maize, contaminated with tetracycline, and fertilized with compost. 25—the date of the analysis, 25 days, 50—the date of the analysis, 50 days. Homogeneous groups denoted with letters (*a–e*) were calculated separately for each group of microorganisms.

The mean OTU number of the bacterial phylum reached 76,782 in the soil sown with Zm and 69,042 in the non-sown soil (Figure 3). In the non-sown soil, the value of the index of Tc effect on OTU reached 1.53, that of compost effect reached 2.17, and that of the cumulative Tc and G effect reached 1.34, whereas the respective values noted in the soil sown with Zm were at 0.07, −0.08, and −0.40. The IF value describing the effect of Zm cultivation on the soil tested reached 1.80.

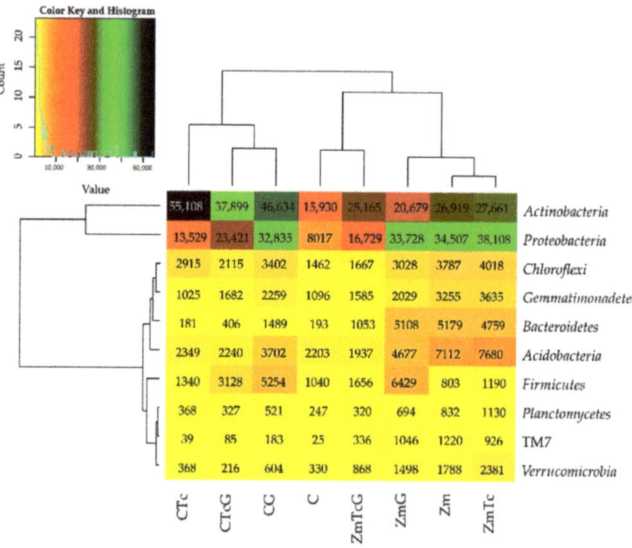

Figure 3. The abundance of bacterial types, OTU ≥ 1% presented by the heat map. C—non-sown soil (control), CTc—non-sown soil contaminated with tetracycline, CG—non-sown soil fertilized with compost, CTcG—non-sown soil contaminated with tetracycline and fertilized with compost, Zm—soil sown with maize (*Zea mays*), ZmTc—soil sown with maize and contaminated with tetracycline, ZmG—soil sown with maize and fertilized with compost, and ZmTcG—soil sown with maize, contaminated with tetracycline, and fertilized with compost.

Both in the soil sown with *Zea mays* and in the non-sown soil, the prevailing bacteria were those belonging to the *Actinobacteria* and *Proteobacteria* phyla (Figure 4).

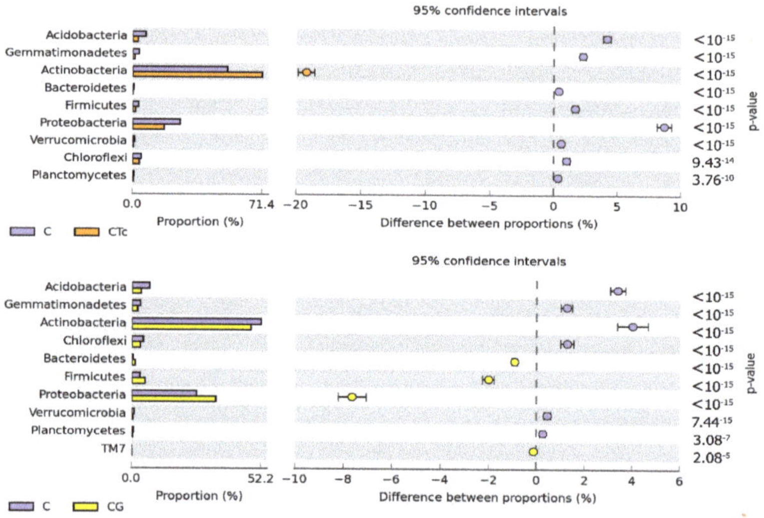

Figure 4. *Cont.*

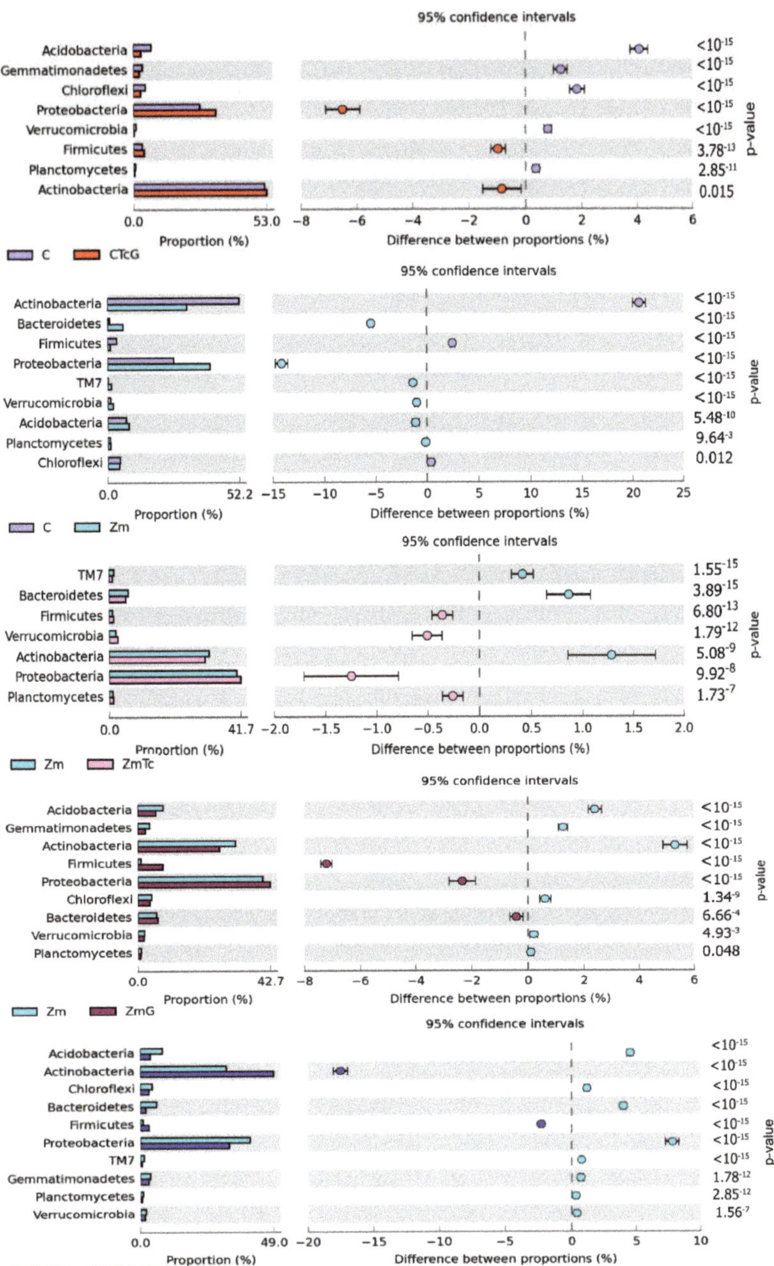

Figure 4. Structure of the abundance of bacterial phyla in the soil computed based on OTU ≥ 1%. C—non-sown soil (control), CTc—non-sown soil contaminated with tetracycline, CG—non-sown soil fertilized with compost, CTcG—non-sown soil contaminated with tetracycline and fertilized with compost, Zm—soil sown with maize (*Zea mays*), ZmTc—soil sown with maize and contaminated with tetracycline, ZmG—soil sown with maize and fertilized with compost, and ZmTcG—soil sown with maize, contaminated with tetracycline, and fertilized with compost.

The greatest changes in the structure of both phyla were triggered by Zm cultivation, which contributed to a 20.64% decrease in the relative abundance of *Actinobacteria* and a 14.16% increase in the relative abundance of *Proteobacteria*. In this experimental variant, soil contamination with Tc decreased the relative abundance of *Actinobacteria* by barely 1.29% and increased that of *Proteobacteria* by 1.25%. The structure of the remaining bacterial phyla remained unaffected by Tc effect. In turn, soil fertilization with G diminished *Actinobacteria* contribution in the phylum structure by 5.32%, but increased that of *Proteobacteria* and *Firmicutes* by 2.33% and 7.21%, respectively. The use of this fertilizer in the Tc-contaminated soil caused a 17.52% increase in the contribution of *Actinobacteria* in the structure as well as decreased contributions of *Proteobacteria*, *Acidobacteria*, and *Bacteroides* by 7.81%, 4.55%, and 4.01%, respectively. Nine bacterial genera were identified to belong to the phylum *Proteobacteria*, eight to the phylum *Actinobacteria*, and one to each of the following phyla: *Acidobacteria*, *Firmicutes*, and *Verrucomicrobia* (Figure 5).

The prevailing genera of the phylum *Proteobacteria* included: *Kaistobacter* and *Rhodoplanes*, and these prevailing in the phylum *Actinobacteria* included: *Cellulosimicrobium*, *Nocardioides*, *Streptomyces*, and *Terracoccus*. *Kaistobacter* was the most abundant in CTc and ZmTc variants; *Cellulosimicrobium* in CTc as well as ZmTcG and ZmG; *Luteibacter* in ZmG; *Sphingomonas*, *Sphingobium*, and *Burkholderia* in the soil sown with Zm; *Rhodanobacter* in ZmTc; *Nocardioides* in CTc, CTcG, and CG; *Rhodoplanes* in CTcG; *Arthrobacter* and *Streptomyces* in ZmTcG; and *Terracoccus* in CTcG.

The core bacteriome identified in all variants with non-sown soil comprised the following genera: *Kaistobacter*, *Cellulosimicrobium*, *Nocardioides*, *Rhodoplanes*, and *Terracoccus* (Figure 6a). Among the aforementioned genera, only *Kaistobacter*, *Cellulosimicrobium*, and *Rhodoplanes* represented the core bacteriobiome of the soil sown with Zm (Figure 6b). These three common genera, colonizing soils regardless of their contamination with Tc, fertilization with G, Zm cultivation, and specific genera appearing exclusive in the soil contaminated with Tc (*Kutzneria*, DA101, *Ralstonia*), should be perceived as providers of species potent to degrade Tc.

Considering the above findings, it may be concluded that the changes observed in the genetic diversity of bacteria were reflected in the values of the Shannon-Wiener index (Figure 7). The contamination of non-sown soil with Tc diminished the diversity of bacteria at the phylum, class, order, and family levels. Soil fertilization with G decreased the diversity at the phylum and class levels, and increased that at the genus level. This increased bacterial diversity at the genus level was also observed in the soil not contaminated with Tc but fertilized with G. The greatest positive impact on bacterial diversity, found at all taxonomic levels, was elicited by Zm cultivation. In this experimental variant, Tc did not diminish the bacterial diversity. Soil fertilization with compost had little effect on the bacterial diversity, increasing it only at the genus level. Fertilization of the soil contaminated with Tc, likewise of the soil non-sown with Zm, diminished the bacterial diversity at the phylum, class, and order levels.

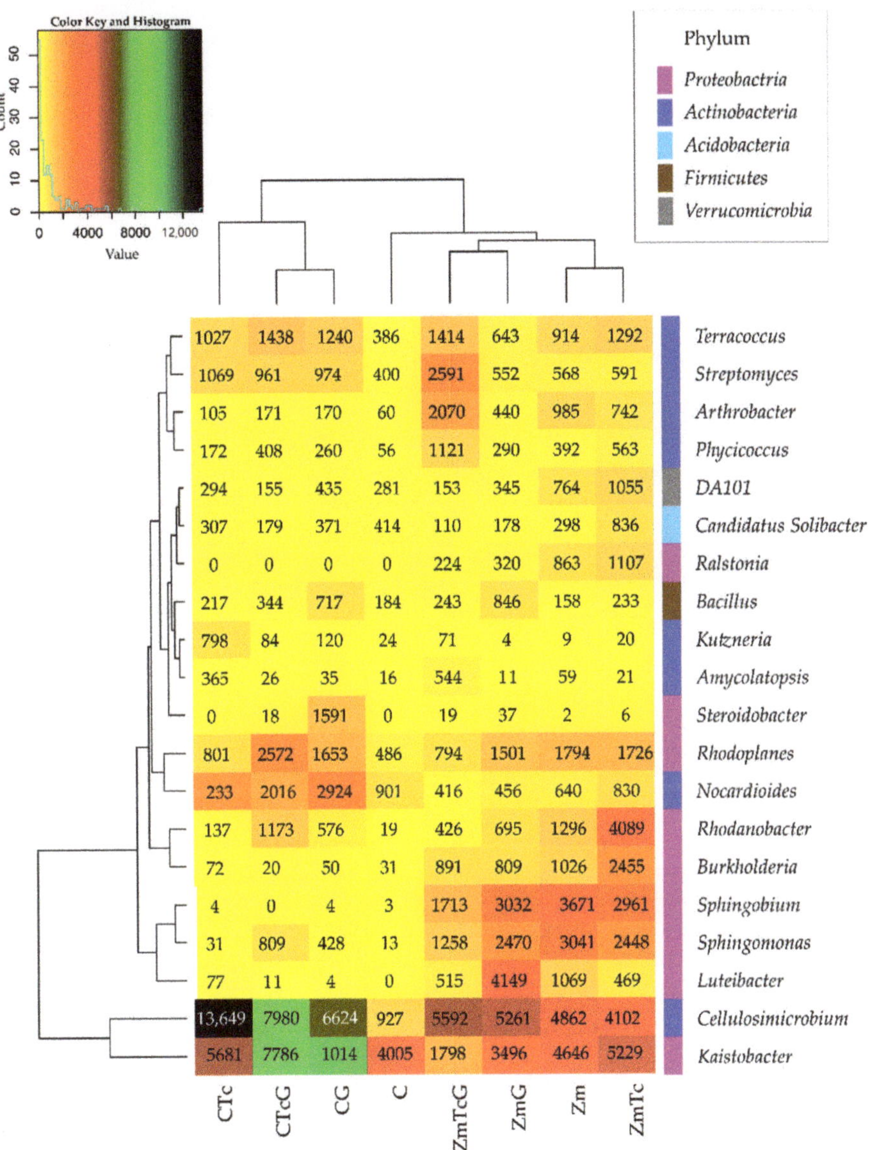

Figure 5. Diversity of bacterial species, presented based on OTU $\geq$ 1%. C—non-sown soil (control), CTc—non-sown soil contaminated with tetracycline, CG—non-sown soil fertilized with compost, CTcG—non-sown soil contaminated with tetracycline and fertilized with compost, Zm—soil sown with maize (*Zea mays*), ZmTc—soil sown with maize and contaminated with tetracycline, ZmG—soil sown with maize and fertilized with compost, and ZmTcG—soil sown with maize, contaminated with tetracycline, and fertilized with compost.

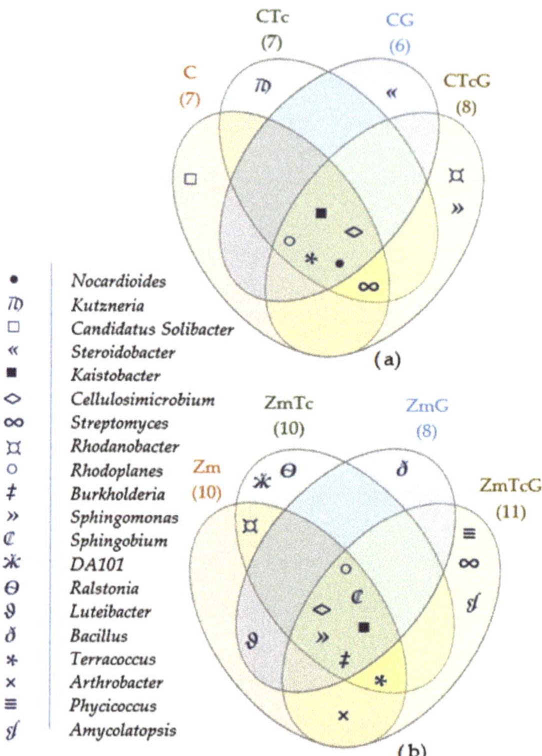

•	*Nocardioides*
ᛏᛆ	*Kutzneria*
□	*Candidatus Solibacter*
«	*Steroidobacter*
■	*Kaistobacter*
◇	*Cellulosimicrobium*
∞	*Streptomyces*
¤	*Rhodanobacter*
○	*Rhodoplanes*
‡	*Burkholderia*
»	*Sphingomonas*
℃	*Sphingobium*
⁂	*DA101*
ϴ	*Ralstonia*
♨	*Luteibacter*
ᚧ	*Bacillus*
✳	*Terracoccus*
×	*Arthrobacter*
≡	*Phycicoccus*
♫	*Amycolatopsis*

Figure 6. Venn diagram depicting unique and core bacterial genera in (**a**) non-sown soil and (**b**) soil sown with *Zea mays*, plotted based on OTU ≥ 1%. C—non-sown soil (control), CTc—non-sown soil contaminated with tetracycline, CG—non-sown soil fertilized with compost, CTcG—non-sown soil contaminated with tetracycline and fertilized with compost, Zm—soil sown with maize (*Zea mays*), ZmTc—soil sown with maize and contaminated with tetracycline, ZmG—soil sown with maize and fertilized with compost, and ZmTcG—soil sown with maize, contaminated with tetracycline, and fertilized with compost.

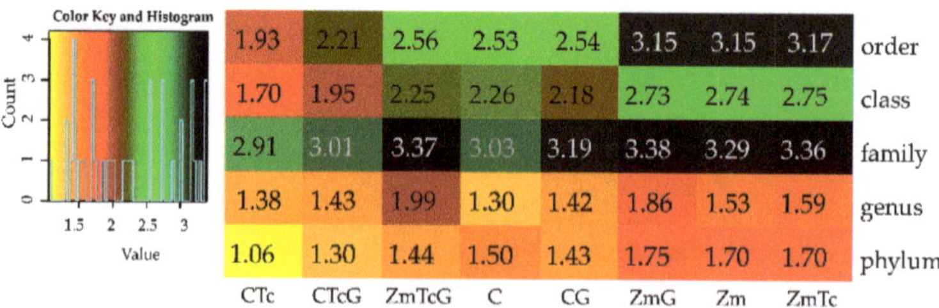

	CTc	CTcG	ZmTcG	C	CG	ZmG	Zm	ZmTc	
	1.93	2.21	2.56	2.53	2.54	3.15	3.15	3.17	order
	1.70	1.95	2.25	2.26	2.18	2.73	2.74	2.75	class
	2.91	3.01	3.37	3.03	3.19	3.38	3.29	3.36	family
	1.38	1.43	1.99	1.30	1.42	1.86	1.53	1.59	genus
	1.06	1.30	1.44	1.50	1.43	1.75	1.70	1.70	phylum

Figure 7. Diversity of bacterial taxa calculated from OTU according to Shannon-Wiener. C—non-sown soil (control), CTc—non-sown soil contaminated with tetracycline, CG—non-sown soil fertilized with compost, CTcG—non-sown soil contaminated with tetracycline and fertilized with compost, Zm—soil sown with maize (*Zea mays*), ZmTc—soil sown with maize and contaminated with tetracycline, ZmG—soil sown with maize and fertilized with compost, and ZmTcG—soil sown with maize, contaminated with tetracycline, and fertilized with compost.

3.3. Response of Oxidoreductases to Soil Contamination with Tetracycline

Tetracycline was found to elicit a less pronounced effect on oxidoreductases than on soil bacteria (Table S2). IF_{Tc} of catalase was low (Table 7), ranging from -0.089 to 0.051, which indicates that Tc inhibited its activity by barely 8.9% or stimulated it by 5.1%. In turn, the IF_{Tc} values computed for dehydrogenases were always negative (ranging from -0.004 to -0.245), which proves a tendency for the inhibiting effect of Tc on these enzymes. The lowest IF_{Tc} values were noted for dehydrogenases on day 50 of the experiment in the soil sown with *Zea mays*. They reached -0.245 in the non-fertilized soil and -0.170 in the soil fertilized with compost.

Table 7. The index of tetracycline effect on the activity of soil enzymes (IF_{Tc}).

The Dose of Grass Compost in g C kg^{-1} DM Soil	$-$Zm		$+$Zm	
	Analysis Day			
	25	50	25	50
	Dehydrogenases			
0	-0.087 [b] ± 0.016	-0.040 [ab] ± 0.008	-0.074 [b] ± 0.018	-0.245 [d] ± 0.021
4	-0.004 [a] ± 0.001	-0.061 [ab] ± 0.009	-0.078 [b] ± 0.016	-0.170 [c] ± 0.013
	Catalase			
0	-0.042 [b] ± 0.010	0.047 [a] ± 0.010	-0.089 [c] ± 0.014	0.051 [a] ± 0.002
4	0.016 [a] ± 0.004	-0.054 [bc] ± 0.012	-0.037 [b] ± 0.010	-0.026 [b] ± 0.004

$-$Zm—unsown soil; $+$Zm—soil sown with *Zea mays*. Homogeneous groups denoted with letters (*a–d*) were calculated separately for each enzyme.

The values of the IF_G index (Table 8) prove that fertilization with G promoted activities of oxidoreductases to a various extent depending on the analytical term, soil contamination with Tc, and cultivation of *Zea mays*. The strongest dehydrogenase stimulation was noted on day 50 of the experiment in the soil sown with Zm and contaminated with Tc (IF_G 0.742) and in the soil not contaminated with this antibiotic (IF_G 0.583), whereas the poorest one was in the non-sown soil (IF_G 0.022—0.046). The catalase activity was most strongly promoted by G fertilization in the non-sown soil not contaminated with Tc on day 50 of the experiment (IF_G 0.156) and in the Tc-contaminated soil on day 25 regardless of its cultivation (IF_G 0.141).

Table 8. The index of compost effect on the activity of soil enzymes (IF_G).

Tc Content (mg kg^{-1} DM of Soil)	$-$Zm		$+$Zm	
	Analysis Day			
	25	50	25	50
	Dehydrogenases			
0	0.061 [d] ± 0.053	0.046 [d] ± 0.005	0.043 [d] ± 0.011	0.583 [b] ± 0.014
100	0.158 [c] ± 0.008	0.022 [d] ± 0.033	0.038 [d] ± 0.018	0.742 [a] ± 0.028
	Catalase			
0	0.076 [bc] ± 0.005	0.156 [a] ± 0.005	0.081 [bc] ± 0.028	0.094 [b] ± 0.005
100	0.141 [a] ± 0.015	0.045 [cd] ± 0.013	0.141 [a] ± 0.015	0.014 [d] ± 0.012

Tc—tetracycline; $-$Zm—unsown soil; $+$Zm—soil sown with *Zea mays*. Homogeneous groups denoted with letters (*a–d*) were calculated separately for each enzyme.

The cultivation of *Zea mays* elicited a significantly positive effect on the activity of dehydrogenases (Table 9), with a stronger stimulating effect observed on day 50 than 25 of the experiment. This effect was, however, significantly suppressed by Tc, as the IF_{Zm} value determined on day 50 of the experiment reached 0.982 in the non-contaminated soil not fertilized with G and 2.000 in the fertilized soil. Soil contamination with Tc decreased the IF_{Zm} value to 0.558 and 1.654, respectively. In contrast to dehydrogenases, the activity of catalase was modified by Zm cultivation to a little extent. In the case of this enzyme, apart from the null IF_{Zm} values noted on day 25 of the experiment in the soil contaminated with Tc, the maximal IF_{Zm} values were noted on day 50; however they were relatively low and ranged from 0.050 to 0.082.

Table 9. The index of *Zea mays* effect on the activity of soil enzymes (IF_{Zm}).

Tc Content (mg kg^{-1} DM of Soil)	Dehydrogenases		Catalase	
	Analysis Day			
	25	50	25	50
	−G			
0	0.237 [ef] ± 0.007	0.982 [c] ± 0.055	0.051 [ab] ± 0.003	0.078 [a] ± 0.014
100	0.254 [e] ± 0.025	0.558 [d] ± 0.040	0.000 [c] ± 0.003	0.082 [a] ± 0.003
	+G			
0	0.215 [ef] ± 0.012	2.000 [a] ± 0.026	0.055 [ab] ± 0.007	0.020 [bc] ± 0.003
100	0.124 [f] ± 0.019	1.654 [b] ± 0.043	0.000 [c] ± 0.003	0.050 [ab] ± 0.005

Tc—tetracycline; −G—soil without grass compost; +G—soil with grass compost. Homogeneous groups denoted with letters (*a–f*) were calculated separately for each enzyme.

4. Discussion

4.1. Effect of Tetracycline and Soil Fertilization on Plants

The continuous release of tetracyclines to the natural environment makes them ubiquitous in soil [36]. The growing concentrations of tetracycline group antibiotics in arable fields worldwide are especially alarming due to their potential impact on crops [70–72]. Antibiotics may exert direct phytotoxic effects contributing to, e.g., a decreased rate of seed germination [71], plant growth and biomass growth [73,74] as well as a decreased rate of respiration or chlorophyll synthesis [70]. In turn, in our study, Tc not only did not decrease the SPAD value but increased it at the 8th leaf stage. The effect of antibiotics is considerably stronger in hydroponic cultures than in soil. According to Liu et al. [71], in cultures of this type, the EC50 reached 57 mg dm^{-3} for oats, 69 mg dm^{-3} for rice, and as much as 203 mg dm^{-3} for cucumber. In turn, in soil cultures, the EC50 determined for these plants exceeded 300 mg kg^{-1} DM soil. This Tc dose analyzed in Phytotoxkits tests reduced the length of *Sinapis alba* L. roots by 40% [75]. The direct phytotoxic effect of antibiotics on plant growth can, undoubtedly, be estimated using phytotoxic kits [70,71,75], but determination of Tc impact on plants in pot and field experiments seems to be crucial. Our vegetation experiment demonstrated that Tc applied in a dose of 100 mg kg^{-1} soil did not inhibit the growth of both aerial parts and roots of *Zea mays*. This is consistent with the findings from the study by Chen et al. [76], where oxytetracycline doses of 15 and 200 mg kg^{-1} had no significant effect on the cultivation of *Amarantus mangestnus* L. and *Trifolium repens* L. The weaker inhibiting effect of tetracycline in the soil than in water may be attributed to its strong absorption in soil [36,58,77].

4.2. Effect of Tetracycline, Fertilization with Compost, and Cultivation of Zea Mays on Soil Bacteria Community

The release of increasing amounts of antibiotics to soil poses potential threats to all microorganisms colonizing this environment [18,24,78]. They can modify the structure and activity of soil bacteria communities [17]. The present study results demonstrated that soil contamination with Tc exerted selective pressure on soil microorganisms. The impact of antibiotics on soil microorganisms is strongly dependent on the extent of soil contamination [79,80]. The antibiotic tested in the present study (100 mg Tc kg^{-1} soil) inhibited the proliferation of organotrophic bacteria and actinobacteria, and promoted that of copiotrophic bacteria. Despite these changes, Tc had little effect upon the colony development index (CD) and ecophysiological diversity index (EP) of organotrophic bacteria and actinobacteria, probably due to its very high affinity to soil components. The percentage of Tc adsorption approximates 100%, whereas its desorption percentage falls under 10% [53].

Antibiotics, i.e., cyprofloxacin, oxytetracycline, sulfamethoxazol, and tylosin, applied in a dose of 1 mg kg^{-1} soil during cabbage, endive, and spinach cultivation had no significant effect on the population numbers of microorganisms [77]. In turn, Santás-Miguel et al. [79] demonstrated that soil contamination with a Tc dose of 2000 mg kg $^{-1}$ DM soil induced tolerance of the bacterial community to this antibiotic, which however diminished with time. A temporary effect of Tc applied in doses of 100 and 500 mg per 1 kg of soil was also demonstrated by Chessa et al. [58,81]. Its adverse effect on the activity and structure of microorganisms was observed to decline after 7 days and to disappear within 60 days. In the present study the effect of Tc on soil bacteriobiome was also observed to vary in time. On day 50 of the experiment, it was completely neutralized in the case of oligotrophic bacteria.

Antibiotics may modify the structure and composition of bacterial communities [74,78,82]. In the current study, Tc presence in the soil had a significant but varying effect on the abundance, diversity, and structure of bacterial communities colonizing it. At the phylum level, the bacterial community was mainly constituted by *Proteobacteria* and *Actinobacteria*. The increased OTU numbers of bacteria from the following genera: *Cellulosimicrobium, Nocardioides, Candidatus Solibacter, Streptomyces, Terracoccus, Arthrobacter, Phycicoccus, Kutzneria,* and *Amycolatopsis* belonging to the phylum *Actinobacteria* and bacteria from the genera: *Rhodoplanes, Burkholderia,* and *Rhodanobacter* belonging to the phylum *Proteobacteria* in the soils contaminated with Tc can be associated with a growing ability of these microorganisms to degrade this antibiotic [54,83] and use it as a source of carbon and energy [24,58]. Potentially, the adverse Tc effect on the bacteria from the phylum *Proteobacteria* represented by *Kaistobacter, Sphingomonas, Sphingobium, Luteibacter,* and *Steroidobacter* genera could have been masked by enhanced activities of other microorganisms capable of developing in the presence of Tc, probably, by using compounds released from lysed cells of microorganisms [58,79,84,85].

Chessa et al. [58] and Chen et al. [52] point to the fact that tetracyclines are antibiotic residues most often reported in manure and sewage sludge. Increased concentrations of antibiotics in soil cause the number of antibiotic-resistant bacteria to increase, which modifies the sensitivity of entire populations of microorganisms to the antibiotic [86,87] and increases the number of genes resistant to tetracyclines in the soil [88,89]. The prevailing bacterial phyla in the soil fertilized with manure containing 171.07–660.20 µg Tc kg^1 were *Proteobacteria, Acidobacteria, Actinobacteria, Chloroflexi,* and *Bacteroidetes,* accounting for 85.2–92.4% of the total soil bacteria population [80]. The long-term use of sewage sludge and hen droppings increased the counts of *Proteobacteria, Acidobacteria, Actinobacteria,* and *Chloroflexi.* In addition, five bacterial phyla (*Chloroflexi, Planctomycetes, Firmicutes, Gemmatimonadetes,* and *Bacteroidetes*) were significantly correlated with antibiotic-resistance genes (ARG) in the soil [88]. The changes observed in the present study in the diversity of microbial communities upon the influence of Tc may also be reflected in the modified functions of soil microorganisms, which probably translates into soil metabolism. The present study results highlighted that, regardless of stress induced by soil contamination with Tc, fertilization with compost, and sowing the soil with *Zea mays*, the stability of the soil microbiome was very well described by the core microbiome being common for all soil types tested and represented by *Kaistobacter, Cellulosimicrobium,* and *Rhodoplanes.* In the non-sown soil contaminated with Tc, the core microbiome comprised *Kutzneria,* whereas in the soil sown with *Zea mays*—it was represented by *Ralstonia* and DA101. This is very precious information considering the paucity of data related to basic changes in the community of microorganisms involved in accelerated removal of tetracyclines. These microorganisms can be used for bioaugmentation of antibiotic-contaminated areas.

Soil supplementation with grass compost, which provides an easily available pool of organic compounds, has turned out to be of key importance during ecosystem adaptation to adverse environmental conditions caused by soil contamination with tetracycline. Compost was found to significantly stimulate the development of autochthonous microorganisms and alleviate the adverse effect of Tc on bacteria proliferation. Also other bioadsorbents seem to be a fine alternative for minimizing the adverse effect of tetracycline on soil microbiomes due to their low cost and their ability to adsorb toxic substances [23,53]. These include: pine bark and crushed clam shell [17], manure [23], and biocharcoal [90]. Definitely lower negative IF_{Tc} values obtained in the soil fertilized using G than in the non-fertilized soil prove that organic matter provided to the soil with compost mitigates the adverse effects of tetracycline on the soil microbiome. The simultaneous use of Tc and compost caused lesser changes in the bacterial phylum structure. Also other studies corroborated the usability of organic matter in restoring the stability of Tc-contaminated soil. Bovine manure modified the bacterial structure in the soil, enhanced microbiological activity, and contributed to the restoration of the microbiological structure in the soil with Tc addition [58]. According to Yue et al. [90], biocharcoal from manure accelerated the removal of tetracyclines and promoted the growth of bacteria potentially degrading tetracyclines (*Acidothermus*, *Sphingomonas*, and *Blastococcus*), which may be used for in situ remediation of tetracycline-contaminated soils. The above considerations allow for the conclusion that the use of compost, likewise the use of manure or biocharcoal, is advisable as—being the source of nutrients and microorganisms—it may accelerate microbiological degradation of tetracycline. In addition, due to its structure, compost may potentially improve the physical properties of soil by modifying its pH, water capacity, and structure. The usability of compost in accelerating the microbiological degradation of tetracycline was also confirmed by the Shannon index values. They prove that compost mitigated the adverse Tc effect on bacterial diversity at all taxonomic levels in the non-sown soil. In the soil sown with *Zea mays*, its alleviating impact was confirmed only at the genus level.

4.3. Effect of Tetracycline, Fertilization with Compost, and Cultivation of Zea Mays on Activities of Soil Oxidoreductases

Both the microorganisms and enzymes they produce are very good indicators of soil health [59,91–94], because they are sensitive to various changes proceeding in the environment [34,59,61]. The present study results demonstrate that activity of oxidoreductases depended on all variables tested, i.e., soil contamination with tetracycline, soil fertilization with compost, soil sowing with *Zea mays*, and experiment duration. The sensitivity of dehydrogenases and catalase to tetracycline was relatively low. This is consistent with sparse and inexplicit reports related to the effect of this group of antibiotics on the activity of soil enzymes [85,95]. Wei et al. [72] and Kessler et al. [95] emphasized that soil contamination with Tc impaired the enzymatic activity of soil, particularly in the case of dehydrogenases. According to Chen et al. [76], oxytetracycline doses of 15 and 200 mg kg^{-1} adversely affected activity of dehydrogenases, whereas cultivation of *Amarantus mangestnus* L. and Trifolium repens L. played an insignificant role in alleviating the adverse effect of this antibiotic. In turn, Liu [96] demonstrated a temporary enhancement of dehydrogenase activity in the soil contaminated with chlorotetracycline. In the present study, Tc only slightly enhanced catalase activity on day 50 of the experiment. Generally, dehydrogenases were more responsive to Tc compared to catalase, probably due to growth inhibition or death of sensitive microorganisms [18,83,97]. In turn, the stimulating effect of Tc on catalase activity, observed in the present study, might have been caused by enhanced proliferation of copiotrophic bacteria in the Tc-contaminated soils. Probably, these bacteria are capable of surviving in the presence of antibiotics, using them as sources of carbon [29].

The enhanced activities of dehydrogenases and catalase upon grass compost application results most likely from the response of organotrophic bacteria, copiotrophic bacteria, and actinobacteria to the nutrient loads supplied to their communities. Compost application to soil provides microorganisms with easily available substrates, e.g., carbohydrates [98–100]. This is indicative of a feedback between bacterial community and activity of oxidoreductases, because these are the microorganisms that trigger the enhancement in the activity of dehydrogenases [100,101]. The latter directly reflects the activity of soil microorganisms [102,103] because dehydrogenases are endogenous enzymes responsible for bio-oxidation of soil organic matter [104,105].

The results of our study demonstrate that tetracycline disturbed the stability of a soil ecosystem and confirm the hypothesis that *Zea mays* cultivation and soil fertilization with good-quality compost are utile in restoring the biological homeostasis of soil contaminated with this antibiotic. This finding suggests the permanent need for searching, developing, and implementing strategies for the bioremediation of soils contaminated with antibiotics.

5. Conclusions

Tetracycline (Tc) present in the soil in a dose of 100 mg kg^{-1} did not inhibit the yield of *Zea mays* (Zm) and did not decrease its leaf greenness index (SPAD). Even though it impaired the proliferation of culture bacteria, it did not affect the values of the colony development index (CD) and the ecophysiological diversity index (EP). Tetracycline exerted a positive effect on copiotrophic bacteria and an adverse effect on organotrophic bacteria, actinobacteria, oligotrophic bacteria as well as soil oxidoreductases. The bacteria belonging to *Acidiobacteria* and *Proteobacteria* phyla were found to be prevailing soil bacteria. In the non-sown soil, tetracycline increased the relative abundance of *Acidiobacteria* by 19% and reduced that of *Proteobacteria* by 9%. In the soil sown with Zm, the Tc effect on the relative abundance of all bacterial phyla was minor and changes observed ranged from −1.29% to 1.25%, which was mainly due to the significant positive impact of Zm on bacteria of the identified phyla. The greatest changes in the structure of *Actinobacteria* and *Proteobacteria* were caused by Zm cultivation, as it decreased the relative OTU number of *Actinobacteria* by 21% and increased that of *Proteobacteria* by 14%. Fertilization with grass compost (G) and Zm cultivation alleviated its adverse effect on the mentioned groups of bacteria and activities of soil dehydrogenases and catalase. The metagenomic analysis demonstrated that Tc application and fertilization of non-sown soil with G as well as Zm cultivation significantly increased the OTU numbers of bacteria. Soil fertilization with G reduced the relative abundance of *Acidiobacteria* and *Actinobacteria* as well as increased *Proteobacteria* abundance in both soil variants. The *Kaistobacter* and *Rhodoplanes* bacteria belonging to the phylum *Proteobacteria* and *Cellulosimicrobium* belonging to the phylum *Actinobacterium* constituted the core microbiome in both the soil sown with Zm and in the non-sown soil. In turn, specific genera identified exclusively in the Tc-contaminated soil turned out to be *Kutzneria (p_Actinobacteria), DA101 (p_Verrumicrobe),* and *Ralstonia (p_Actinobacteria).* The mentioned genera should be perceived as sources of species effective in bioaugmentation of soils contaminated with Tc. In the non-sown soil, Tc diminished diversity of bacteria at all taxonomic levels, except for the genus level. This unfavorable phenomenon was mitigated by soil fertilization with compost and by *Zea mays* cultivation.

The present study recommends soil fertilization with grass compost and *Zea mays* cultivation in bioremediation of soils contaminated with tetracycline and indicates the usability of the so-called core bacteria, developing well in the presence of tetracycline, for the bioaugmentation of soils contaminated with this antibiotic.

Supplementary Materials: The following supporting information can be downloaded at: https://www.mdpi.com/article/10.3390/ijerph19127357/s1, Table S1. The number of bacteria, 10^9 cfu kg^{-1} DM of soil; Table S2: Enzymatic activity in soil, kg^{-1} DM of soil h^{-1}.

Author Contributions: Conceptualization and experimental design and methodology, J.W., A.B. and J.K.; investigation, J.W. and A.B.; bioinformatic analysis and visualization of the data, A.B.; statistical analyses, J.W. and A.B.; writing—review and editing, J.W. and A.B.; supervision, J.K.; J.W., corresponding author. All authors have read and agreed to the published version of the manuscript.

Funding: This research was funded by the University of Warmia and Mazury in Olsztyn, Faculty of Agriculture and Forestry, Department of Soil Science and Microbiology (grant No. 30.610.006-110) and the project was financially supported by the Minister of Education and Science in the range of the program entitled "Regional Initiative of Excellence" for the years 2019–2022, Project No. 010/RID/2018/19, amount of funding 12.000.000 PLN.

Institutional Review Board Statement: Not applicable.

Informed Consent Statement: Not applicable.

Data Availability Statement: Sequencing data has been deposited with GenBank NCBI. They are available online: https://www.ncbi.nlm.nih.gov/nuccore/?term=ON042235:ON042332[accn] (accessed on 27 March 2022) under accession numbers ON042235–ON042332.

Acknowledgments: The authors thank Miroslaw Kucharski for bioinformatics consultation.

Conflicts of Interest: The authors declare no conflict of interest. The funders had no role in the design of the study; in the collection, analyses, interpretation of data; in the writing of the manuscript and in the decision to publish the results.

References

1.   Chopra, I.; Roberts, M. Tetracycline antibiotics: Mode of action, applications, molecular biology, and epidemiology of bacterial resistance. *Microbiol. Mol. Biol. Rev.* **2001**, *65*, 232–260. [CrossRef] [PubMed]

2.   Rusu, A.; Buta, E.L. The Development of third-generation tetracycline antibiotics and new perspectives. *Pharmaceutics* **2021**, *13*, 2085. [CrossRef] [PubMed]

3.   Granados-Chinchilla, F.; Rodríguez, C. Tetracyclines in food and feedingstuffs: From regulation to analytical methods, bacterial resistance, and environmental and health implications. *J. Anal. Methods Chem.* **2017**, *2017*, e1315497. [CrossRef] [PubMed]

4.   Fuoco, D. Classification framework and chemical biology of tetracycline-structure-based drugs. *Antibiotics* **2012**, *1*, 1–13. [CrossRef]

5.   OIE. *OIE Annual Report on Antimicrobial Agents Intended for Use in Animals: Better Understanding of the Global Situation*; Fifth Report; OIE: Paris, France, 2021; p. 136.

6.   EMA. *Sales of Veterinary Antimicrobial Agents in 31 European Countries in 2019 and 2020. EMA*; Publications Office of the European Union: Luxembourg, Germany, 2021; p. 130.

7.   WHO. *Critically Important Antimicrobials for Human Medicine, 6th Revision*; 6th Revision; World Health Organization: Geneva, Switzerland, 2019; p. 52.

8.   Ahn, J.G.; Cho, H.-K.; Li, D.; Choi, M.; Lee, J.; Eun, B.-W.; Jo, D.S.; Park, S.E.; Choi, E.H.; Yang, H.-J.; et al. Efficacy of tetracyclines and fluoroquinolones for the treatment of macrolide-refractory mycoplasma pneumoniae pneumonia in children: A systematic review and meta-analysis. *BMC Infect. Dis.* **2021**, *21*, 1003. [CrossRef]

9.   Furlong-Silva, J.; Cross, S.D.; Marriott, A.E.; Pionnier, N.; Archer, J.; Steven, A.; Merker, S.S.; Mack, M.; Hong, Y.-K.; Taylor, M.J.; et al. Tetracyclines improve experimental lymphatic filariasis pathology by disrupting interleukin-4 receptor–mediated lymphangiogenesis. *J. Clin. Investig.* **2021**, *131*, e140853. [CrossRef]

10. Mdegela, R.H.; Mwakapeje, E.R.; Rubegwa, B.; Gebeyehu, D.T.; Niyigena, S.; Msambichaka, V.; Nonga, H.E.; Antoine-Moussiaux, N.; Fasina, F.O. Antimicrobial use, residues, resistance and governance in the food and agriculture sectors, Tanzania. *Antibiotics* **2021**, *10*, 454. [CrossRef]

11. Van, T.T.H.; Yidana, Z.; Smooker, P.M.; Coloe, P.J. Antibiotic use in food animals worldwide, with a focus on africa: Pluses and minuses. *J. Glob. Antimicrob. Resist.* **2020**, *20*, 170–177. [CrossRef]

12. Albernaz-Gonçalves, R.; Olmos Antillón, G.; Hötzel, M.J. Linking animal welfare and antibiotic use in pig farming—A review. *Animals* **2022**, *12*, 216. [CrossRef]

13. Doidge, C.; West, H.; Kaler, J. Antimicrobial resistance patterns of escherichia coli isolated from sheep and beef farms in england and wales: A Comparison of disk diffusion interpretation methods. *Antibiotics* **2021**, *10*, 453. [CrossRef]

14. Hao, H.; Cheng, G.; Iqbal, Z.; Ai, X.; Hussain, H.I.; Huang, L.; Dai, M.; Wang, Y.; Liu, Z.; Yuan, Z. Benefits and risks of antimicrobial use in food-producing animals. *Front. Microbiol.* **2014**, *5*, 288. [CrossRef] [PubMed]

15. Tasho, R.P.; Cho, J.Y. Veterinary antibiotics in animal waste, its distribution in soil and uptake by plants: A review. *Sci. Total Environ.* **2016**, *563–564*, 366–376. [CrossRef] [PubMed]

16. Daghrir, R.; Drogui, P. Tetracycline antibiotics in the environment: A review. *Environ. Chem. Lett.* **2013**, *11*, 209–227. [CrossRef]

17. Santás-Miguel, V.; Fernández-Sanjurjo, M.J.; Núñez-Delgado, A.; Álvarez-Rodríguez, E.; Díaz-Raviña, M.; Arias-Estévez, M.; Fernández-Calviño, D. Use of waste materials to prevent tetracycline antibiotics toxicity on the growth of soil bacterial communities. *Environ. Res.* **2021**, *193*, 110404. [CrossRef] [PubMed]

18. Sun, Y.; Guo, Y.; Shi, M.; Qiu, T.; Gao, M.; Tian, S.; Wang, X. Effect of antibiotic type and vegetable species on antibiotic accumulation in soil-vegetable system, soil microbiota, and resistance genes. *Chemosphere* **2021**, *263*, 128099. [CrossRef]

19. Awad, Y.M.; Kim, S.-C.; Abd El-Azeem, S.A.M.; Kim, K.-H.; Kim, K.-R.; Kim, K.; Jeon, C.; Lee, S.S.; Ok, Y.S. Veterinary antibiotics contamination in water, sediment, and soil near a swine manure composting facility. *Environ. Earth Sci.* **2014**, *71*, 1433–1440. [CrossRef]

20. An, J.; Chen, H.; Wei, S.; Gu, J. Antibiotic contamination in animal manure, soil, and sewage sludge in Shenyang, Northeast China. *Environ. Earth Sci.* **2015**, *74*, 5077–5086. [CrossRef]

21. Li, C.; Chen, J.; Wang, J.; Ma, Z.; Han, P.; Luan, Y.; Lu, A. Occurrence of antibiotics in soils and manures from greenhouse vegetable production bases of Beijing, China and an associated risk assessment. *Sci. Total Environ.* **2015**, *521–522*, 101–107. [CrossRef]

22. Andreu, V.; Vazquez-Roig, P.; Blasco, C.; Picó, Y. Determination of tetracycline residues in soil by pressurized liquid extraction and liquid chromatography tandem mass spectrometry. *Anal. Bioanal. Chem.* **2009**, *394*, 1329–1339. [CrossRef]

23. Conde-Cid, M.; Álvarez-Esmorís, C.; Paradelo-Núñez, R.; Nóvoa-Muñoz, J.C.; Arias-Estévez, M.; Álvarez-Rodríguez, E.; Fernández-Sanjurjo, M.J.; Núñez-Delgado, A. Occurrence of tetracyclines and sulfonamides in manures, agricultural soils and crops from different areas in Galicia (NW Spain). *J. Clean. Prod.* **2018**, *197*, 491–500. [CrossRef]

24. Hamscher, G.; Sczesny, S.; Höper, H.; Nau, H. Determination of persistent tetracycline residues in soil fertilized with liquid manure by high-performance liquid chromatography with electrospray ionization tandem mass spectrometry. *Anal. Chem.* **2002**, *74*, 1509–1518. [CrossRef] [PubMed]

25. Łukaszewicz, P.; Białk-Bielińska, A.; Dołżonek, J.; Kumirska, J.; Caban, M.; Stepnowski, P. A New approach for the extraction of tetracyclines from soil matrices: Application of the microwave-extraction technique. *Anal. Bioanal. Chem.* **2018**, *410*, 1697–1707. [CrossRef] [PubMed]

26. Yang, Z.; Nie, G.; Feng, G.; Han, J.; Huang, L.; Zhang, X. Genome-wide identification, characterization, and expression analysis of the NAC transcription factor family in orchardgrass (*Dactylis Glomerata* L.). *BMC Genom.* **2021**, *22*, 178. [CrossRef] [PubMed]

27. Lyu, J.; Yang, L.; Zhang, L.; Ye, B.; Wang, L. Antibiotics in soil and water in china–a systematic review and source analysis. *Environ. Pollut.* **2020**, *266*, 115147. [CrossRef]

28. Wei, R.; Ge, F.; Zhang, L.; Hou, X.; Cao, Y.; Gong, L.; Chen, M.; Wang, R.; Bao, E. Occurrence of 13 veterinary drugs in animal manure-amended soils in Eastern China. *Chemosphere* **2016**, *144*, 2377–2383. [CrossRef]

29. Xu, L.; Wang, W.; Xu, W. Effects of tetracycline antibiotics in chicken manure on soil microbes and antibiotic resistance genes (ARGs). *Environ. Geochem. Health* **2022**, *44*, 273–284. [CrossRef]

30. Chen, C.; Li, J.; Chen, P.; Ding, R.; Zhang, P.; Li, X. Occurrence of antibiotics and antibiotic resistances in soils from wastewater irrigation areas in Beijing and Tianjin, China. *Environ. Pollut.* **2014**, *193*, 94–101. [CrossRef]

31. He, L.-Y.; He, L.-K.; Gao, F.-Z.; Wu, D.-L.; Zou, H.-Y.; Bai, H.; Zhang, M.; Ying, G.-G. Dissipation of antibiotic resistance genes in manure-amended agricultural soil. *Sci. Total Environ.* **2021**, *787*, 147582. [CrossRef]

32. Huang, R.; Guo, Z.; Gao, S.; Ma, L.; Xu, J.; Yu, Z.; Bu, D. Assessment of veterinary antibiotics from animal manure-amended soil to growing alfalfa, alfalfa silage, and milk. *Ecotoxicol. Environ. Saf.* **2021**, *224*, 112699. [CrossRef]

33. Haenni, M.; Dagot, C.; Chesneau, O.; Bibbal, D.; Labanowski, J.; Vialette, M.; Bouchard, D.; Martin-Laurent, F.; Calsat, L.; Nazaret, S.; et al. Environmental contamination in a high-income country (france) by antibiotics, antibiotic-resistant bacteria, and antibiotic resistance genes: Status and possible causes. *Environ. Int.* **2022**, *159*, 107047. [CrossRef]

34. Stachurová, T.; Sýkorová, N.; Semerád, J.; Malachová, K. Resistant genes and multidrug-resistant bacteria in wastewater: A study of their transfer to the water reservoir in the Czech Republic. *Life* **2022**, *12*, 147. [CrossRef] [PubMed]

35. Zhi, S.; Zhou, J.; Yang, F.; Tian, L.; Zhang, K. Systematic analysis of occurrence and variation tendency about 58 typical veterinary antibiotics during animal wastewater disposal processes in Tianjin, China. *Ecotoxicol. Environ. Saf.* **2018**, *165*, 376–385. [CrossRef] [PubMed]

36. Scaria, J.; Anupama, K.V.; Nidheesh, P.V. Tetracyclines in the environment: An overview on the occurrence, fate, toxicity, detection, removal methods, and sludge management. *Sci. Total Environ.* **2021**, *771*, 145291. [CrossRef] [PubMed]

37. Lu, W.; Wang, M.; Wu, J.; Jiang, Q.; Jin, J.; Jin, Q.; Yang, W.; Chen, J.; Wang, Y.; Xiao, M. Spread of chloramphenicol and tetracycline resistance genes by plasmid mobilization in agricultural soil. *Environ. Pollut.* **2020**, *260*, 113998. [CrossRef] [PubMed]

38. IUSS Working Group WRB. *World Reference Base for Soil Resources 2014, Update 2015. International Soil Classification System for Naming Soils and Creating Legends for Soil Maps*; World Soil Resources Reports No. 106; FAO: Rome, Italy, 2015.

39. *ISO 11261*; Soil Quality—Determination of Total Nitrogen—Modified Kjeldahl Method. ISO: Geneva, Switzerland, 1995.

40. Nelson, D.; Sommers, L. Total carbon, organic carbon, and organic matter. In *Method of Soil Analysis: Chemical Methods*; Sparks, D.L., Ed.; American Society of Agronomy: Madison, WI, USA, 1996; pp. 1201–1229.

41. Egner, H.; Riehm, H.; Domingo, W. Untersuchungen Über Die Chemische Bodenanalyse Als Grundlage Für Die Beurteilung Des Nährstoffzustandes Der böden. II. Chemische extractionsmethoden zur phospor- und kaliumbestimmung. *Ann R. Agric. Coll. Swed.* **1960**, *26*, 199–215.

42. Schlichting, E.; Blume, H.; Stahr, K. *Bodenkundliches Praktikum. Pareys Studientexte 81*; Blackwell Wissenschafts-Verlag: Berlin, Germany, 1995.

43. *ISO 10390*; Soil Quality—Determination of PH. International Organization for Standardization: Geneva, Switzerland, 2005.

44. Klute, A. *Methods of Soil Analysis*; American Society of Agronomy, Agronomy Monograph 9: Madison, WI, USA, 1996.

45. Thomas, G.W. Exchangeable Cations. In *Methods of Soil Analysis*; Soil Science Society of America, American Society of Agronomy, Inc.: Madison, WI, USA, 1982; pp. 159–165.

46. Bunt, J.S.; Rovira, A.D. Microbiological studies of some subantarctic soils. *J. Soil Sci.* **1955**, *6*, 119–128. [CrossRef]

47. Ohta, H.; Hattori, T. Oligotrophic bacteria on organic debris and plant roots in a paddy field soil. *Soil Biol. Bioch.* **1983**, *15*, 1–8.

48. Parkinson, D.; Gray, T.R.G.; Williams, S.T. *Methods for Studying the Ecology of Soil Microorganisms*; IBP Handbook 19; Blackwell Scientific Publication: Oxford, UK, 1971.

49. Öhlinger, R. Dehydrogenase activity with the substrate TTC. In *Methods in Soil Biology*; Schinner, F., Ohlinger, R., Kandler, E., Margesin, R., Eds.; Springer: Berlin/Heidelberg, Germany, 1996; pp. 241–243.

50. Johnson, J.L.; Temple, K.L. Some variables affecting the measurement of "catalase activity" in soil. *Soil Sci. Soc. Am. J.* **1964**, *28*, 207–209. [CrossRef]

51. Babić, S.; Horvat, A.J.M.; Mutavdžić Pavlović, D.; Kaštelan-Macan, M. Determination of PKa Values of active pharmaceutical ingredients. *Trends Anal. Chem.* **2007**, *26*, 1043–1061. [CrossRef]

52. Chen, M.; Yi, Q.; Hong, J.; Zhang, L.; Lin, K.; Yuan, D. Simultaneous Determination of 32 antibiotics and 12 pesticides in sediment using ultrasonic-assisted extraction and high performance liquid chromatography-tandem mass spectrometry. *Anal. Methods* **2015**, *7*, 1896–1905. [CrossRef]

53. Conde-Cid, M.; Núñez-Delgado, A.; Fernández-Sanjurjo, M.J.; Álvarez-Rodríguez, E.; Fernández-Calviño, D.; Arias-Estévez, M. Tetracycline and sulfonamide antibiotics in soils: Presence, fate and environmental risks. *Processes* **2020**, *8*, 1479. [CrossRef]

54. Cycoń, M.; Mrozik, A.; Piotrowska-Seget, Z. Antibiotics in the soil environment—degradation and their impact on microbial activity and diversity. *Front. Microbiol.* **2019**, *10*, 338. [CrossRef] [PubMed]

55. Erenstein, O.; Chamberlin, J.; Sonder, K. Estimating the global number and distribution of maize and wheat farms. *Glob. Food Sec.* **2021**, *30*, 100558. [CrossRef]

56. OECD-FAO. *OECD-FAO Agricultural Outlook 2021–2030*; OECD Publishing: Paris, France, 2021; p. 337. Available online: https://www.oecd.org/development/oecd-fao-agricultural-outlook-19991142.htm (accessed on 12 February 2022).

57. Chao, H.; Zheng, X.; Xia, R.; Sun, M.; Hu, F. Incubation Trial indicated the earthworm intestinal bacteria as promising biodigestor for mitigating tetracycline resistance risk in anthropogenic disturbed forest soil. *Sci. Total Environ.* **2021**, *798*, 149337. [CrossRef] [PubMed]

58. Chessa, L.; Pusino, A.; Garau, G.; Mangia, N.P.; Pinna, M.V. Soil microbial response to tetracycline in two different soils amended with cow manure. *Environ. Sci. Pollut. Res.* **2016**, *23*, 5807–5817. [CrossRef] [PubMed]

59. Borowik, A.; Wyszkowska, J.; Kucharski, M.; Kucharski, J. Implications of soil pollution with diesel oil and bp petroleum with ACTIVE Technology for soil health. *Int. J. Environ. Res. Public Health* **2019**, *16*, 2474. [CrossRef]

60. Ferris, M.J.; Muyzer, G.; Ward, D.M. Denaturing gradient gel electrophoresis profiles of 16s rrna-defined populations inhabiting a hot spring microbial mat community. *Appl. Environ. Microbiol.* **1996**, *62*, 340–346. [CrossRef]

61. Zaborowska, M.; Wyszkowska, J.; Borowik, A.; Kucharski, J. Bisphenol A—A dangerous pollutant distorting the biological properties of soil. *Int. J. Mol. Sci.* **2021**, *22*, 12753. [CrossRef]

62. Borowik, A.; Wyszkowska, J.; Wyszkowski, M. Resistance of aerobic microorganisms and soil enzyme response to soil contamination with Ekodiesel Ultra Fuel. *Environ. Sci. Pollut. Res.* **2017**, *24*, 24346–24363. [CrossRef]

63. De Leij, F.A.A.M.; Whipps, J.M.; Lynch, J.M. The Use of colony development for the characterization of bacterial communities in soil and on roots. *Microb. Ecol.* **1993**, *27*, 81–97. [CrossRef]

64. TIBCO Software Inc. Statistica (Data Analysis Software System), Version 13. Available online: https://www.tibco.com/products/data-science (accessed on 12 February 2022).

65. Parks, D.H.; Tyson, G.W.; Hugenholtz, P.; Beiko, R.G. STAMP: Statistical analysis of taxonomic and functional profiles. *Bioinformatics* **2014**, *30*, 3123–3124. [CrossRef] [PubMed]

66. RStudio Team. *R Studio: Integrated Development*; RStudio, Inc.: Boston, MA, USA, 2019; Available online: http://www.rstudio.com/ (accessed on 5 March 2022).

67. R Core Team. *R: A Language and Environment for Statistical Computing*; R Foundation for Statistical Computing: Vienna, Austri, 2019; Available online: https://www.gbif.org/tool/81287/r-a-language-and-environment-for-statistical-computing (accessed on 5 March 2022).

68. Warnes, G.R.; Bolker, B.; Bonebakker, L.; Gentleman, R.; Huber, W.; Liaw, A.; Lumley, T.; Maechler, M.; Magnusson, M.; Moeller, S.; et al. Gplots: Various R Programming Tools for Plotting Data. R Package Version 2.17.0. 2020. Available online: https://CRAN.R-Project.org/package=gplots (accessed on 5 March 2022).

69. Heberle, H.; Meirelles, G.V.; da Silva, F.R.; Telles, G.P.; Minghim, R. InteractiVenn: A web-based tool for the analysis of sets through venn diagrams. *BMC Bioinform.* **2015**, *16*, 169. [CrossRef] [PubMed]

70. Carballo, M.; Rodríguez, A.; de la Torre, A. Phytotoxic effects of antibiotics on terrestrial crop plants and wild plants: A systematic review. *Arch. Environ. Contam. Toxicol.* **2022**, *82*, 48–61. [CrossRef] [PubMed]

71. Liu, F.; Ying, G.-G.; Tao, R.; Zhao, J.-L.; Yang, J.-F.; Zhao, L.-F. Effects of six selected antibiotics on plant growth and soil microbial and enzymatic activities. *Environ. Pollut.* **2009**, *157*, 1636–1642. [CrossRef]

72. Wei, X.; Wu, S.C.; Nie, X.P.; Yediler, A.; Wong, M.H. The effects of residual tetracycline on soil enzymatic activities and plant growth. *J. Environ. Sci. Health B* **2009**, *44*, 461–471. [CrossRef]

73. Pan, M.; Chu, L.M. Phytotoxicity of veterinary antibiotics to seed germination and root elongation of crops. *Ecotoxicol. Environ. Saf.* **2016**, *126*, 228–237. [CrossRef]

74. Wang, R.; Wang, J.; Wang, J.; Zhu, L.; Zhang, W.; Zhao, X.; Ahmad, Z. Growth inhibiting effects of four antibiotics on cucumber, rape and chinese cabbage. *Bull. Environ. Contam. Toxicol.* **2019**, *103*, 187–192. [CrossRef]

75. Timmerer, U.; Lehmann, L.; Schnug, E.; Bloem, E. Toxic effects of single antibiotics and antibiotics in combination on germination and growth of *Sinapis alba* L. Plants **2020**, *9*, 107. [CrossRef]

76. Chen, W.; Liu, W.; Pan, N.; Jiao, W.; Wang, M. Oxytetracycline on functions and structure of soil microbial community. *J. Soil Sci. Plant Nutr.* **2013**, *13*, 967–975. [CrossRef]

77. Sun, Y.; Lyu, H.; Cheng, Z.; Wang, Y.; Tang, J. Insight into the mechanisms of ball-milled biochar addition on soil tetracycline degradation enhancement: Physicochemical properties and microbial community structure. *Chemosphere* **2022**, *291*, 132691. [CrossRef]

78. Zheng, J.; Zhang, J.; Gao, L.; Kong, F.; Shen, G.; Wang, R.; Gao, J.; Zhang, J. The Effects of tetracycline residues on the microbial community structure of tobacco soil in pot experiment. *Sci. Rep.* **2020**, *10*, 8804. [CrossRef] [PubMed]

79. Santás-Miguel, V.; Rodríguez-González, L.; Núñez-Delgado, A.; Álvarez-Rodríguez, E.; Díaz-Raviña, M.; Arias-Estévez, M.; Fernández-Calviño, D. Time-course evolution of bacterial community tolerance to tetracycline antibiotics in agricultural soils: A laboratory experiment. *Chemosphere* **2022**, *291*, 132758. [CrossRef] [PubMed]

80. Xu, L.S.; Wang, W.Z.; Deng, J.B.; Xu, W.H. The residue of tetracycline antibiotics in soil and brassica juncea var. gemmifera, and the diversity of soil bacterial community under different livestock manure treatments. *Environ. Geochem. Health*, 2022; *Online ahead of print.* [CrossRef]

81. Santás-Miguel, V.; Arias-Estévez, M.; Díaz-Raviña, M.; Fernández-Sanjurjo, M.J.; Álvarez-Rodríguez, E.; Núñez-Delgado, A.; Fernández-Calviño, D. Interactions between soil properties and tetracycline toxicity affecting to bacterial community growth in agricultural soil. *Appl. Soil Ecol.* **2020**, *147*, 103437. [CrossRef]

82. Tong, X.; Wang, X.; He, X.; Wang, Z.; Li, W. Effects of Antibiotics on microbial community structure and microbial functions in constructed wetlands treated with artificial root exudates. *Environ. Sci. Processes Impacts* **2020**, *22*, 217–226. [CrossRef] [PubMed]

83. Shi, Y.; Lin, H.; Ma, J.; Zhu, R.; Sun, W.; Lin, X.; Zhang, J.; Zheng, H.; Zhang, X. Degradation of tetracycline antibiotics by *Arthrobacter nicotianae* OTC-16. *J. Hazard. Mater.* **2021**, *403*, 123996. [CrossRef]

84. Ma, J.; Lin, H.; Sun, W.; Wang, Q.; Yu, Q.; Zhao, Y.; Fu, J. Soil microbial systems respond differentially to tetracycline, sulfamonomethoxine, and ciprofloxacin entering soil under pot experimental conditions alone and in combination. *Environ. Sci. Pollut. Res.* **2014**, *21*, 7436–7448. [CrossRef]

85. Santás-Miguel, V.; Díaz-Raviña, M.; Martín, A.; García-Campos, E.; Barreiro, A.; Núñez-Delgado, A.; Álvarez-Rodríguez, E.; Arias-Estévez, M.; Fernández-Calviño, D. Soil enzymatic activities and microbial community structure in soils polluted with tetracycline antibiotics. *Agronomy* **2021**, *11*, 906. [CrossRef]

86. Kumar, A.; Kumar, A.; MMS., C.-P.; Chaturvedi, A.K.; Shabnam, A.A.; Subrahmanyam, G.; Mondal, R.; Gupta, D.K.; Malyan, S.K.; Kumar, S.S.; et al. Lead toxicity: Health hazards, influence on food chain, and sustainable remediation approaches. *Int. J. Environ. Res. Public Health* **2020**, *17*, 2179. [CrossRef]

87. Torres-Cortés, G.; Millán, V.; Ramírez-Saad, H.C.; Nisa-Martínez, R.; Toro, N.; Martínez-Abarca, F. Characterization of novel antibiotic resistance genes identified by functional metagenomics on soil samples. *Environ. Microbiol.* **2011**, *13*, 1101–1114. [CrossRef]

88. Chen, Q.; An, X.; Li, H.; Su, J.; Ma, Y.; Zhu, Y.-G. Long-term field application of sewage sludge increases the abundance of antibiotic resistance genes in soil. *Environ. Int.* **2016**, *92–93*, 1–10. [CrossRef]

89. Kyselková, M.; Jirout, J.; Vrchotová, N.; Schmitt, H.; Elhottová, D. Spread of tetracycline resistance genes at a conventional dairy farm. *Front. Microbiol.* **2015**, *6*, 536. [CrossRef] [PubMed]

90. Yue, Y.; Liu, Y.-J.; Wang, J.; Vukanti, R.; Ge, Y. Enrichment of potential degrading bacteria accelerates removal of tetracyclines and their epimers from cow manure biochar amended soil. *Chemosphere* **2021**, *278*, 130358. [CrossRef] [PubMed]

91. Bhowmik, A.; Kukal, S.S.; Saha, D.; Sharma, H.; Kalia, A.; Sharma, S. Potential Indicators of soil health degradation in different land use-based ecosystems in the Shiwaliks of Northwestern India. *Sustainability* **2019**, *11*, 3908. [CrossRef]

92. Kucharski, J.; Tomkiel, M.; Baćmaga, M.; Borowik, A.; Wyszkowska, J. Enzyme activity and microorganisms diversity in soil contaminated with the Boreal 58 WG Herbicide. *J. Environ. Sci. Health B* **2016**, *51*, 446–454. [CrossRef] [PubMed]

93. Steffan, J.J.; Brevik, E.C.; Burgess, L.C.; Cerdà, A. The effect of soil on human health: An overview. *Eur. J. Soil Sci.* **2018**, *69*, 159–171. [CrossRef]

94. Wyszkowska, J.; Borowik, A.; Olszewski, J.; Kucharski, J. Soil Bacterial community and soil enzyme activity depending on the cultivation of *Triticum aestivum, Brassica napus, and Pisum sativum* ssp. *arvense*. *Diversity* **2019**, *11*, 246. [CrossRef]

95. Kessler, N.C.H.; Sampaio, S.C.; do Prado, N.V.; Remor, M.B.; dos Reis, R.R.; Cordovil, C.M. Effects of tetracyclines on enzymatic activity and soil nutrient availability. *J. Soil Sci. Plant Nutr.* **2020**, *20*, 2657–2670. [CrossRef]

96. Liu, B.; Li, Y.; Zhang, X.; Wang, J.; Gao, M. Effects of chlortetracycline on soil microbial communities: Comparisons of enzyme activities to the functional diversity via Biolog EcoPlates™. *Eur. J. Soil Biol.* **2015**, *68*, 69–76. [CrossRef]

97. Zhao, X.; Li, X.; Li, Y.; Zhang, X.; Zhai, F.; Ren, T.; Li, Y. Metagenomic analysis reveals functional genes in soil microbial electrochemical removal of tetracycline. *J. Hazard. Mater.* **2021**, *408*, 124880. [CrossRef]

98. Hammerschmiedt, T.; Holatko, J.; Kucerik, J.; Mustafa, A.; Radziemska, M.; Kintl, A.; Malicek, O.; Baltazar, T.; Latal, O.; Brtnicky, M. Manure maturation with biochar: Effects on plant biomass, manure quality and soil microbiological characteristics. *Agriculture* **2022**, *12*, 314. [CrossRef]

99. Khalifa, T.H.; Mariey, S.A.; Ghareeb, Z.E.; Khatab, I.A.; Alyamani, A. Effect of organic amendments and nano-zinc foliar application on alleviation of water stress in some soil properties and water productivity of barley yield. *Agronomy* **2022**, *12*, 585. [CrossRef]

100. Wojewódzki, P.; Lemanowicz, J.; Debska, B.; Haddad, S.A. Soil enzyme activity response under the amendment of different types of biochar. *Agronomy* **2022**, *12*, 569. [CrossRef]

101. Kaczyńska, G.; Borowik, A.; Wyszkowska, J. Soil dehydrogenases as an indicator of contamination of the environment with petroleum products. *Water Air Soil Pollut.* **2015**, *226*, 372. [CrossRef] [PubMed]

102. Kucharski, J.; Wieczorek, K.; Wyszkowska, J. Changes in the enzymatic activity in sandy loam soil exposed to zinc pressure. *J. Elem.* **2011**, *16*, 577–589. [CrossRef]

103. Wyszkowska, J.; Boros-Lajszner, E.; Lajszner, W.; Kucharski, J. Reaction of soil enzymes and spring barley to copper chloride and copper sulphate. *Environ. Earth Sci.* **2017**, *76*, 403. [CrossRef]
104. Singh, M.; Sarkar, B.; Bolan, N.S.; Ok, Y.S.; Churchman, G.J. Decomposition of soil organic matter as affected by clay types, pedogenic oxides and plant residue addition rates. *J. Hazard. Mater.* **2019**, *374*, 11–19. [CrossRef]
105. Zhang, N.; He, X.-D.; Gao, Y.-B.; Li, Y.-H.; Wang, H.-T.; Ma, D.; Zhang, R.; Yang, S. Pedogenic Carbonate and soil dehydrogenase activity in response to soil organic matter in *Artemisia ordosica* community. *Pedosphere* **2010**, *20*, 229–235. [CrossRef]

International Journal of
Environmental Research and Public Health

Article

Effect Mechanism of Land Consolidation on Soil Bacterial Community: A Case Study in Eastern China

Yaoben Lin [1], Yanmei Ye [2], Shuchang Liu [2], Jiahao Wen [3] and Danling Chen [4],*

[1] School of Law and Politics, Nanjing Tech University, Nanjing 211816, China; lyb@njtech.edu.cn
[2] Land Academy for National Development (LAND), Zhejiang University, Hangzhou 310058, China; yymzjuedu@163.com (Y.Y.); 13395775305@163.com (S.L.)
[3] Department of Watershed Sciences, Utah State University, Logan, UT 84321, USA; X1525866@163.com
[4] Department of Land Management, College of Public Administration, Huazhong Agricultural University, Wuhan 430070, China
* Correspondence: danling1235@yeah.net

Abstract: Farmland consolidation is an effective tool to improve farmland infrastructures, soil quality, and sustain a healthy farmland ecosystem and rural population, generating contributions to food security and regional sustainable development. Previous studies showed that farmland consolidation regulates soil physical and chemical properties. Soil microorganisms also play an important role in soil health and crop performance; however, few studies reported how farmland consolidation influence soil microecology. Here, we used DNA sequencing technology to compare bacterial community structure in farmlands with and without consolidation. DNA sequencing technology is the most advanced technology used to obtain biological information in the world, and it has been widely used in the research of soil micro-ecological environment. In September 2018, we collected soil samples in Jiashan County, Zhejiang Province, China, and used DNA sequence technology to compare the bacterial community structure in farmlands with and without consolidation. Our results found that (1) farmland consolidation had significant impacts on soil microbial characteristics, which were mainly manifested as changes in microbial biomass, microbial diversity and community structure. Farmland consolidation can increase the relative abundance of the three dominant bacteria phyla and the three fungal dominant phyla, but it also negatively affects the relative abundance of the six dominant bacteria phyla and the three fungal dominant phyla. (2) Farmland consolidation had an indirect impact on soil bacterial community structure by adjusting the soil physical and chemical properties. (3) The impact of heavy metals on bacterial community structure varied significantly under different levels of heavy metal pollution in farmland consolidation areas. There were 6, 3, 3, and 5 bacterial genera that had significant correlations with heavy metal content in cultivated land with low pollution, light pollution, medium pollution, and heavy pollution, respectively. The number of heavy metal-tolerant bacteria in the soil generally increased first and then decreased under heavy metal polluted conditions. Our study untangled the relationship between varied farmland consolidation strategies and bacteria through soil physcicochemical properties and metal pollution conditions. Our results can guide farmland consolidation strategies and sustain soil health and ecological balance in agriculture.

Keywords: farmland consolidation; cultivated land quality; soil basic physical and chemical properties; heavy metals; microorganisms

Citation: Lin, Y.; Ye, Y.; Liu, S.; Wen, J.; Chen, D. Effect Mechanism of Land Consolidation on Soil Bacterial Community: A Case Study in Eastern China. *Int. J. Environ. Res. Public Health* **2022**, *19*, 845. https://doi.org/10.3390/ijerph19020845

Academic Editor: Paul B. Tchounwou

Received: 29 November 2021
Accepted: 28 December 2021
Published: 13 January 2022

Publisher's Note: MDPI stays neutral with regard to jurisdictional claims in published maps and institutional affiliations.

1. Introduction

Farmland consolidation is an effective means to improve farmland infrastructure, improve farmland quality, and protect farmland ecology [1,2]. It is a complex system engineering that plays an important regulatory role in multiple fields such as land ecology, land economy, and land ownership [3–5]. Therefore, farmland consolidation has an important contribution to ensuring global food security and regional sustainable development and rural depopulation on a global scale [6,7]. The methods under farmland consolidation mainly include building ditches, merging plots, land levelling, applying organic fertilizers, and comprehensive improvement. A large number of studies have shown that farmland consolidation regulates soil physical and chemical properties and concentrations of heavy metals [6,8,9]. As the most abundant microorganism in the soil environment, soil bacteria are extremely susceptible to varied soil properties [10]. As a key indicator of soil quality and ecological status, soil bacteria play an important role in soil pollution restoration and promotion of crop growth [11,12]. Therefore, it is important to explore how farmland consolidation affects the relative abundance and community structure of bacteria by adjusting soil physical and chemical properties and the content of heavy metals, which can provide information on sustainable development in farmlands.

The farmland conditions can not only affect crop growth and food security, but also have a significant impact on regional climate, hydrology, and soil [13–16]. Soil bacteria are the key driver of nutrient cyclings and energy flow in the soil environment, contributing to a healthy soil environment, enhanced regional ecosystem service values and ecosystem functions [17]. Soil bacteria are an important part of the earth's ecosystem, and they affect global climate through the regulation of CO_2, CH_4, N_2O, and other greenhouse gases [18]. Many studies reported the relationship between soil bacteria and climate change, and it has also been confirmed that the regulation of carbon, nitrogen, phosphorus, and other element cycles by soil microorganisms plays a vital role in the feedback of climate change [19,20]. In addition, soil bacteria can promote or inhibit the plant performance and further affect the plant community structure and functions [21,22]. Nitrogen-fixing bacteria in the soil can promote the uptakes of nitrogen (N), phosphorus (P), potassium (K), and other nutrients by plants in the nitrogen-fixing symbiotic relationship. Since N, P, K, and other elements are key nutrients that influence plant growth and reproduction, nitrogen-fixing bacteria can influence the expansion of the growth range of plant communities [23]. Hence, the types and functions of soil bacteria regulate plant community structure and function, as well as evolutionary processes in the ecosystem.

The structure of soil bacterial community is highly susceptible to different soil conditions. As a human activity, farmland consolidation has a strong interference on the physical and chemical properties of the soil, which will directly or indirectly affect the soil bacteria and change its ecological function [8,10,24]. Therefore, the impact of farmland consolidation on soil bacteria will be an important content in the future research fields of land use, environmental management, and microbial diversity [25,26]. Farmland consolidation is one of the most effective land management measures to improve agricultural production and ecological environment, including merging scattered land, improving agricultural facilities and soil quality, which has been widely used in most countries in the world [1,2,16]. Nowadays, in order to cope with the increasing risks of farmland, researchers are exploring the multifunctional potential of farmland consolidation to solve development problems such as agriculture, nature, landscape, economy, and tourism [27,28]. However, farmland consolidation also affects the physical and chemical properties of farmland soil and bacterial communities, especially in terms of soil bacterial diversity [6,27]. In order to promote the strategic deployment of China's ecological civilization construction, the farmland consolidation is also developing in a more ecological direction. However, the soil microorganisms that are profoundly affected by farmland consolidation have not received sufficient attention and need to be paid enough attention in future research and practice [29]. Under the current situation, it is possible to improve soil quality and increase the productivity of cultivated land by adopting appropriate farmland consolidation methods [30,31]. On this basis, we call to scientifically assess the impact of farmland consolidation on the structure of soil microbial communities, and extract microbial indicators of farmland quality [32,33]. This will be an important topic in the field of global farmland consolidation and ecological environmental protection in farmlands.

However, there were few studies on how farmland consolidation affects soil bacterial characteristics currently, and the mechanism behind them was not clear, which was a blank in the field of farmland consolidation and soil microecology research. Therefore, this study attempted to use experimental methods to explore the influence mechanism of farmland consolidation on soil bacteria in a county of the east coast of China. Specifically, we explored how farmland consolidation affected the relative abundance of bacteria and the community structure by regulating the soil properties and heavy metal contents, established a new research framework for this type of research, and provided information on farmland consolidation measures and ecological restoration in soils.

2. Framework and Data Collection

2.1. Research Framework

As an engineering method that directly affects arable land, farmland consolidation regulates arable land morphology, land landscape, soil properties, and fertility through measures such as building ditches, merging plots, land levelling, and applying organic fertilizers, combined or respectively [34]. Specifically, the influences of farmland consolidation on soil characteristics include: (1) Increasing the intensity of tillage activities to reduce soil bulk density, increase soil porosity, and improve organic fertilizer uniformity; (2) Applying organic fertilizer to increase soil nutrients availability, optimize the structure of soil microbial community, and improve the structure of soil aggregates and soil fertility; (3) improving farmland irrigation and drainage facilities to ensure the water requirement for crop growth and implement soil moisture management; (4) strengthening the management and protection, standardizing the technical process, and implementing supervision, of farmland consolidation, to sustain soil and ecosystem health [35].

Soil microorganisms are mainly affected by the dual effects of natural environment and human disturbance, including natural factors such as soil nutrients, pH, moisture, soil pollutants, climate change, biological invasion, plant species, as well as farming methods, irrigation management, land use changes and farmland consolidation methods and other human activities [36,37]. This study area is a small-scale farmland soil microbial study with a single farming method, and it has the characteristics of close to natural and local conditions. Therefore, this study limits the influencing factors of soil bacteria to soil properties and farmland consolidation methods to help reveal the true mechanism of action and find effective management strategies.

Farmland consolidation is an engineering method that directly affects the soil, including multiple methods such as building ditches, merging plots, land levelling, and applying organic fertilizers. This has a direct effect on the physical and chemical properties of the soil, and it will also directly or indirectly affect the quantity and community structure of soil microorganisms (Figure 1). In addition, through the implementation of mechanical engineering, farmland consolidation can reclaim non-agricultural land into cultivated land, and the infrastructure conditions of agricultural land can also be improved through consolidation projects. The former can lead to heavy metals residue from mines, factories and other lands, and enter the soil to cause pollution. The latter can adjust heavy metal concentrations in the soil through engineering measures, both of which will affect soil bacterial community structure.

In the field of farmland consolidation, few studies have focused on the impact of heavy metal content and basic physical and chemical properties of soil on soil bacteria, and the situation is more complicated in different levels of heavy metal pollution, so further exploration and discussion are needed. Based on the existing research results, this paper puts forward the analytical ideas and empirical research on the changes in soil heavy metal content and the basic physical and chemical properties of soil and the indirect effects of soil microbial characteristics by farmland consolidation, as shown in Figure 1.

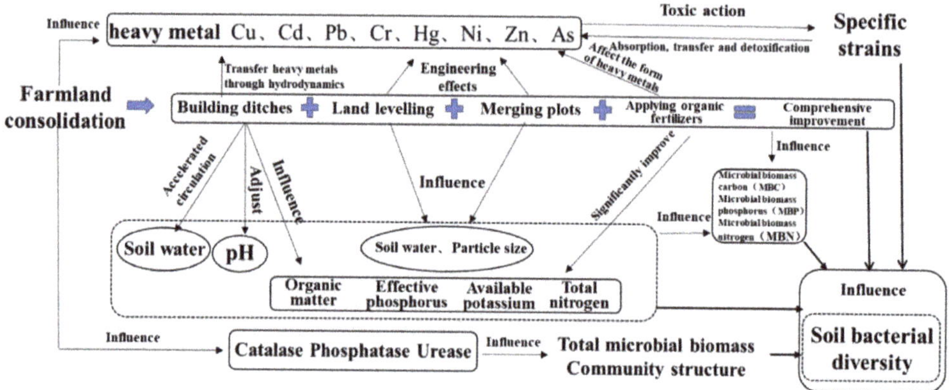

Figure 1. Conceptual diagram illustrating how farmland consolidation influence soil bacteria.

2.2. Study Area

The study area is located in a county of the east coast of China. The total area of the county is about 500 km², of which the water area accounts for about 15% and the land area accounts for about 85% (Data come from Jiashan County Natural Resources Bureau). The area belongs to the Southeast Asian monsoon region, with four distinct seasons and a mild climate. The annual average temperature is 15.6 °C, the average annual rainfall is 1155.7 mm, the annual average relative humidity is 65%, the annual average insolation duration is about 2000 h, and the frost-free period is 236 days (Data come from Jiashan County Natural Resources Bureau). The county has an advantageous geographical location and pleasant climate, making it an important grain producing area in China. However, the sustainable development of agriculture in this region is being challenged by fragmentation of cultivated land, incomplete agricultural facilities, degradation of soil fertility, and soil pollution. Therefore, farmland consolidation is an important strategy in recent years to improve the agricultural production and soil health in this area.

Through data collection, field investigations, and household interviews, we found that the county's farmland consolidation methods mainly included four types: building ditches, merging plots, land levelling, and applying organic fertilizers. In the farmland where these four measures were applied, we defined as comprehensive improvement. Collectively, this study divided the county's farmland into six types: building ditches, merging plots, land levelling, applying organic fertilizers, comprehensive improvement, and non-agricultural land consolidation areas.

2.3. Soil Collection and Analysis

2.3.1. Soil Sampling

In September 2018, 40 farmland consolidation areas were randomly selected in the study area (Figure 2). All targeted sites have a rice planting history from 4 to 8 years, after farmland consolidation, numbered A01–A40. Ten plots of cultivated land without farmland consolidation history were randomly selected around the farmland consolidation area. These sites had decades of rice planting history, numbered B01–B10. The farmland in the study area were dominated by green-purple mud fields and yellow-spot fields. The sampling points in the non-agricultural land consolidation area were next to the farmland consolidation area, and had similar soil properties to the farmland consolidation area, which could be used as an effective control group versus the farmland consolidation area. The specific sampling point distribution is shown in Supplementary S1.

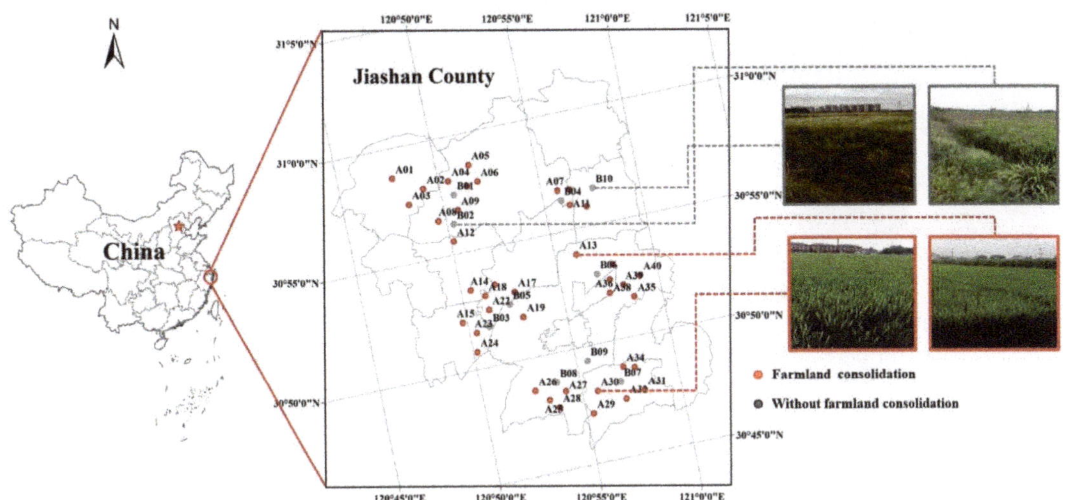

Figure 2. Location map of soil samples.

At each site, we divided a plot of 10×10 m from the area away from the road. We collected surface soil (0–15 cm) samples using five-point method and mixed into one soil sample at each plot [6,38]. Soil samples from all 50 sampling locations were filtered in sterilized stainless steel vessels to remove tree roots, rocks, plant, and animal debris with tweezers. We weighted about 20 g from each soil sample and stored them in a refrigerator at -80 °C for DNA extraction [39]. Five hundred grams of soil samples were stored in a sealed bag and transported back to the laboratory for soil characteristics and heavy metal analysis. After homogeneity, we extracted soil DNA 3 times from each sample and for further analysis.

2.3.2. Soil Basic Physical and Chemical Properties Test

We measured soil pH in the soil solution at 1:2.5 (soil:water) with a digital pH meter. The soil water content (SW) was calculated by using the oven-drying method at 105 °C for 12 h [40]. Soil organic matter (OM) was calculated by measuring the total organic carbon content using a total organic carbon analyzer (BOCS301, Shimadzu, Kyoto, Japan). Soil total nitrogen (TN), soil available phosphorus (AP), soil total phosphorus (TP), and soil available potassium (AK) were respectively determined by automatic Kjeldahl nitrogen analyzer (K9860, Hai Energy, Qingdao, China), flame photometer and spectrophotometer photometric measurement [41]. Soil catalase, urease, and phosphatase activities were determined by sodium phenate, sodium phenol-sodium hypochlorite colorimetry, and phenyl disodium phosphate colorimetry [42]. Each measurement was repeated at least three times for each sample. All data are detailed in Supplementary S2.

2.3.3. Soil Heavy Metal Content Test

We extracted soil samples using HCl-HNO_3-HF-$HClO_4$, the concentrations of Cu, Zn, Cr, Cd, Pb, and Ni in the soil was measured by an inductively coupled plasma source mass spectrometer (Agilent 7800, Palo Alto, CA, USA) [43]. Specifically, we digested 0.1 g soil samples using 3 mL 37% HCl, 1 mL 65% HNO_3, 6 mL 65% HF, and 0.5 mL 65% $HClO_4$. The digestion solution was evaporated to near dryness and dissolved in 1.0 mL of 65% HNO_3, and then 20 mL of deionized water was added. The concentrations of Hg and As in the soil was pretreated with aqua regia according to the China National Standard (GB 22105-2008) and then measured by an atomic fluorescence spectrophotometer (AF-630, BFRL, Beijing,

China) [43]. In this experiment, we repeated measurement 3 times, and used a blank control to ensure the measurement quality.

2.3.4. Soil Microbial Properties Determination

(1)   Soil microbial biomass determination

Soil microbial biomass usually includes soil microbial biomass carbon (MBC), microbial biomass nitrogen (MBN), microbial biomass phosphorus (MBP), etc., and it is mainly determined by chloroform fumigation extraction method [44]. We fumigated 10 g of sieved soil and the blank control group with de-ethanol chloroform, added K_2SO_4 for shaking extraction and filtered after 24 h. For microbial biomass carbon, after adding $K_2Cr_2O_7$ and H_2SO_4 solution to the extract to boil, $FeSO_4 \cdot 7H_2O$ was added and the coefficient conversion titrated. The microbial biomass nitrogen was processed by adding $CuSO_4$ and concentrated H_2SO_4 to the extract, then adding NaOH and connecting a distillation nitrogen analyzer to absorb the released NH_3, and the result was calculated by coefficient conversion. The microbial biomass phosphorus was measured in a spectrophotometer by using $NaHCO_3$ and KH_2PO_4.

(2)   DNA extraction and sequencing analysis

According to the instructions of the FastDNA SPIN kit (MP Biomedicals, Santa Ana, CA, USA), the microbial DNA for PCR amplification was extracted from 0.5 g of soil sample. The Nanodrop 2000 spectrophotometer (Thermo Scientific, Waltham, MA, USA) was used to measure the concentration and quality of the extracted DNA, and the extracted DNA was stored in a refrigerator at -20 °C for later analysis. For bacteria, primers 338F and 806R were used for PCR amplification of the V3–V4 region of the 16S rDNA gene [45].

PCR used 20 µL reaction system, including MdNTPs, FastPfu Buffer, FastPfu Polymerase, primers, BSA, DNA template, and finally added pure water to 20 µL. We carried out the following amplification procedure with ABI GeneAmp® 9700 PCR machine: predenaturation at 95 °C for 3 min; 95 °C for 30 s, 60 °C for 30 s, 72 °C for 45 s, repeat 10 cycles; 72 °C extend down for 10 min; perform 30 min at 10 °C. After the amplified products were electrophoresed on a 2% agarose gel, the PCR mixed products were recovered with a gel extraction kit (Omega, Norcross, GA, USA).

We used the Illumina HiSeq4000 platform for sequencing and microbial community analysis [11]. The sequencing data was analyzed on Illumina HiSeq4000, and the original DNA sequencing data was processed by the Quantitative Insights Into Microbial Ecology 2 platform (QIIME 2, University of California, San Diego, CA, USA) [46]. We used Usearch 7.1 to divide high-quality sequences with 97% similarity into operational taxonomic units (OTUs). Before clustering the sequences in OTUs, sequences that only appear once were removed to improve the accuracy of diversity assessment.

2.4. Statistic Analysis

The microbial α diversity index could reflect the richness and diversity of microbial communities, including Shannon, Simpsoneven, Simpson and Chao 1, which were all evaluated by the mothur software package (version v.1.30.1, University of Michigan, Ann Arbor, MI, USA) [47]. The composition of the microbial community was mainly displayed in the form of bar graphs and Heatmap graphs through the "vegan" package. The β diversity index was calculated by QIIME to analyze the differences of microbial communities. The drivers of the differences in soil microbial community composition could be processed and discriminated by PLS-DA (Partial Least Squares Discriminant Analysis). The correlations between environmental factors and soil microbial species were analyzed by Spearman method and displayed in the form of Heatmap. The functional composition of soil microorganisms was used to predict the function of amplicon sequencing data of bacteria through PICRUSt2 software (Harvard University, Cambridge, Massachusetts, USA). These were the scientific and effective bioinformatics statistical analysis methods commonly used in the world [11].

We used geological accumulation index (*GI*), proposed by Muller, to assess heavy metal pollution levels, which had been widely used in paddy soils [48]. The Nerome Comprehensive Index (*NI*) is another common method based on *GI* to reflect the level of heavy metal pollution [49]. Collectively, the level of heavy metal pollution in the soil can be effectively assessed by combining the two methods of *GI* and *NI*. *GI* is a geochemical standard for determining the pollution level of a single heavy metal in the soil by comparing it with the pre-industrial level. The formula is as follows:

$$GI_i = log_2\left(\frac{C_i}{1.5B_i}\right) \tag{1}$$

where C_i represents the concentration of a single heavy metal in the soil collected from sample i; B_i represents the geochemical background value of a single heavy metal in the sample in this area [50]. In addition, one study reported the influence of the external environment and human activities on the fluctuation of metal content can be expressed by a coefficient of 1.5 [51]. *NI* was determined based on the results of *GI*, and a reasonable and comprehensive assessment of pollution could be obtained. The formula is as follows:

$$NI_i = \sqrt{\frac{GI_{i_{ave}}^2 + GI_{i_{max}}^2}{2}} \tag{2}$$

where GI_{iave} and GI_{imax} show the average and maximum GI_i values of eight heavy metals, respectively. According to the *NI* value, the level of heavy metal pollution can be divided into severe level (SL, $NI > 3$), moderate level (ML, $2 < NI \leq 3$), light level (LL, $1 < NI \leq 2$), and clean level (CL, $NI \leq 1$) [52].

3. Results and Discussion

3.1. Effect of Land Consolidation on Soil Bacterial Community

3.1.1. Changes in Soil Bacterial Diversity

Soil microorganisms include bacteria, fungi, archaea, actinomycetes, viruses, algae, and protozoa. Among them, bacteria have an absolute advantage in quantity and volume play an important role in the operation of soil ecosystems [53,54]. Therefore, this article selected bacteria to represent soil microorganisms for analysis to determine the degree of impact of different farmland consolidation measures on soil microbial diversity. α diversity can characterize community diversity characteristics such as the number of species, uniformity, and relative abundance of microbial communities in the habitat, while β diversity can reflect the differences between different groups of microbial communities to analyze changes in space. These two microbial diversity indexes can reveal the structure and stability of microbial ecosystem functions, which have been widely used for diversity analysis.

Analysis of α Diversity of Bacterial Community

After 16S rDNA amplification and quality control optimization of soil samples in the study area, a total of 1,249,000 valid sequences were obtained, with an average length of about 439 bp, and aggregated into 11,295 OTUs. By making the Shannon dilution curve (Figure 3), a curve that tends to be flat was obtained. This showed that the amount of sequencing data in this experiment was appropriate and sufficient to reflect most of the bacterial diversity information in the soil sample. It was impossible to generate more strains if the amount of sequencing data continues to increase. Therefore, the 16S rDNA sequencing data obtained in this experiment had been able to accurately reflect the needs of bacterial diversity.

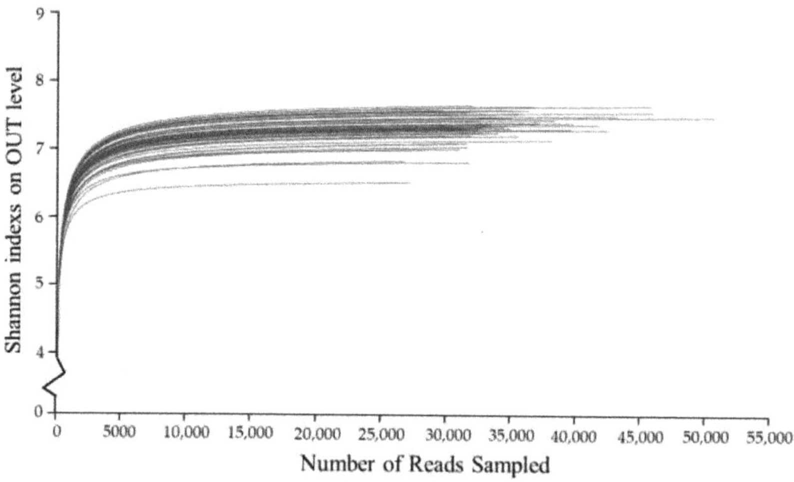

Figure 3. Shannon dilution curve of different samples.

After calculating the α diversity index of bacterial communities in soil samples, the correlation and significant relationship between the α diversity indexes of different groups of soil samples were calculated through Spearman correlation analysis (Table 1). The α diversity index mainly included Sobs and Chao reflecting the community richness index, Simpsoneven and Shannoneven reflecting the community evenness index, Shannon and Invsimpson reflecting the community diversity index, and Coverage reflecting the community coverage. The community richness index refers to the sum of the abundance of all microbial species in the soil sample. The larger the value, the richer the community species. In the study area, the soil bacterial community richness indexes Sobs and Chao of the farmland consolidation were significantly higher than those in the non-agricultural land consolidation area at the $p < 0.05$ level. Simpsoneven and Shannoneven, reflecting the community evenness index, could be used to measure the consistency of the relative abundance of microbial community species. This species index did not show the significant difference among the groups, but the value of farmland consolidation areas was higher than that in non-agricultural land consolidation areas. The results indicated, after farmland consolidation, the bacterial community species abundance in cultivated soil will be more uniform. The community diversity index Shannon and Invsimpson were widely used indexes to reflect the community diversity. The higher the value, the higher the community diversity. In the study area, the soil bacterial community diversity index in the farmland consolidation area was significantly higher than that in the non-agricultural land consolidation area, especially in the comprehensive consolidation area, the soil bacterial community diversity index was the highest. The community coverage index could reflect the depth of sequencing, and the value close to 1 indicating that the depth of sequencing was sufficient to cover most of the microbial community species in the soil sample. The Coverage index of soil bacteria in the study area was all greater than 0.95, indicating that the experimental data had high credibility and operability.

Our results showed that the analysis of Shannon dilution curve and Coverage index indicated that the data of this experiment was of high quality and strong operability. The measurement and calculation of the bacterial community α diversity index confirmed that farmland consolidation had a stronger effect on improving the diversity of soil bacterial communities, which was mainly reflected in the community richness index and community diversity index.

Table 1. Table of soil bacterial diversity indicators under different farmland consolidation.

Land Consolidation Measures	The Community Richness Index		The Community Evenness Index		The Community Diversity Index		The Community Coverage
	Sobs	Chao	Shannoneven	Simpsoneven	Shannon	Invsimpson	Coverage
Comprehensive improvement	4312.00 ± 452.33a	5866.69 ± 523.61a	0.89 ± 0.01a	0.17 ± 0.02a	7.44 ± 0.12a	730.73 ± 118.61a	0.95 ± 0.01a
Applying organic fertilizers	4318.92 ± 447.21a	5768.07 ± 530.29a	0.88 ± 0.01a	0.15 ± 0.04a	7.36 ± 0.16a	658.19 ± 189.31a	0.96 ± 0.01a
Building ditches	4370.70 ± 443.19a	5808.02 ± 493.61a	0.88 ± 0.01a	0.16 ± 0.03a	7.39 ± 0.14a	693.77 ± 161.98a	0.95 ± 0.01a
Merging plots	4349.68 ± 440.11a	5823.19 ± 528.11a	0.88 ± 0.01a	0.15 ± 0.04a	7.38 ± 0.15a	675.83 ± 176.20a	0.96 ± 0.01a
Land levelling	4405.47 ± 396.52a	5963.16 ± 483.81a	0.88 ± 0.01a	0.16 ± 0.04a	7.39 ± 0.17a	696.92 ± 202.67a	0.96 ± 0.01a
Non-agricultural land consolidation	3338.20 ± 675.30b	4348.85 ± 922.49b	0.87 ± 0.01a	0.13 ± 0.04a	7.01 ± 0.25b	439.17 ± 166.72b	0.96 ± 0.01a

Note: The average value in the table is presented in the form of mean ± standard deviation. Different lowercase letters in the same column represent significant differences at the $p < 0.05$ level.

Analysis of β Diversity of Bacterial Community

β diversity analysis is usually used to compare and analyze the diversity of microbial communities between different sample groups, that is, the difference analysis between samples, mainly by evaluating the abundance information of community species and the evolutionary relationship between samples to measure the distance between samples—so as to effectively reflect the significant differences in microbial communities between different sample groups. PCoA analysis is a commonly used method for β diversity analysis. The R language is used to sort the eigenvectors and eigenvalues, and the most important eigenvalues are selected for representation in the coordinate system. Identify the principal components that have an important influence on the composition of the sample microbial community by way of dimensionality reduction.

In this study, all samples were divided into two groups: farmland consolidation area and non-agricultural land consolidation area for PCoA analysis. Our results showed that the bacterial communities in the two areas had different clusters, which indicates that there were obvious differences in the soil bacterial community structure between the groups (Figure 4). Two components explained more than 50% for the difference in bacterial community composition of the samples. Hence, we concluded that farmland consolidation affects the change of bacterial community structure by changing the physical and chemical properties of the soil.

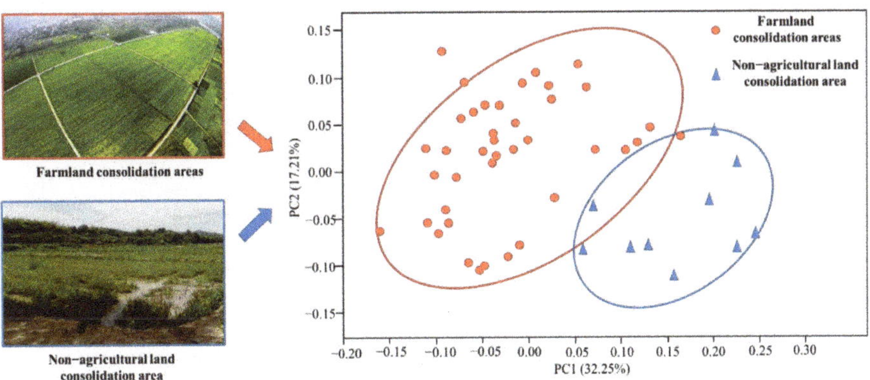

Figure 4. PCoA diagrams of bacterial communities with OTU levels of samples between different groups.

3.1.2. Variations in Soil Bacterial Community Structure

Analysis on the Changes of Bacterial Community at the Phylum Level

A total 1,249,000 high-quality soil bacterial sequences were obtained from 50 samples in this study, which were aggregated into 11,295 bacterial OTUs. These high-quality sequences could be divided into 56 bacterial phyla. The dominant phyla with relative abundance greater than 2% included Proteobacteria, Chloroflexi, Acidobacteria, Actinomycetes, Actinobacteria, Bacteroidetes, Nitrospirae, Gemmatimonadetes, Cyanobacteria. The number of effective sequences of these dominant bacteria phyla was 1,112,282, accounting for about 89% of the total; the number of OTUs is 7861, accounting for about 70% of the total.

Among them, the relative abundance was absolutely dominant in three phyla. Among them, the sequence number and OTU number of Proteobacteria were 399,929 and 2830 respectively, accounting for about 32% and 25% respectively. The sequence number and OTU number of Chloroflexi were 239,428 and 1972, respectively, accounting for 19% and 17%, respectively. The sequence number and OTU number of Acidobacteria were 182,952 and 973, respectively, accounting for about 15% and 9% respectively. The sequence numbers and OTU numbers of Actinobacteria were 100,050 and 541, respectively, accounting for about 8% and 5%, respectively. The sequence numbers and OTU numbers of Bacteroidetes were 63,609 and 760, respectively, accounting for about 5% and 7%, respectively. The sequence numbers and OTU numbers of Nitrospirae were 53,756 and 170, respectively, accounting for about 4% and 2%, respectively. The sequence numbers and OTU numbers of Gemmatimonadetes were 45,718 and 305, respectively, accounting for about 4% and 3%, respectively. The sequence numbers and OTU numbers of Cyanobacteria were 26,840 and 310, respectively, accounting for approximately 2% and 3%, respectively (Figure 5).

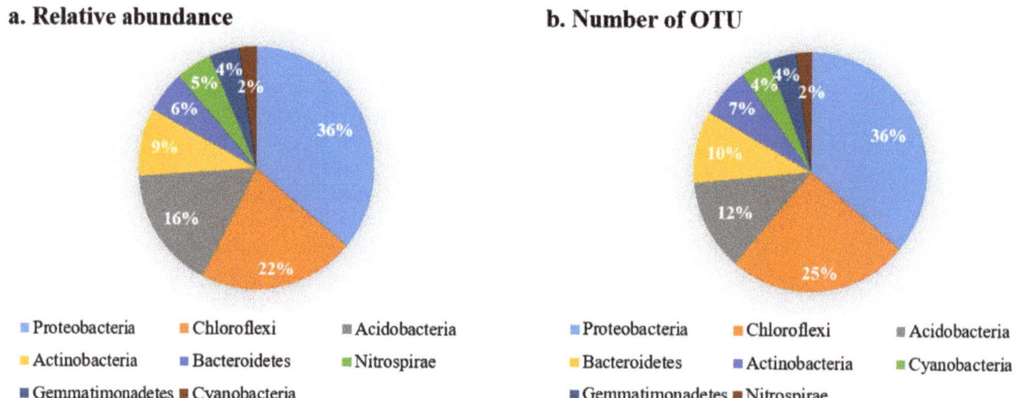

Figure 5. The relative abundance (**a**) and OTU (**b**)composition of the dominant bacteria in the soil samples in the study area.

The relative abundance of the soil samples in the top 20 dominant bacterial phyla was extracted for comparison (Figure 6). The abscissa in the Figure 6 represented the sample number, which was introduced in Supplementary S1. The results showed that the average relative abundance of Chloroflexi, Gemmatimonadetes, and Saccharibacteria in all grouped soils in the farmland consolidation area was higher than that in the non-agricultural land consolidation area. However, the average relative abundance of Bacteroidetes, Nitrospirae, Firmicutes, Ignavibacteriae, Spirochaetae, and Nitrospinae was lower than that of non-agricultural land consolidation area. Our results indicated that the farmland consolidation in the study area could increase the relative abundance of the three dominant bacteria phyla, but it also had a negative impact on the relative abundance of the six dominant bacteria phyla.

Analysis on the Changes of Bacterial Community at the Genus Level

We divided the sampling site type into six groups: building ditches, merging plots, land levelling, applying organic fertilizers, comprehensive improvement, and non-agricultural land consolidation areas, and extracted the top 20 dominant bacterial genera with relative abundance in each group of soil samples for comparative analysis (Figure 7).

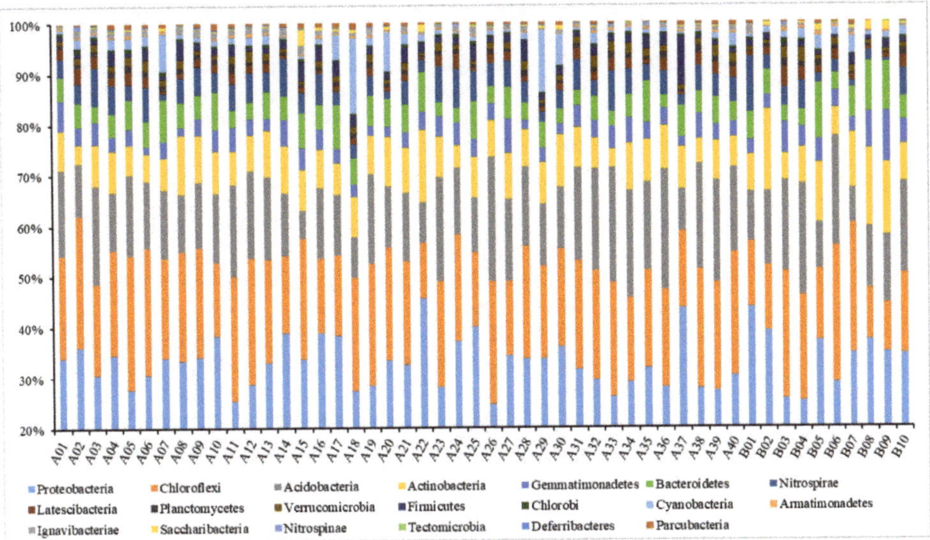

Figure 6. Relative abundance of dominant bacterial phyla in all samples of the study area.

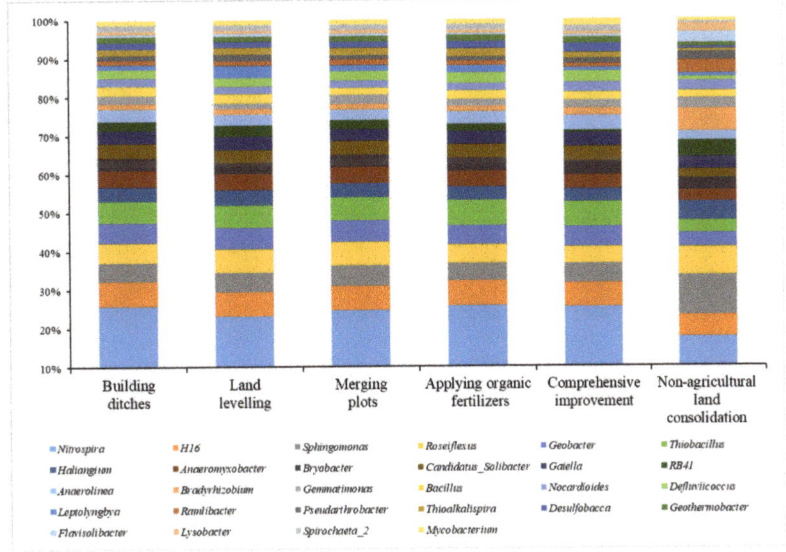

Figure 7. Relative abundance of dominant bacterial genera in all soil samples in the study area.

A total of 1,249,000 high-quality soil bacterial sequences from 50 samples in the study area could be divided into 1103 bacterial genera. Specifically, the dominant bacterial genera with relative abundance greater than 1% included *Nitrospira, H16, Sphingomonas,* and *Roseiflexus*. The number of effective sequences of these dominant bacteria was 98,949, accounting for about 8% of the total. Among the collected soil samples, the average relative abundance of *Nitrospira* was the highest, about 4.30%, which was significantly higher than that of other bacterial genera. This genus accounted for the largest proportion of 4.97% of the soil in the plots where comprehensive improvement measures were implemented, and the lowest proportion of soil in the non-agricultural land consolidation area was 3.66%. The average relative abundance of *H16* is 1.23%, which was relatively close in all soil groups, and its proportion ranges from 1.16 to 1.25%. The average relative abundance of *Sphingomonas* was 1.21%, the largest proportion of soils in non-agricultural land consolidation areas was 2.14%, and the lowest proportion in the soil with applying organic fertilizers was 0.84%. The average relative abundance of *Roseiflexus* was 1.18%, the largest proportion of soils in non-agricultural land consolidation areas was 1.52%, and the lowest proportion of soils under comprehensive improvement was 0.82%.

Analysis of Changes in Bacterial Community Function

In order to analyze the differences in soil bacterial functions between different groups, COG family information was obtained through the PICRUSt2 software to calculate the corresponding functional abundance and perform a significant analysis (Figure 8). According to the prediction results of soil bacterial community function, the most important functions included energy production and conversion, amino acid transport and metabolism, ribosome structure and biogenesis, and cell wall/membrane biogenesis. The relative abundance of bacterial community functions such as energy production and transformation, amino acid transport and metabolism, cell wall/membrane biogenesis in the soil of farmland consolidation areas was significantly higher than that in non-agricultural land consolidation areas ($p < 0.01$). This was caused by the significant differences in the composition and structure of the corresponding bacterial communities.

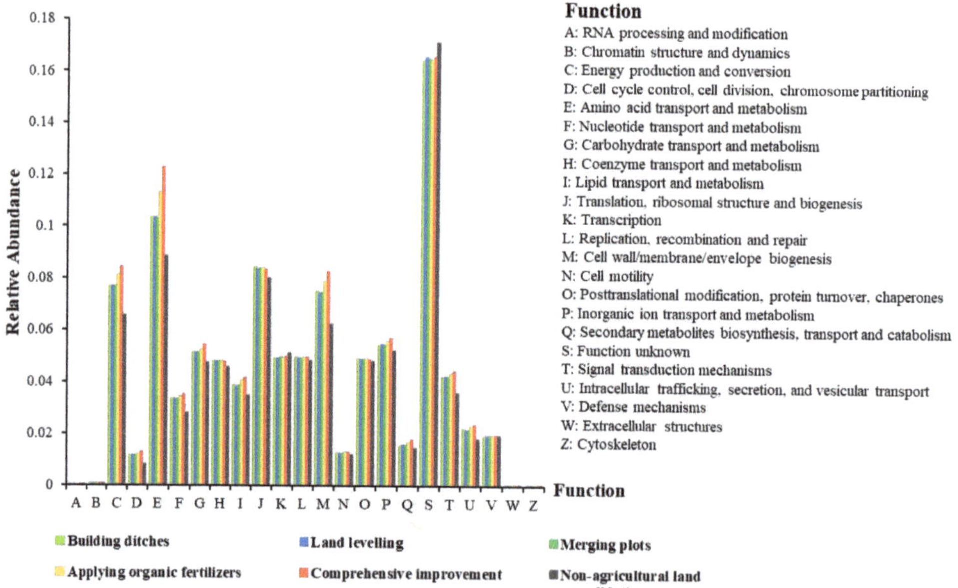

Figure 8. The relative abundance of bacterial community functions in soil samples between different groups.

3.2. Farmland Consolidation Regulates the Basic Physical and Chemical Properties of Soil and Its Mechanism of Action on Bacteria

3.2.1. Farmland Consolidation Promotes Changes in Basic Physical and Chemical Properties of Soil

Soil Physical Properties

The targeted soil physical properties in this study consisted of soil particle size and soil water content, which were the core content of current soil physics research. Soil particle size can directly affect the distribution of soil pores and have a significant impact on the aeration and water holding capacity of the soil. A reasonable particle size distribution is conducive to the healthy growth of crop roots and effectively increases food production [55]. Soil moisture is an important driver for soil physical processes, which can be affected by natural factors such as vegetation, terrain, climate, and human activities at different scales [56], but it can also affect the nutrient cycles and energy flow in the soil environment and directly affect the growth and development of crops.

The soil particle size in the study area ranged from 10.16 to 30.42 μm, and the average soil particle size in the non-agricultural land consolidation area was the smallest, which was significantly lower than that in the farmland consolidation area (Figure 9). Studies had shown that when the soil particle size in paddy soil was less than 20 μm, the content of soil organic matter could be significantly reduced, and when the soil particle size ranged from 20 to 250 μm, it would play a good carbon sink function [57,58]. This was probably because soil particles with a larger particle size have better permeability, the positive charge carried by themselves was easy to combine with the humus carrying negative charges, and it could be decomposed by soil microorganisms easily to increase the organic matter content in the soil [59]. In this study area, the average soil particle size in each farmland consolidation grouping was greater than 20 μm and the distribution was relatively uniform, while the average soil particle size in non-agricultural land consolidation was only 17.48 μm. Our results suggested that farmland consolidation could change soil particle size distribution through measures such as project implementation and soil fertilization, which had a certain driving effect on the change of soil properties. The soil water content in the study area ranged from 20.70% to 54.18%. The comprehensive improvement area had the highest average soil water content (47.00%), followed by the average soil water content of the plots where building ditches were implemented (43.87%). The average soil water content in non-agricultural land consolidation areas was the lowest (30.39%). Our results indicated that the improvement of farmland irrigation facilities and the optimization of soil mechanical structure could effectively increase soil water content and provide sufficient water supply for crop growth.

Soil Chemistry

Soil chemistry is an important branch of soil science, including soil pH, organic matter, available potassium, available phosphorus, and other indicators. It plays a key role in the process of soil productivity, self-purification capacity, carbon emissions, and nutrient balance [60]. The current frontier research field of soil chemistry is a cross-discipline of soil chemistry and microbiology. Soil microbes play an important role in the development of soil chemistry research.

The soil pH value in the study area ranged from 6.15 to 8.30. The average soil pH value in the non-agricultural land consolidation area was the highest and the value fluctuates the most, which was significantly higher than areas where the soil pH was close to neutral, such as building ditches, merging plots, applying organic fertilizers, and comprehensive improvement (Table 2). This was because measures such as building ditches and merging plots could effectively promote the flow of water in the farmland, while applying organic fertilizers could reduce the input of inorganic fertilizers, which could effectively regulate the pH of the soil [61]. For soil nutrients, the content ranges of organic matter, available phosphorus, available potassium, and total nitrogen are 16.00~70.40 g/kg, 8.56~330.79 mg/kg, 14.45~80.25 μg/mL and 0.93~3.75 g/kg, respectively. The average soil nutrient of non-agricultural land consolidation areas was the lowest, and was significantly lower than that of farmland consolidation areas, especially the areas where organic fertilizers and comprehensive improvement were applied have higher soil nutrient content. Our results showed that the application of organic fertilizers in farmland consolidation areas could effectively improve soil nutrients, and the construction of ditches accelerates the flow of water to promote nutrient cycling, thereby creating a healthy soil environment to accelerate the promotion of microorganisms in soil nutrient cycling, and achieve benign cycle.

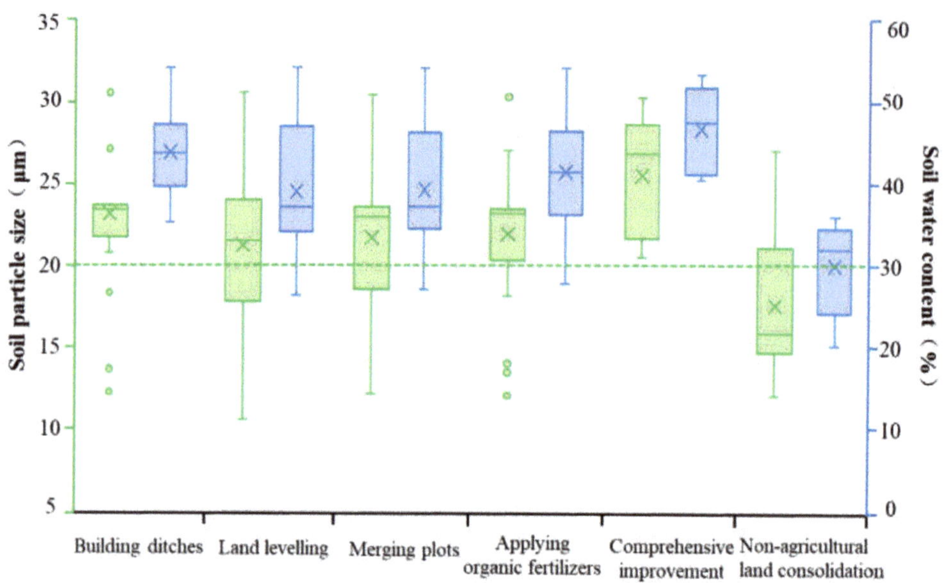

Figure 9. Distribution of soil particle size (green) and soil water content (blue) in the study area.

Table 2. Soil chemical properties among different groups in the study area.

Group	pH	Organic Matter (g/kg)	Available Phosphorus (mg/kg)	Available Potassium (μg/mL)	Total Nitrogen (g/kg)
Building ditches	7.09 ± 0.63b	46.63 ± 14.22b	90.78 ± 83.78a	29.65 ± 13.31a	2.41 ± 0.73b
Land levelling	7.23 ± 0.63a	41.97 ± 15.33b	72.85 ± 54.01b	28.99 ± 12.00a	2.20 ± 0.77b
Merging plots	7.11 ± 0.57b	41.52 ± 15.30b	70.24 ± 57.77b	28.60 ± 15.25a	2.23 ± 0.77b
Applying organic fertilizers	7.11 ± 0.64b	52.49 ± 11.59a	93.67 ± 84.50a	31.38 ± 14.57a	2.51 ± 0.64b
Comprehensive improvement	6.93 ± 0.51b	57.98 ± 11.66a	79.80 ± 58.37b	37.78 ± 7.61a	3.08 ± 0.45a
Non-agricultural land consolidation	7.38 ± 0.71a	30.21 ± 9.80c	53.43 ± 37.13c	24.00 ± 14.93b	1.69 ± 0.45c

Note: The average value in the table is presented in the form of mean ± standard deviation. Different lowercase letters in the same column represent significant differences at the $p < 0.05$ level.

Soil Enzyme Activity

Soil enzymes are an important driving factor for soil biogeochemical cycles. They can not only sense changes in soil properties, but their activity represents the capacity of soil nutrient supply, which is a key indicator of soil quality [62]. Among them, catalase is mainly involved in the chemical process of soil redox, which can effectively characterize the content of soil organic matter and the degree of soil decay [63]. Phosphatase and urease participate in the soil nitrogen and phosphorus cycles respectively, and are important indicators to characterize the conversion capacity of soil nitrogen and phosphorus [64]. Additionally, soil enzyme activities can also interact with soil microbial communities. For example, soil enzymes can participate in the degradation of microbial residues, and changes in the structure of microbial communities will also affect soil enzyme activities, thereby affecting the process of soil organic matter decomposition and nutrient cycling [65]. In addition, fertilization and engineering measures will also have a certain impact on soil enzyme activity. Studies have shown that applying nitrogen fertilizer and increasing soil moisture can significantly increase soil enzyme activity [66,67]. Therefore, this study selected catalase, phosphatase, and urease for determination and comparison between groups to try to analyze the effect of farmland improvement on soil enzymes.

According to the results of laboratory tests, in the study area, the soil catalase activity ranged from 125.45 to 276.90 mg/g, and there was no significant difference between the groups (Table 3). However, the soil catalase activity in the non-agricultural land consolidation area was the lowest, with an average of 183.58 mg/g. The soil phosphatase activity ranged from 2.80 to 45.20 mg/g. Among them, the soil phosphatase activity of comprehensive improvement was the highest, and the soil phosphatase activity of farmland consolidation areas was significantly higher than that of non-agricultural land consolidation areas. The soil urease activity ranged from 0.01 to 0.98 mg/g, and there was no significant difference between the groups. However, the soil urease activity in the non-agricultural land consolidation area was the lowest, and the soil urease activity in the comprehensive improvement was the highest. Farmland consolidation had effectively improved the soil environment of farmland by adjusting soil pH, improving soil nutrients, accelerating soil water circulation, and had a greater promotion effect on the increase of soil enzyme activity [68,69].

Table 3. Three soil enzyme activities in the study area.

Group	Catalase (mg/g)	Phosphatase (mg/g)	Urease (mg/g)
Building ditches	207.66 ± 38.30a	16.017 ± 8.44a	0.25 ± 0.16a
Land levelling	203.62 ± 38.48a	17.35 ± 10.65a	0.28 ± 0.22a
Merging plots	201.61 ± 32.37a	15.28 ± 9.97b	0.23 ± 0.15a
Applying organic fertilizers	198.61 ± 41.72a	17.54 ± 9.68a	0.25 ± 0.16a
Comprehensive improvement	206.72 ± 28.98a	20.88 ± 10.58a	0.33 ± 0.16a
Non-agricultural land consolidation	183.58 ± 50.72a	12.32 ± 11.87c	0.22 ± 0.30a

Note: The average value in the table is presented in the form of mean ± standard deviation. Different lowercase letters in the same column represent significant differences at the $p < 0.05$ level.

3.2.2. The Mechanism of Basic Physical and Chemical Properties of Soil on Bacterial Community

Soil Physical Properties and Bacterial Community

Soil particle size (SPD) and soil water content (SW) are important indicators for evaluating soil physical properties. They have a strong influence on other soil physical and chemical properties and microbial characteristics, and can directly or indirectly affect crop growth and food quality [70]. Soil particle size is an effective index that can characterize soil porosity and aeration. The size of soil particle size determines the amount of humus that it adheres to, and also reflects the ability to provide nutrients for microorganisms. The soil water content can reflect the water supply capacity and nutrient retention capacity of the soil. As the soil water content increases, it will promote the reproduction of soil microorganisms to increase the number and relative abundance of microorganisms [71]. Therefore, soil particle size and soil water content are key indicators of soil physical properties that affect the structure and diversity of soil microbial communities. However, soil particle size and soil water content are easily affected by farmland consolidation projects, so the correlation analysis of soil particle size, water content, and soil microorganisms in farmland consolidation areas has certain research significance.

According to Spearman correlation measurement, the soil particle size in the study area was significantly correlated with the relative abundance of 11 bacterial phyla, and the soil water content was significantly correlated with the relative abundance of 7 bacterial phyla (Figure 10). In farmland consolidation areas where construction of ditches was implemented, the soil water content and the relative abundance of bacteria were mainly positively correlated, but not significant. This was probably because the soil water content of all plots was higher after the construction of ditches in the farmland, so there was no significant difference. The soil particle size in this type of farmland was significantly negatively correlated with Cyanobacteria and Elusimicrobia. This was because these two bacteria tend to be synergetic in a eutrophic environment. However, the area with larger soil particle size would accelerate the water cycle and inhibit the growth of the relative abundance of these two bacteria. In farmland consolidation areas where land levelling was implemented, the soil water content was significantly positively correlated with the relative abundance of Aminicenantes, Firmicutes, Ignavibacteriae, and Spirochaetae. This was because the farmland soil that implemented the land levelling was subjected to certain mechanical compaction, which would affect the physical characteristics of the soil's water circulation and air permeability, which led to the increase in the relative abundance of some bacteria phyla with the increase of soil water content. However, the Cyanobacteria and the soil particle size were significantly negatively correlated at the level of 0.001. This was due to the mechanical compaction of the soil particle morphology, which resulted in the deterioration of soil voids and water ventilation performance, which restricted the reproduction of the cyanobacteria, and the larger the soil particle size, the greater the impact it would bear. In farmland where combined plots, application of organic fertilizers, and comprehensive improvement were implemented, soil water content and particle size affected the relative abundance of Cyanobacteria, Firmicutes, Elusimicrobia, and Ignavibacteriae, which were probably due to the combined effects of soil nutrients, pH and other environmental factors, and had a certain universality. Different from farmland consolidation areas, the influence of soil water content and particle size on the relative abundance of bacterial communities in non-agricultural land consolidation areas was mostly significantly positively correlated. This was probably due to the uneven drainage and nutrient supply in the non-agricultural land consolidation area. The bacterial community could obtain a better living environment in the soil with high water content and large particle size, thereby significantly increasing its relative abundance. These conclusions were consistent with the previous studies [72].

Soil Chemical Properties and Bacterial Communities

Soil pH, organic matter (SOM), available phosphorus (AP), available potassium (AK), and total nitrogen (TN) are representative indicators of soil chemical properties, which can effectively characterize the basic chemical properties and fertility status of the soil, and are also important factors affecting the structure of bacterial communities [73]. Previous studies have shown that the pH value in the black soil is the primary influencing factor that affects the bacterial community structure, while the total nitrogen in the soil under continuous soybean cropping conditions is a key environmental factor that affects soil bacteria [74,75]. Because bacterial communities are more sensitive to changes in soil nutrients and pH [76], and farmland consolidation measures have a strong influence on soil chemical properties. Therefore, the analysis of the correlation between soil chemical properties and bacterial communities in farmland consolidation areas has a certain reference value for the research on farmland microecology.

Spearman correlation analysis was carried out on the relative abundance of bacterial communities and soil chemical properties in different groups of farmland in the study area. The results showed that a total of 18 bacterial phyla were significantly correlated with soil chemical indicators such as soil pH, SOM, AP, AK, and TN (Figure 11). The 8 bacterial phyla that were significantly related to pH include Bacteroidetes, Aminicenantes, Ignavibacteriae, and Nitrospirae, which mainly appeared in farmland consolidation areas. This was probably because the engineering measures in the farmland consolidation area had a certain adjustment effect on the soil pH, but they also interfered with the soil environment, resulting in uneven pH distribution and a significant impact on the relative abundance of soil bacteria in local plots. There were 7 bacterial phyla that were significantly affected by SOM, and the relative abundance of soil bacteria in the plots where comprehensive improvement was implemented showed a strong positive correlation with SOM, including Latescibacteria, Nitrospinae, Planctomycetes, and other phyla. Our results suggested that increasing soil organic matter content and improving nutrient cycling conditions through farmland consolidation had a greater promotion effect on the survival and reproduction of soil bacteria. AP, AK, and TN, as important soil nutrients, also had a greater impact on the structural changes of soil bacteria. A total of 17 bacterial phyla were significantly related to these three nutrients. Among them, there were 9 bacterial phyla in non-agricultural land consolidation areas that were significantly correlated with soil AP, AK, and TN, but most were negatively correlated. Our results indicated that, due to the massive application of inorganic fertilizers in non-agricultural land consolidation areas, the soil environment was polluted and soil nutrient dynamic balance was destroyed, thereby reducing soil bacterial activity [77].

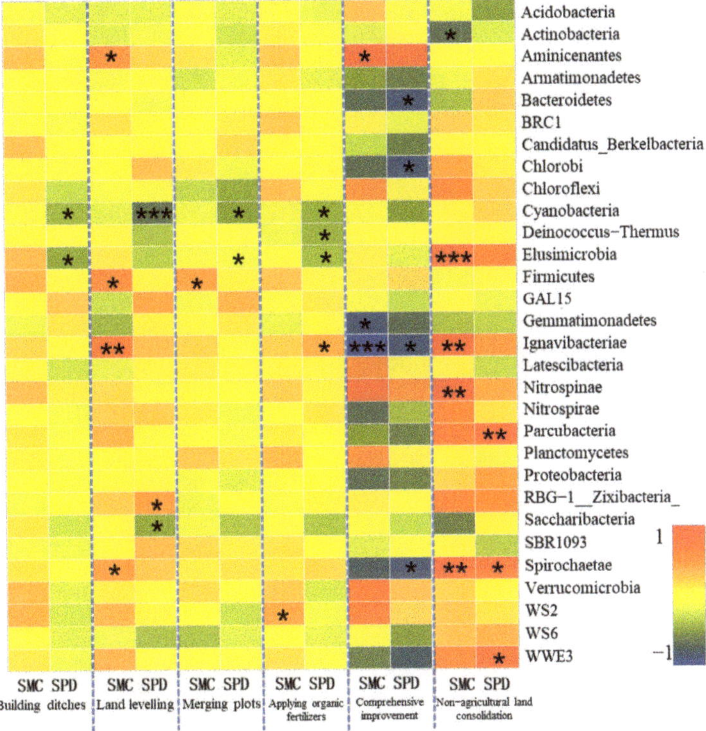

Figure 10. Correlation between soil particle size, moisture content and bacterial community structure in different groups. (Note: Warm colors indicate positive correlation, and cool colors indicate negative correlation. *, **, *** indicate significant correlation at the levels of 0.05, 0.01, and 0.001, respectively).

Soil Enzyme Activity and Bacterial Community

Soil microbes and crop secretions are important sources of soil enzymes, so soil enzyme activity can be used as an indicator of soil microbial activity and has a close relationship with the structure of soil bacterial community [78]. Existing studies have shown that soil urease (URE), phosphatase (PDXP), and catalase (CAT) were significantly positively correlated with the total amount of soil bacteria ($p < 0.05$) [79]. There are also many studies on the relationship between soil enzyme activity and bacterial community structure, but they only stay at the level of bacterial diversity analysis without performing correlation analysis on specific bacterial species [79,80].

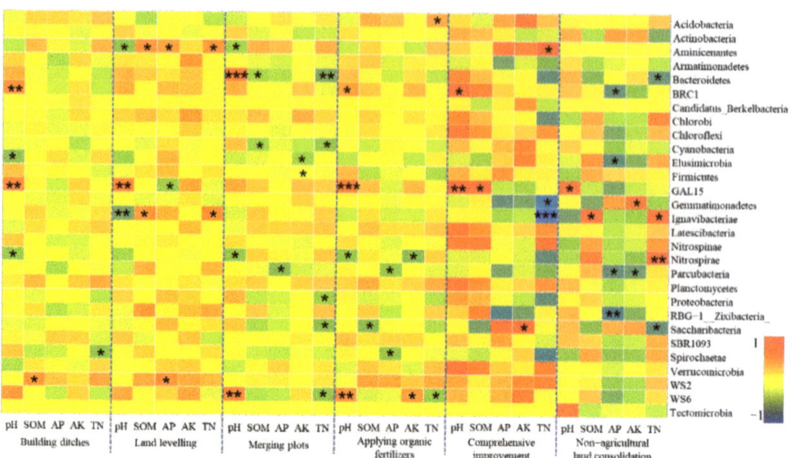

Figure 11. Correlation between soil chemical properties and bacterial community structure in different groups. (Note: Warm colors indicate positive correlation, and cool colors indicate negative correlation. *, **, *** indicate significant correlation at the levels of 0.05, 0.01, and 0.001, respectively).

A total of 14 bacterial phyla in farmland soil samples in the study area were significantly correlated with soil enzyme activity indicators such as URE, PDXP, and CAT (Figure 12). In the farmland consolidation area, the soil enzymes in the farmland where the organic fertilizer was applied and the comprehensive improvement was carried out had a significant positive correlation with the more dominant bacteria phyla. This was because the two groups of soils had a higher content of soil enzymes, and the bacteria obtained soil nutrients such as carbon, phosphorus, and nitrogen decomposed by enzymes could achieve rapid reproduction. The phylum Chloroflexi and Planctomycetes were significantly positively correlated with catalase (CAT) and phosphatase (PDXP) in farmland soils under comprehensive improvement. This was because many bacteria in these two phyla could produce energy and nutrient elements through photosynthesis and oxidation, which could help the microorganisms and enzyme activities in the soil to a certain extent. However, in farmland soils in non-farmland consolidation areas, most bacterial phyla had a negative correlation with soil enzymes. This was also because there were fewer nutrients available for soil enzyme decomposition in this group, and it was difficult to promote the development of the bacterial community, which was caused by the microecological imbalance.

3.3. Farmland Consolidation Regulates Soil Heavy Metal Content and Its Mechanism of Action on Bacteria

3.3.1. Effects of Farmland Consolidation on Soil Heavy Metal Content

Heavy Metal Content of Farmland Soil

The levels of heavy metals in farmland soils of different groups in the study area are shown in Table 4. Among the eight heavy metals tested in farmland soils in the study area, the average content of seven heavy metals was greater than the background value of the soil, and only the average content of the heavy metal As was slightly lower than the background value. Among them, the average contents of Cu, Cd, Pb, Cr, Hg, Ni, and Zn were 45.17, 2.5, 41.94, 223.79, 0.55, 56.77, 123.27 mg/kg, respectively, which were 2.00, 14.71, 1.17, 4.00, 3.24, 2.38, 1.48 times of the soil background value in the region. This showed that there was a relatively serious accumulation of heavy metals in farmland soils in this study area.

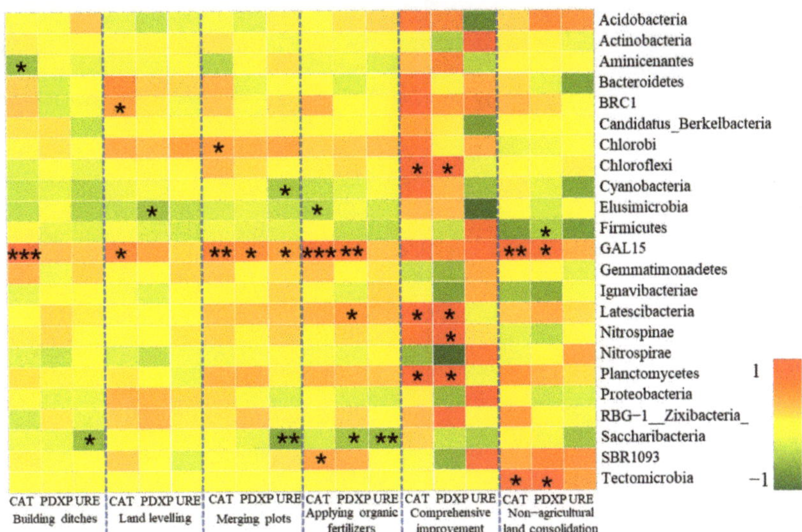

Figure 12. The correlation between soil enzyme activity and bacterial community structure in different groups. (Note: Warm colors indicate positive correlation, and cool colors indicate negative correlation. *, **, *** indicate significant correlation at the levels of 0.05, 0.01, and 0.001, respectively).

Table 4. Soil heavy metal content in the study area.

	Cu	Cd	Pb	Cr	As	Hg	Ni	Zn
				Unit: mg/kg				
Building ditches	45.95 ± 12.74	2.28 ± 0.23	43.39 ± 13.23	214.65 ± 11.67	4.66 ± 3.83	0.57 ± 0.28	53.15 ± 5.60	125.75 ± 26.70
Land levelling	39.44 ± 7.92	2.23 ± 0.27	37.54 ± 12.81	214.61 ± 11.34	4.21 ± 3.51	0.45 ± 0.15	54.06 ± 7.63	110.47 ± 18.39
Merging plots	41.72 ± 8.83	2.27 ± 0.33	37.62 ± 12.06	212.64 ± 11.22	4.99 ± 3.91	0.53 ± 0.25	54.55 ± 7.44	112.51 ± 21.58
Applying organic fertilizers	46.55 ± 12.25	2.27 ± 0.31	42.98 ± 10.38	216.45 ± 11.12	3.89 ± 2.90	0.53 ± 0.21	55.00 ± 6.86	128.36 ± 24.72
Comprehensive improvement	39.90 ± 9.17	2.28 ± 0.13	40.29 ± 14.98	208.70 ± 6.59	2.93 ± 3.94	0.47 ± 0.14	51.10 ± 6.93	117.82 ± 20.45
Non-agricultural land consolidation	48.97 ± 6.71	3.30 ± 0.55	46.83 ± 8.81	252.96 ± 11.05	7.88 ± 5.29	0.65 ± 0.37	64 ± 10.85	136.82 ± 21.17
Maximum	84	3.98	77.5	268.8	14.4	1.38	80.4	198
Minimum	25	1.57	20	195	0.81	0.1	39	70.6
Average value	45.17	2.5	41.94	223.79	5.29	0.55	56.77	123.27
Variation coefficient (%)	22.38	22.86	26.4	8.72	77.91	49.99	15.96	20.7
Background value	22.6	0.17	35.7	56	6.9	0.17	23.9	83.1

Note: The content of heavy metals in the table is presented in the form of average ± standard deviation.

The order of the coefficient of variation of the eight heavy metals in the study area was: As > Hg > Pb > Cd > Cu > Zn > Ni > Cr. Among them, the coefficients of variation of As, Hg, Pb, Cd, Cu and Zn were all greater than 20%, belonging to moderate intensity variation, indicating that these heavy metals were significantly affected by external interference. The specific manifestation was that the content of heavy metals in the soil varies greatly in space, which was mainly attributed to the influence of human disturbance factors such as farming methods, fertilizer application, management measures, and pollutant emissions. The small coefficient of variation of Cr and Ni indicated that the spatial distribution of these two heavy metal elements was relatively uniform, and there might be a certain degree of homology [81]. As an artificial measure that strongly disturbed the soil environment, farmland consolidation was an important factor affecting the spatial distribution of soil heavy metal content. There were large differences in the content of heavy metals between the areas where different agricultural land consolidation measures were implemented and the non-agricultural land consolidation areas, and the highest average values of various heavy metal content were in the non-agricultural land consolidation areas. The content of heavy metals in farmland where construction of ditches, land levelling, combined plots, application of organic fertilizers, and comprehensive improvement were implemented was lower than that of non-agricultural land consolidation areas. This was probably because the construction of ditches could significantly improve the transfer of heavy metals by water in the soil, and the application of organic fertilizers could provide nutrients for the growth and reproduction of microorganisms to play the function of transferring and absorbing related heavy metals, which was consistent with the existing research results [82].

Heavy Metal Pollution Level of Farmland Soil

According to the results of the Geological Accumulation Index (*GI*) (Figure 13), the pollution degree of the heavy metals in the study area was in descending order: Cd > Cr > Hg > Ni > Cu > Zn > Pb > As. Among them, Cd was the heavy metal with the highest geological accumulation index and the most polluted heavy metal in the study area. The geological accumulation index of the three heavy metals Pb, As, and Zn was mostly negative, indicating that the soil in the study area was not polluted by them [83]. At the same time, there were big differences in the geological accumulation level of heavy metals among different groups. In the study area, the cumulative geological index of heavy metals in the farmland under comprehensive improvement was generally low, while the cumulative index of heavy metals in the non-agricultural land consolidation area had the highest value.

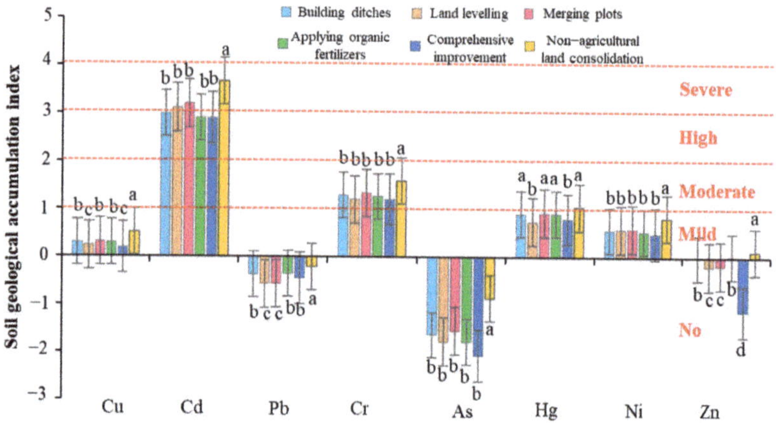

Figure 13. Evaluation results of soil geological accumulation index in the study area. Note: Different lowercase letters represent significant differences at the $p < 0.05$ level.

The comprehensive pollution level of heavy metals in farmland soil could be determined by measuring the Nerome comprehensive index (*NI*). The average range of the soil Nemerow comprehensive index of each group in the study area was 0.64 to 3.68. According to the soil heavy metal pollution index classification standard [52], when the *NI* is less than 1, the soil is low pollution; when the *NI* is 1 to 2, the soil is lightly polluted; When the *NI* is 2 to 3, the soil is moderately polluted, which will pose a toxic threat to rice; when the *NI* is 3 to 4, the soil is heavily polluted, which will seriously affect the growth and development of crops. According to the calculation results of Nerome comprehensive index, the farmland soil in the non-agricultural land consolidation area was at a heavily polluted level, the farmland soil in the combined plot was at a moderately polluted level, and the farmland soil in other farmland consolidation areas was at a lightly polluted level (Figure 14). This was because the implementation of construction ditches could promote soil water circulation to improve the effective transfer of heavy metals, and the application of organic fertilizers could increase the abundance of heavy metal-tolerant bacteria to adsorb, migrate and detoxify related heavy metals. All of these could effectively reduce the level of heavy metal pollution in farmland consolidation areas. This conclusion was consistent with the existing research results [82]. At the same time, this also explained why the farmland soil that implements comprehensive improvement had the lowest level of heavy metal pollution.

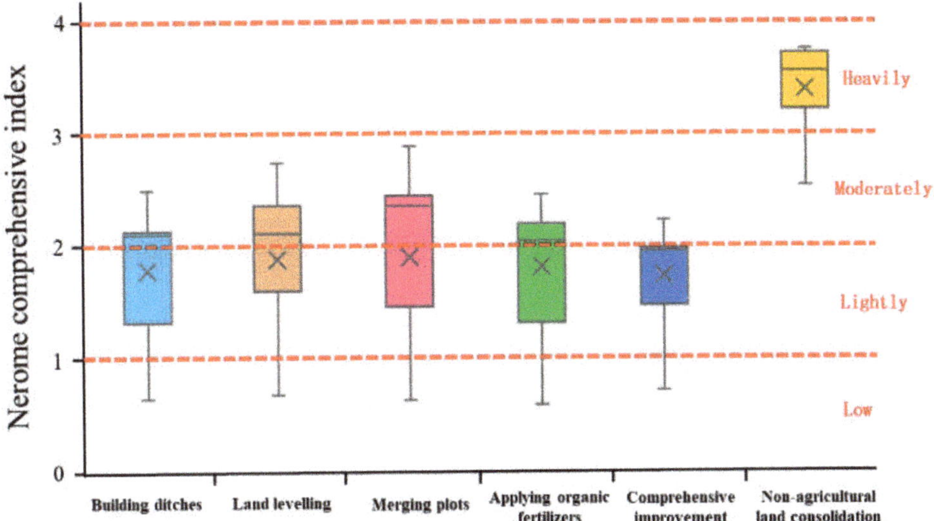

Figure 14. Evaluation results of Nerome comprehensive index in the soil of the study area.

Through the above calculation analysis and field investigation, our results showed that the reasons for the large differences in the level of heavy metal pollution in farmland soils in the study area were as follows:

(1) In recent years, the area where the study area was located has continuously strengthened the development and utilization of the industrial area, causing industrial pollutants to continue to flow into the surrounding farmland, causing serious soil heavy metal pollution. (2) Many oily pollutants, such as fertilizer bags, pesticide bottles, and herbicides, which were likely to cause cadmium (Cd) pollution, were scattered in the farmland, but had not been effectively removed. (3) Pollutants caused by fertilizer application, poultry breeding and domestic garbage in the study area were also sources of heavy metal pollution in the soil [84]. (4) Measures such as building ditches and applying organic fertilizers in farmland consolidation areas could accelerate water circulation to transfer heavy metals, and provide nutrients for bacteria with heavy metal repair functions to achieve the adsorption and detoxification of heavy metals, which effectively reduced the level of heavy metal pollution in farmland [85].

3.3.2. The Mechanism of Soil Heavy Metal Pollution on Bacterial Communities
Bacterial Communities at Low Pollution Levels

A total of 10 samples in the study area belonged to the low level of heavy metal pollution, all located in the farmland consolidation area. According to CCA analysis, in these 10 samples, heavy metals Cr ($r^2 = 0.30$, $p = 0.027$), As ($r^2 = 0.27$, $p = 0.032$), Hg ($r^2 = 0.28$, $p = 0.031$), Ni ($r^2 = 0.45$, $p = 0.013$) were significantly related to bacterial community structure. According to Spearman correlation analysis, the content of heavy metals was significantly related to 11 bacterial phyla and 8 bacterial genera ($p < 0.05$) (Figure 15).

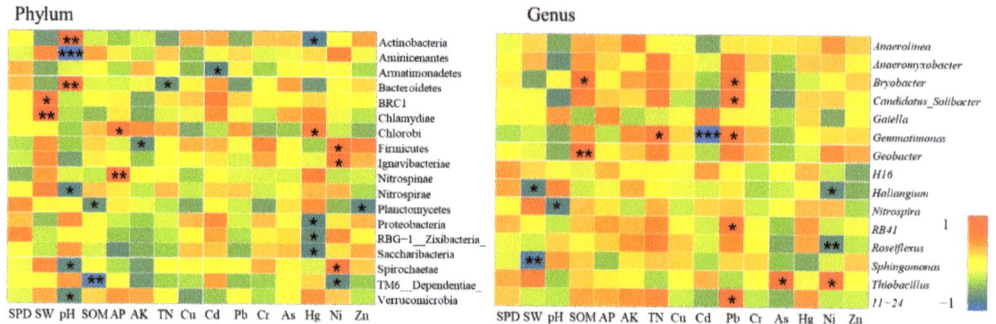

Figure 15. Correlation between soil environmental factors and bacterial community structure with low heavy metal pollution levels. (Note: Warm colors indicate positive correlation, and cool colors indicate negative correlation. *, **, *** indicate significant correlation at the levels of 0.05, 0.01, and 0.001, respectively).

At the bacterial phyla level, the relative abundance of soil bacteria in farmland with low soil pollution levels of heavy metals was significantly related to the physical and chemical properties of the soil. In particular, Actinobacteria, Aminicenantes, Bacteroidetes, Chlamydiae, Nitrospinae, TM6__Dependentiae_were significantly correlated with SW, pH, SOM, and AP at the level of 0.01. The relative abundance of soil bacteria in this area was also significantly related to the content of heavy metals such as Cd, Hg, Ni, Zn, and the relative abundance of 9 bacterial phyla were significantly related to the content of Hg and Ni. Among them, Chlorobi had a significant positive correlation with Hg, while Firmicutes, Ignavibacteriae, and Spirochaetae had a significant positive correlation with Ni. Our results suggested that these four types of bacteria have strong adaptability under low pollution levels of heavy metals, and could absorb and transfer heavy metals Hg and Ni to a certain extent. Studies have shown that pH could further change the microbial community structure by affecting the leaching of heavy metals and adjusting nutrients such as AP and SOM, which was consistent with the results of this research [86].

At the level of bacterial genera, *Bryobacter, Gemmatimonas, Geobacter, Haliangium, Nitrospira, Sphingomonas* were significantly related to the basic physical and chemical properties of soil such as SW, pH, SOM, and TN. The heavy metals Cd, Pb, As, and Ni were the main influencing factors affecting the genus structure of bacteria. Among them, *Bryobacter, Candidatus_Solibacter, Gemmatimonas, RB41, 11–24* had a significant positive correlation with Pb content, while *Thiobacillus* had a significant positive correlation with As and Ni. This showed that these bacteria had a certain tolerance to heavy metals such as Pb, As, and Ni in the low pollution level of heavy metals, and had a strong heavy metal degradation function, which had a great effect on improving the soil environment.

Bacterial Communities at Light Pollution Levels

A total of 14 samples in the study area belonged to the level of light heavy metal pollution. One of the samples was located in the non-agricultural land consolidation area, and the rest were in the farmland consolidation area. According to CCA analysis, in these 14 samples, heavy metals Hg ($r^2 = 0.53$, $p = 0.029$), Cd ($r^2 = 0.38$, $p = 0.081$), Pb ($r^2 = 0.26$, $p = 0.094$), As ($r^2 = 0.24$, $p = 0.023$), Cr ($r^2 = 0.19$, $p = 0.032$) were significantly related to the bacterial community structure. According to Spearman correlation analysis, the content of heavy metals was significantly related to 5 bacterial phyla and 4 bacterial genera ($p < 0.05$) (Figure 16).

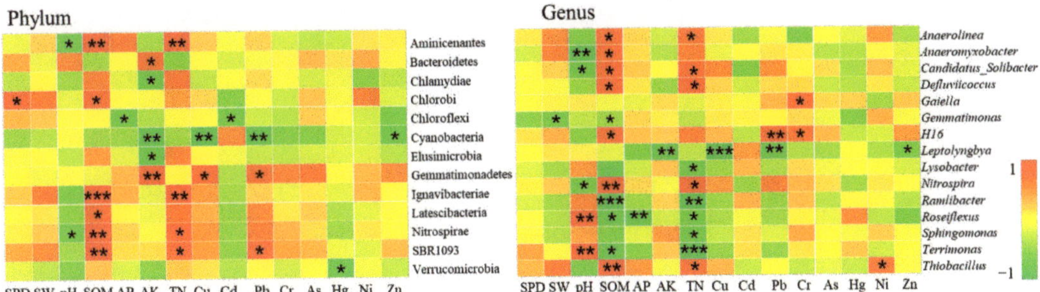

Figure 16. Correlation between soil environmental factors and bacterial community structure at light level of heavy metal pollution. (Note: Warm colors indicate positive correlation, and cool colors indicate negative correlation. *, **, *** indicate significant correlation at the levels of 0.05, 0.01, and 0.001, respectively).

At the phylum level, the relative abundance of farmland soil bacteria at this pollution level was significantly related to the physical and chemical properties of soils such as SOM, AK, TN, especially Aminicenantes, Ignavibacteriae, Nitrospirae, and SBR1093. Heavy metals Cu and Pb had a strong influence on the structure of bacterial community, and were significantly positively correlated with Gemmatimonadetes and SBR1093. Our results indicated that Gemmatimonadetes and SBR1093 could still have strong adsorption and degradation effects on heavy metal elements such as Cu and Pb under conditions of sufficient nutrients and in an environment with a light heavy metal pollution level.

At the genus level, the basic physical and chemical properties of soil pH, SOM, TN were important influencing factors that affected the structure of bacterial community, and had a significant impact on the relative abundance of *Candidatus_Solibacter*, *Nitrospira*, *Roseiflexus*, and *Terrimonas*. Cu, Pb, Cr, Ni, and Zn were the main heavy metal elements that affected the structure of bacterial genera. Among them, bacterial genera of *Gaiella*, *H16*, *Thiobacillus* were significantly positively correlated with the contents of Pb, Cr, and Ni. This showed that these bacterial genera had a certain tolerance to heavy metals such as Pb, Cr, Ni in the light pollution level of heavy metals, and had a strong function of heavy metal degradation, which could effectively improve the soil environmental conditions.

Bacterial Communities at Moderate Pollution Levels

A total of 15 samples in the study area belonged to the moderate level of heavy metal pollution, of which two samples were located in the non-agricultural land consolidation area, and the rest were located in the farmland consolidation area. According to CCA analysis, among these 15 samples, heavy metals Cu ($r^2 = 0.46$, $p = 0.043$), Pb ($r^2 = 0.45$, $p = 0.042$), Hg ($r^2 = 0.35$, $p = 0.045$), Zn($r^2 = 0.38$, $p = 0.035$) were significantly related to bacterial community structure. According to Spearman's correlation analysis, the content of heavy metals was significantly related to 11 bacterial phyla and 6 bacterial genera ($p < 0.05$) (Figure 17).

At the phylum level, the correlation between the relative abundance of farmland soil bacteria and the basic physical and chemical properties of the soil at the moderate pollution level was not significant, only two phyla were significantly related to pH and one phyla was significantly related to SPD. Cd, Pb, Cr, As were the main heavy metal elements that affected bacteria. Among them, Pb was significantly positively correlated with GAL15 and SBR1093; Cr was significantly positively correlated with BRC1, Chloroflexi, RBG-1__Zixibacteria_, Verrucomicrobia; As was significantly positively correlated with GAL15; and Cd was significantly negatively correlated with Chloroflexi, Latescibacteria, and Planctomycetes. Our results showed that in a moderately polluted environment with heavy metals, the basic physical and chemical properties of soil were no longer the main environmental factors affecting the soil bacterial community. At this pollution level, some of the bacteria still had good tolerance to heavy metals such as Pb, Cr, As, and Cd, and had a strong absorption effect on them.

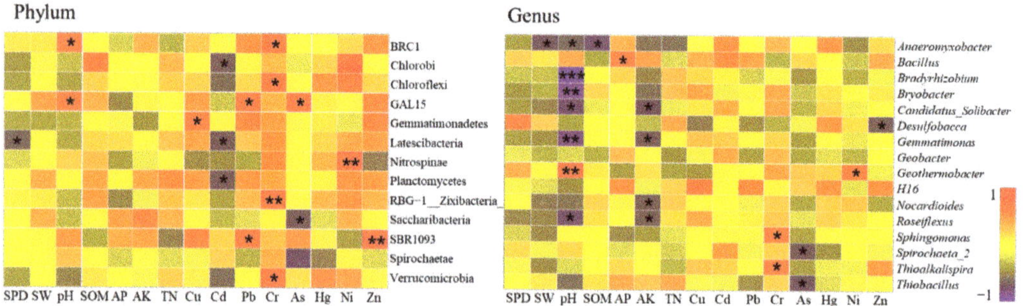

Figure 17. Correlation between soil environmental factors and bacterial community structure at moderate levels of heavy metal pollution. (Note: Warm colors indicate positive correlation, and cool colors indicate negative correlation. *, **, *** indicate significant correlation at the levels of 0.05, 0.01, and 0.001, respectively).

At the genus level, the soil pH and AK were important influencing factors on the structure of bacterial communities, and had a significant impact on the relative abundance of bacterial genera *Candidatus_Solibacter, Gemmatimonas, Roseiflexus*. The influence of heavy metals on the relative abundance of bacterial genera was limited. Only Cr and Ni had a significant positive correlation with bacterial genera *Sphingomonas, Thioalkalispira, Geothermobacter*. This also showed that under this pollution level, the basic physical and chemical properties of the soil had a greater impact on the community structure of bacterial genera, and some bacterial genera still had a certain degrading effect on heavy metals such as Cr and Ni. This conclusion was consistent with the results of existing literatures [87].

Bacterial Communities under Heavy Pollution Levels

A total of 11 samples in the study area were in the heavily polluted area, of which 7 samples were in the non-agricultural land consolidation area, and the remaining 4 samples were in the farmland consolidation area. According to CCA analysis, in these 11 samples, heavy metals Cd (r^2 = 0.34, p = 0.020), Cr (r^2 = 0.28, p = 0.030), Ni (r^2 = 0.41, p = 0.017) were significantly related to the bacterial community structure. According to the Spearman correlation analysis, the content of heavy metals was significantly related to 7 bacterial phyla, and 6 bacterial genera (p < 0.05) (Figure 18).

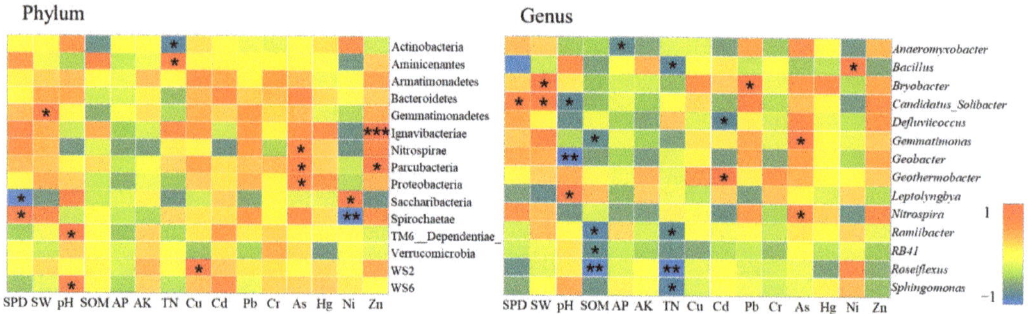

Figure 18. Correlation between soil environmental factors and bacterial community structure at heavy pollution levels. (Note: Warm colors indicate positive correlation, and cool colors indicate negative correlation. *, **, *** indicate significant correlation at the levels of 0.05, 0.01, and 0.001, respectively).

At the phyla level, the relative abundance of farmland soil bacteria at the heavily polluted level was not significantly correlated with SW, pH, and TN. Only SPD, SW, and pH were significantly related to some bacteria phyla. Cu, As, Ni, Zn were the main heavy metal elements that affect bacteria. Among them, Cu had a significant positive correlation with WS2; As had a significant positive correlation with Nitrospirae, Parcubacteria, Proteobacteria; Ni had a significant positive correlation with Saccharibacteria; and Zn had a significant positive correlation with Ignavibacteriae and Parcubacteria. Our results indicated that there were still a certain number of heavy metal-tolerant bacteria phyla in heavily polluted environment, which played an important role in stabilizing the quality of the soil environment.

At the genus level, nutrient indicators such as SOM and TN had a greater impact on the structure of bacterial community. There was a significant negative correlation with the relative abundance of *Roseiflexus* and *Ramlibacter*. This might be that in this environment, the genus with heavy metal tolerance could significantly inhibit the toxic effects of heavy metals, creating favorable conditions for other microorganisms to absorb nutrients to increase their survival rate, resulting in a rapid decline in nutrients. This was consistent with the research results of Li [88]. The heavy metals Cd and As were the main influencing factors, which had a significant effect on the relative abundance of some bacterial genera. Among them, Cd and *Geothermobacter*, Pb and *Bryobacter*, As and *Gemmatimonas* and *Nitrospira*, Ni and *Bacillus*, all showed a significant positive correlation. Under heavy pollution levels, soil nutrients were particularly important for the survival of microorganisms, and heavy metal-tolerant bacteria provided a more favorable growth environment for other bacteria, which was consistent with existing research conclusions [89].

4. Conclusions

This study used soil experimental analysis and high-throughput sequencing technology of gene amplicons to measure the basic physical and chemical properties of soil, soil heavy metal content, and soil microbial characteristics. We used the basic physical and chemical properties of the soil and heavy metal content as an intermediary, to study the impact mechanism of farmland consolidation on microorganisms, to explore the impact of different farmland consolidation measures on the soil micro-ecological environment, and drew the following main conclusions:

(1) Farmland consolidation had a significant impact on soil microbial characteristics, which was mainly manifested in changes in soil microbial biomass, microbial diversity and community structure. The soil microbial biomass carbon and nitrogen in farmland consolidation areas were significantly higher than those in non-agricultural land consolidation areas, and the microbial biomass phosphorus in soil samples from most farmland consolidation areas was significantly higher than that in non-agricultural land consolidation areas. In the study area, the soil bacterial and fungal community richness indexes Sobs and Chao of the cultivated land that had implemented farmland consolidation were significantly higher than the non-agricultural land consolidation areas at the $p < 0.05$ level. The soil bacterial community diversity indexes Shannon and Invsimpson in farmland consolidation areas were significantly higher than those in non-agricultural land consolidation areas, especially the soil bacterial community diversity index in comprehensive improvement areas was the highest.

(2) Farmland consolidation could have a significant impact on the basic physical and chemical properties of the soil. In the study area, the soil particle size and water content of the agricultural land consolidation area were significantly higher than those on the non-agricultural land consolidation area, and the soil pH value of the non-agricultural land consolidation area was significantly higher than that of the construction ditches, combined plots, application of organic fertilizer, and comprehensive improvement areas where the soil pH was close to neutral. Regarding soil nutrients, the content of organic matter, available phosphorus, available potassium, and total nitrogen in non-agricultural land consolidation areas was also significantly lower than

that in farmland consolidation areas, especially the application of organic fertilizer and comprehensive improvement areas had higher soil nutrients. In addition, the soil catalase, phosphatase, and urease activities in farmland consolidation areas were significantly higher than those in non-agricultural land consolidation areas. Our results showed that farmland consolidation had effectively improved the soil environment of farmland by adjusting soil pH, improving soil nutrients, accelerating soil water circulation, improving soil enzyme activity, and creating favorable conditions for the survival and reproduction of soil microorganisms.

(3) Farmland consolidation had an indirect impact on soil bacteria by adjusting the basic physical and chemical properties of the soil. Studies have shown that the effects of different farmland consolidation measures on the relative abundance of soil bacteria were quite different. The soil water content in the farmland through the implementation of construction ditches was significantly improved, and the area with larger soil particle size could accelerate the water cycle, thereby effectively inhibiting the increase in the relative abundance of Cyanobacteria and Elusimicrobia. However, in areas with land levelling, Cyanobacteria was significantly negatively correlated with soil particle size. This was due to the mechanical compaction of soil particle morphology, which resulted in soil voids and poor water ventilation performance, which restricted the reproduction of Cyanobacteria. The larger the soil particle size, the greater the impact it would bear. Important soil nutrients such as SOM, AP, AK, TN also had a greater impact on the structural changes of soil bacteria, but there was a significant negative correlation between soil bacteria and soil nutrients in non-agricultural land consolidation areas. This was probably due to the large-scale application of inorganic fertilizers in non-agricultural land consolidation areas, resulting in soil environmental pollution, destroying the dynamic balance of soil nutrients, and reducing soil bacterial activity. In addition, in farmland consolidation areas, there was a significant positive correlation between soil enzymes and more dominant bacteria in farmland where organic fertilizers were applied and comprehensive improvement was implemented. This was because these two groups of soils had a high content of soil enzymes, and bacteria could reproduce quickly by obtaining nutrients such as carbon, phosphorus, and nitrogen that were decomposed by soil enzymes.

(4) Farmland consolidation had a significant effect on the content of heavy metals in the soil. Among the eight heavy metals tested in the farmland soil of the study area, the average content of seven heavy metals was greater than the background value of the soil, and only the average value of As was slightly lower than the background value. This showed that there was a relatively serious accumulation of heavy metals in farmland soils in this study area. As an artificial measure that strongly disturbed the soil environment, farmland consolidation was an important factor affecting the spatial distribution of soil heavy metal content. There were large differences in the content of heavy metals between farmland with different farmland consolidation measures and farmland in non-agricultural land consolidation areas, and the highest average values of various heavy metal content were in non-agricultural land consolidation areas. The content of heavy metals in farmland where building ditches, merging plots, land levelling, applying organic fertilizers, and comprehensive improvement were implemented was lower than that of non-agricultural land consolidation areas.

(5) The impact of heavy metals on bacterial community structure varied greatly under different levels of heavy metal pollution. Cultivated lands with low pollution levels were all located in farmland consolidation areas. A total of 4 bacterial phyla exhibited strong absorption and transfer functions for heavy metals such as Hg and Ni, and 6 bacterial genera showed a significant positive correlation with heavy metals such as Pb, As, and Ni. Most of the soil samples at the lightly polluted level were located in farmland consolidation areas. Among them, the bacteria Gemmatimonadetes and SBR1093 had strong adsorption and degradation functions on the heavy metals Cu

and Pb. The bacteria genera *Gaiella, H16, Thiobacillus* and the heavy metals Pb, Cr, Ni content were significantly positively correlated. Among the soil samples with moderate pollution levels, 13 samples were located in farmland consolidation areas. A total of 11 bacterial phyla were significantly correlated with the heavy metals Pb, Cr, As, and Cd, respectively. Bacterial genera such as *Sphingomonas, Thioalkalispira,* and *Geothermobacter* were significantly positively correlated with the heavy metals Cr and Ni respectively. Among the soil samples with heavily polluted levels, a total of 7 samples were located in non-agricultural land consolidation areas. Among them, the heavy metals Cu, As, Ni, and Zn had significant effects on the 7 bacterial phyla. The bacterial genera *Geothermobacter, Bryobacter, Gemmatimonas, Nitrospira,* and *Bacillus* were significantly positively correlated with Cd, Pb, As, and Ni under the condition of consuming a lot of soil nutrients.

With the continuous increase of heavy metal pollution, the number of heavy metal-tolerant bacteria in the soil generally increased first and then decreased. This was probably because with the increase of pollution level, the vitality of heavy metal-tolerant bacteria was stimulated and better heavy metal adsorption and detoxification functions appear. However, as pollution continues to intensify, some strains were eliminated, leaving dominant strains. These strains, which had strong heavy metal absorption and detoxification functions in heavy metal polluted environments, could be used as effective bioremediation methods to improve the soil environment. They should be added to the cultivated land quality evaluation system to serve the cultivated land quality improvement project in farmland consolidation.

Supplementary Materials: The following supporting information can be downloaded at: https://www.mdpi.com/article/10.3390/ijerph19020845/s1, Supplementary S1: Layout of sample points; Supplementary S2: Soil index data.

Author Contributions: Conceptualization, Y.L. and D.C.; methodology, Y.L.; software, S.L.; validation, Y.Y. and J.W.; formal analysis, Y.L.; investigation, S.L.; resources, D.C.; data curation, Y.L.; writing—original draft preparation, Y.L.; writing—review and editing, Y.L.; visualization, J.W.; supervision, D.C.; project administration, Y.L. and D.C.; funding acquisition, Y.L. and D.C. All authors have read and agreed to the published version of the manuscript.

Funding: This research was supported by National Natural Science Foundation of China (Project No. 42101307), the Humanity and Social Science Research Funds of Ministry of Education of China (Project No. 21YJC790006), China Association for Science and Technology High-end Technology Innovation Think Tank Youth Project(Project No. 2021ZZZLFZB1207130) and University Philosophy and Social Science research Project of Jiangsu Education Department (Project No. 2021SJA0205).

Institutional Review Board Statement: Not applicable.

Informed Consent Statement: Not applicable.

Conflicts of Interest: The authors declare no conflict of interest.

References

1. Djanibekov, U.; Finger, R. Agricultural risks and farm land consolidation process in transition countries: The case of cotton in Uzbekistan. *Agric. Syst.* **2018**, *164*, 223–235. [CrossRef]
2. Demetriou, D. The assessment of land valuation in land consolidation schemes: The need for a new land valuation framework. *Land Use Policy* **2016**, *54*, 487–498. [CrossRef]
3. Wang, J.; Yan, S.; Guo, Y.; Li, J.; Sun, G. The effects of land consolidation on the ecological connectivity based on ecosystem service value: A case study of Da'an land consolidation project in Jilin province. *J. Geogr. Sci.* **2015**, *25*, 603–616. [CrossRef]
4. Akkaya Aslan, S.T.; Gundogdu, K.S.; Yaslioglu, E.; Kirmikil, M.; Arici, I. Personal, physical and socioeconomic factors affecting farmers' adoption of land consolidation. *Span. J. Agric. Res. SJAR* **2007**, *5*, 204–213. [CrossRef]
5. Sklenička, P.; Hladík, J.; Střeleček, F.; Kottová, B.; Lososová, J.; Číhal, L.; Šálek, M. Historical, environmental and socio–economic driving forces on land ownership fragmentation, the land consolidation effect and project costs. *Agric. Econ.* **2009**, *55*, 571–582. [CrossRef]
6. Legrand, F.; Picot, A.; Cobo–Diaz, J.F.; Carof, M.; Chen, W.; Le Floch, G. Effect of tillage and static abiotic soil properties on microbial diversity. *Appl Soil Ecol.* **2018**, *132*, 135–145. [CrossRef]

7. de Paul Obade, V. Integrating management information with soil quality dynamics to monitor agricultural productivity. *Sci. Total Environ.* **2019**, *651*, 2036–2043. [CrossRef]
8. Malik, A.A.; Puissant, J.; Buckeridge, K.M.; Goodall, T.; Jehmlich, N.; Chowdhury, S.; Gweon, H.S.; Peyton, J.M.; Mason, K.E.; van Agtmaal, M.; et al. Land use driven change in soil pH affects microbial carbon cycling processes. *Nat. Commun.* **2018**, *9*, 3591. [CrossRef]
9. Hadzi, G.Y.; Ayoko, G.A.; Essumang, D.K.; Osae, S.K.D. Contamination impact and human health risk assessment of heavy metals in surface soils from selected major mining areas in Ghana. *Environ. Geochem. Health* **2019**, *41*, 2821–2843. [CrossRef]
10. Wang, J.; Deng, H.; Wu, S.; Deng, Y.; Liu, L.; Han, C.; Jiang, Y.; Zhong, W. Assessment of abundance and diversity of exoelectrogenic bacteria in soil under different land use types. *Catena* **2019**, *172*, 572–580. [CrossRef]
11. Jiang, B.; Adebayo, A.; Jia, J.; Xing, Y.; Deng, S.; Guo, L.; Liang, Y.; Zhang, D. Impacts of heavy metals and soil properties at a Nigerian e–waste site on soil microbial community. *J. Hazard. Mater.* **2019**, *362*, 187–195. [CrossRef] [PubMed]
12. Majeed, A.; Muhammad, Z.; Ahmad, H. Plant growth promoting bacteria: Role in soil improvement, abiotic and biotic stress management of crops. *Plant Cell Rep.* **2018**, *37*, 1599–1609. [CrossRef]
13. Bender, S.F.; Wagg, C.; van der Heijden, M.G.A. An Underground Revolution: Biodiversity and Soil Ecological Engineering for Agricultural Sustainability. *Trends Ecol. Evol.* **2016**, *31*, 440–452. [CrossRef] [PubMed]
14. Simonin, M.; Richaume, A. Impact of engineered nanoparticles on the activity, abundance, and diversity of soil microbial communities: A review. *Environ. Sci. Pollut. Res.* **2015**, *22*, 13710–13723. [CrossRef]
15. Anjos, D.C.; Hernandez, F.F.F.; Bañuelos, G.S.; Dangi, S.R.; Tirado–Corbalá, R.; Da Silva, F.N.; Filho, P.F.M. Microbial community and heavy metals content in soils along the Curu River in Ceará, Brazil. *Geoderma Reg.* **2018**, *14*, e173. [CrossRef]
16. Stańczuk–Gałwiaczek, M.; Sobolewska–Mikulska, K.; Ritzema, H.; van Loon–Steensma, J.M. Integration of water management and land consolidation in rural areas to adapt to climate change: Experiences from Poland and the Netherlands. *Land Use Policy* **2018**, *77*, 498–511. [CrossRef]
17. Bardgett, R.D.; van der Putten, W.H. Belowground biodiversity and ecosystem functioning. *Nature* **2014**, *515*, 505–511. [CrossRef]
18. Falkowski, P.G.; Fenchel, T.; Delong, E.F. The microbial engines that drive Earth's biogeochemical cycles. *Science* **2008**, *320*, 1034–1039. [CrossRef]
19. Frey, S.D.; Lee, J.; Melillo, J.M.; Six, J. The temperature response of soil microbial efficiency and its feedback to climate. *Nat. Clim. Change* **2013**, *3*, 395–398. [CrossRef]
20. Karhu, K.; Auffret, M.D.; Dungait, J.A.; Hopkins, D.W.; Prosser, J.I.; Singh, B.K.; Subke, J.A.; Wookey, P.A.; Agren, G.I.; Sebastia, M.T.; et al. Temperature sensitivity of soil respiration rates enhanced by microbial community response. *Nature* **2014**, *513*, 81–84. [CrossRef] [PubMed]
21. Ferreira, D.A.; Da Silva, T.F.; Pylro, V.S.; Salles, J.F.; Andreote, F.D.; Dini–Andreote, F. Soil Microbial Diversity Affects the Plant–Root Colonization by Arbuscular Mycorrhizal Fungi. *Microb. Ecol.* **2021**, *82*, 100–103. [CrossRef]
22. Ke, P.J.; Wan, J. Effects of soil microbes on plant competition: A perspective from modern coexistence theory. *Ecol. Monogr.* **2020**, *90*, e01391. [CrossRef]
23. Wang, C.; Zhou, J.; Jiang, K.; Liu, J.; Du, D. Responses of soil N–fixing bacteria communities to invasive plant species under different types of simulated acid deposition. *Sci. Nat.* **2017**, *104*, 43. [CrossRef] [PubMed]
24. Wang, H.; Wang, S.; Zhang, Y.; Wang, X.; Wang, R.; Li, J. Tillage system change affects soil organic carbon storage and benefits land restoration on loess soil in North China. *Land Degrad. Dev.* **2018**, *29*, 2880–2887. [CrossRef]
25. Beattie, R.E.; Henke, W.; Campa, M.F.; Hazen, T.C.; McAliley, L.R.; Campbell, J.H. Variation in microbial community structure correlates with heavy–metal contamination in soils decades after mining ceased. *Soil Biol. Biochem.* **2018**, *126*, 57–63. [CrossRef]
26. Liu, J.; Yu, Z.; Yao, Q.; Hu, X.; Zhang, W.; Mi, G.; Chen, X.; Wang, G. Distinct soil bacterial communities in response to the cropping system in a Mollisol of northeast China. *Appl. Soil Ecol.* **2017**, *119*, 407–416. [CrossRef]
27. Muchová, Z.; Konc, Ľ.; Petrovič, F. Land plots valuation in land consolidation in slovakia: A need for a new approach. *Int. J. Strateg. Prop. Manag.* **2018**, *22*, 372–380. [CrossRef]
28. Johansen, P.H.; Ejrnæs, R.; Kronvang, B.; Olsen, J.V.; Præstholm, S.; Schou, J.S. Pursuing collective impact: A novel indicator–based approach to assessment of shared measurements when planning for multifunctional land consolidation. *Land Use Policy* **2018**, *73*, 102–114. [CrossRef]
29. Mirzavand, J.; Moradi–Talebbeigi, R. Relationships between field management, soil compaction, and crop productivity. *Arch. Agron Soil Sci.* **2020**, *144*, 1–12. [CrossRef]
30. He, M.; Wang, Y.; Tong, Y.; Zhao, Y.; Qiang, X.; Song, Y.; Wang, L.; Song, Y.; Wang, G.; He, C. Evaluation of the environmental effects of intensive land consolidation: A field–based case study of the Chinese Loess Plateau. *Land Use Policy* **2020**, *94*, 104523. [CrossRef]
31. Wu, C.; Huang, J.; Zhu, H.; Zhang, L.; Minasny, B.; Marchant, B.P.; McBratney, A.B. Spatial changes in soil chemical properties in an agricultural zone in southeastern China due to land consolidation. *Soil Tillage Res.* **2019**, *187*, 152–160. [CrossRef]
32. Li, P.; Chen, Y.; Hu, W.; Li, X.; Yu, Z.; Liu, Y. Possibilities and requirements for introducing agri–environment measures in land consolidation projects in China, evidence from ecosystem services and farmers' attitudes. *Sci. Total Environ.* **2019**, *650*, 3145–3155. [CrossRef] [PubMed]
33. Lin, Y.; Ye, Y.; Wu, C.; Yang, J.; Hu, Y.; Shi, H. Comprehensive assessment of paddy soil quality under land consolidation: A novel perspective of microbiology. *PeerJ* **2019**, *7*, e7351. [CrossRef]

34. Kadlec, V.; Žížala, D.; Novotný, I.; Heřmanovská, D.; Kapička, J.; Tippl, M. Land Consolidations as an Effective Instrument in Soil Conservation. *Ekologia* **2014**, *33*, 188–200. [CrossRef]
35. Zhong, L.; Wang, J.; Zhang, X.; Ying, L.; Zhu, C. Effects of agricultural land consolidation on soil conservation service in the Hilly Region of Southeast China–Implications for land management. *Land Use Policy* **2020**, *95*, 104637. [CrossRef]
36. Zhou, J.; Deng, Y.; Shen, L.; Wen, C.; Yan, Q.; Ning, D.; Qin, Y.; Xue, K.; Wu, L.; He, Z.; et al. Temperature mediates continental-scale diversity of microbes in forest soils. *Nat. Commun.* **2016**, *7*, 12083. [CrossRef] [PubMed]
37. Zhu, X.; Chen, B.; Zhu, L.; Xing, B. Effects and mechanisms of biochar–microbe interactions in soil improvement and pollution remediation: A review. *Environ. Pollut.* **2017**, *227*, 98–115. [CrossRef]
38. Yu, X.; Yang, L.; Fei, S.; Ma, Z.; Hao, R.; Zhao, Z. Effect of Soil Layer and Plant–Soil Interaction on Soil Microbial Diversity and Function after Canopy Gap Disturbance. *Forests* **2018**, *9*, 680. [CrossRef]
39. Lüneberg, K.; Schneider, D.; Siebe, C.; Daniel, R. Drylands soil bacterial community is affected by land use change and different irrigation practices in the Mezquital Valley, Mexico. *Sci. Rep.* **2018**, *8*, 1413. [CrossRef]
40. Baldoncini, M.; Albéri, M.; Bottardi, C.; Chiarelli, E.; Raptis, K.G.C.; Strati, V.; Mantovani, F. Biomass water content effect on soil moisture assessment via proximal gamma–ray spectroscopy. *Geoderma* **2019**, *335*, 69–77. [CrossRef]
41. Song, X.; Yang, F.; Ju, B.; Li, D.; Zhao, Y.; Yang, J.; Zhang, G. The influence of the conversion of grassland to cropland on changes in soil organic carbon and total nitrogen stocks in the Songnen Plain of Northeast China. *Catena* **2018**, *171*, 588–601. [CrossRef]
42. Vourlitis, G.L.; Kirby, K.; Vallejo, I.; Asaeli, J.; Holloway, J.M. Potential soil extracellular enzyme activity is altered by long–term experimental nitrogen deposition in semiarid shrublands. *Appl. Soil Ecol.* **2021**, *158*, 103779. [CrossRef]
43. Wu, S.; Zhou, S.; Bao, H.; Chen, D.; Wang, C.; Li, B.; Tong, G.; Yuan, Y.; Xu, B. Improving risk management by using the spatial interaction relationship of heavy metals and PAHs in urban soil. *J. Hazard. Mater.* **2019**, *364*, 108–116. [CrossRef]
44. Brookes, P.C.; Landman, A.; Pruden, G.; Jenkinson, D.S. Chloroform fumigation and the release of soil nitrogen: A rapid direct extraction method to measure microbial biomass nitrogen in soil. *Soil Biol. Biochem.* **1985**, *17*, 837–842. [CrossRef]
45. Caban, J.R.; Kuppusamy, S.; Kim, J.H.; Yoon, Y.; Kim, S.Y.; Lee, Y.B. Green manure amendment enhances microbial activity and diversity in antibiotic–contaminated soil. *Appl. Soil Ecol.* **2018**, *129*, 72–76. [CrossRef]
46. Nottingham, A.T.; Fierer, N.; Turner, B.L.; Whitaker, J.; Ostle, N.J.; McNamara, N.P.; Bardgett, R.D.; Leff, J.W.; Salinas, N.; Silman, M.R.; et al. Microbes follow Humboldt: Temperature drives plant and soil microbial diversity patterns from the Amazon to the Andes. *Ecology* **2018**, *99*, 2455–2466. [CrossRef] [PubMed]
47. Barnes, A.D.; Allen, K.; Kreft, H.; Corre, M.D.; Jochum, M.; Veldkamp, E.; Clough, Y.; Daniel, R.; Darras, K.; Denmead, L.H.; et al. Direct and cascading impacts of tropical land–use change on multi–trophic biodiversity. *Nat. Ecol. Evol.* **2017**, *1*, 1511–1519. [CrossRef]
48. Wang, D.; Bai, J.; Wang, W.; Zhang, G.; Cui, B.; Liu, X.; Li, X. Comprehensive assessment of soil quality for different wetlands in a Chinese delta. *Land Degrad. Dev.* **2018**, *29*, 3783–3794. [CrossRef]
49. Kong, M.; Zhong, H.; Wu, Y.; Liu, G.; Xu, Y.; Wang, G. Developing and validating intrinsic groundwater vulnerability maps in regions with limited data: A case study from Datong City in China using DRASTIC and Nemerow pollution indices. *Environ. Earth Sci.* **2019**, *78*, 262. [CrossRef]
50. Negahban, S.; Mokarram, M.; Pourghasemi, H.R.; Zhang, H. Ecological risk potential assessment of heavy metal contaminated soils in Ophiolitic formations. *Environ. Res.* **2021**, *192*, 110305. [CrossRef]
51. Keshavarzi, A.; Kumar, V. Ecological risk assessment and source apportionment of heavy metal contamination in agricultural soils of Northeastern Iran. *Int. J. Environ. Health Res.* **2019**, *29*, 544–560. [CrossRef] [PubMed]
52. Liu, C.; HUANG, Y.; LEI, M.; HAO, X.; LI, X.; TIE, B.; XIE, J. Soil Contamination and Assessment of Heavy Metals of Xiangjiang River Basin. *Environ. Sci.* **2012**, *33*, 260–265.
53. Dobrovol Skaya, T.G.; Zvyagintsev, D.G.; Chernov, I.Y.; Golovchenko, A.V.; Zenova, G.M.; Lysak, L.V.; Manucharova, N.A.; Marfenina, O.E.; Polyanskaya, L.M.; Stepanov, A.L.; et al. The role of microorganisms in the ecological functions of soils. *Eurasian Soil Sci.* **2015**, *48*, 959–967. [CrossRef]
54. Liang, J.; Liu, J.; Jia, P.; Yang, T.; Zeng, Q.; Zhang, S.; Liao, B.; Shu, W.; Li, J. Novel phosphate–solubilizing bacteria enhance soil phosphorus cycling following ecological restoration of land degraded by mining. *ISME J.* **2020**, *14*, 1600–1613. [CrossRef]
55. Schneider, F.; Don, A.; Hennings, I.; Schmittmann, O.; Seidel, S.J. The effect of deep tillage on crop yield—What do we really know? *Soil Tillage Res.* **2017**, *174*, 193–204. [CrossRef]
56. Vanderlinden, K.; Vereecken, H.; Hardelauf, H.; Herbst, M.; Martínez, G.; Cosh, M.H.; Pachepsky, Y.A. Temporal Stability of Soil Water Contents: A Review of Data and Analyses. *Vadose Zone J.* **2012**, *11*, j2011–j2178. [CrossRef]
57. Li, M.; Nie, S.; Chen, X.; Luo, L.; Zhu, H.; Shi, H.; Ge, T.; Tong, C.; Wu, J. Istribution Characteristics of Rice Photosynthesized Carbon in Soil Aggregates of Different Size and Density. *Environ. Sci.* **2013**, *34*, 1568–1575.
58. Spohn, M. Phosphorus and carbon in soil particle size fractions: A synthesis. *Biogeochemistry* **2020**, *147*, 225–242. [CrossRef]
59. Zhao, X.; Zhao, L.; Guo, X.; Li, M.; Yu, S.; Wang, M. Particle component and distribution characteristics of organic carbon of sediments in water and shore soils. *J. Soil Water Conserv.* **2014**, *28*, 304–308.
60. LI, F.; Renkou, X.; Wenfeng, T.; Shungui, Z.; Tongxu, L.; Zhenqing, S.; Liping, F.; Chengshuai, L.; Fanghua, L.; Xiaomin, L.; et al. The Frontier and Perspectives of Soil Chemistry in the New Era. *Acta Pedol. Sinica.* **2020**, *57*, 1088–1104.

61. Laurent, C.; Bravin, M.N.; Crouzet, O.; Pelosi, C.; Tillard, E.; Lecomte, P.; Lamy, I. Increased soil pH and dissolved organic matter after a decade of organic fertilizer application mitigates copper and zinc availability despite contamination. *Sci. Total Environ.* **2020**, *709*, 135927. [CrossRef]

62. Ghosh, A.; Singh, A.B.; Kumar, R.V.; Manna, M.C.; Bhattacharyya, R.; Rahman, M.M.; Sharma, P.; Rajput, P.S.; Misra, S. Soil enzymes and microbial elemental stoichiometry as bio-indicators of soil quality in diverse cropping systems and nutrient management practices of Indian Vertisols. *Appl. Soil Ecol.* **2020**, *145*, 103304. [CrossRef]

63. Kuscu, I.S.K. Changing of soil properties and urease–catalase enzyme activity depending on plant type and shading. *Environ. Monit. Assess.* **2019**, *191*, 178. [CrossRef] [PubMed]

64. Adetunji, A.T.; Lewu, F.B.; Mulidzi, R.; Ncube, B. The biological activities of beta–glucosidase, phosphatase and urease as soil quality indicators: A review. *J. Soil Sci. Plant Nutr.* **2017**, *17*, 794–807. [CrossRef]

65. Chen, X.; Hao, B.; Jing, X.; He, J.; Ma, W.; Zhu, B. Minor responses of soil microbial biomass, community structure and enzyme activities to nitrogen and phosphorus addition in three grassland ecosystems. *Plant. Soil.* **2019**, *444*, 21–37. [CrossRef]

66. Song, D.; Chen, L.; Zhang, S.; Zheng, Q.; Ullah, S.; Zhou, W.; Wang, X. Combined biochar and nitrogen fertilizer change soil enzyme and microbial activities in a 2–year field trial. *Eur. J. Soil Biol.* **2020**, *99*, 103212. [CrossRef]

67. Maxwell, T.L.; Augusto, L.; Bon, L.; Courbineau, A.; Altinalmazis–Kondylis, A.; Milin, S.; Bakker, M.R.; Jactel, H.; Fanin, N. Effect of a tree mixture and water availability on soil nutrients and extracellular enzyme activities along the soil profile in an experimental forest. *Soil Biol. Biochem.* **2020**, *148*, 107864. [CrossRef]

68. Gomez, E.J.; Delgado, J.A.; Gonzalez, J.M. Environmental factors affect the response of microbial extracellular enzyme activity in soils when determined as a function of water availability and temperature. *Ecol. Evol.* **2020**, *10*, 10105–10115. [CrossRef] [PubMed]

69. Díaz, F.J.; Sanchez–Hernandez, J.C.; Notario, J.S. Effects of irrigation management on arid soils enzyme activities. *J. Arid Environ.* **2021**, *185*, 104330. [CrossRef]

70. Bogunovic, I.; Pereira, P.; Galic, M.; Bilandzija, D.; Kisic, I. Tillage system and farmyard manure impact on soil physical properties, CO_2 emissions, and crop yield in an organic farm located in a Mediterranean environment (Croatia). *Environ. Earth Sci.* **2020**, *79*, 70. [CrossRef]

71. Mickan, B.S.; Abbott, L.K.; Solaiman, Z.M.; Mathes, F.; Siddique, K.H.M.; Jenkins, S.N. Soil disturbance and water stress interact to influence arbuscular mycorrhizal fungi, rhizosphere bacteria and potential for N and C cycling in an agricultural soil. *Biol. Fertil. Soils* **2019**, *55*, 53–66. [CrossRef]

72. Moyano, F.E.; Manzoni, S.; Chenu, C. Responses of soil heterotrophic respiration to moisture availability: An exploration of processes and models. *Soil Biol. Biochem.* **2013**, *59*, 72–85. [CrossRef]

73. Zornoza, R.; Acosta, J.A.; Martínez–Martínez, S.; Faz, A.; Bååth, E. Main factors controlling microbial community structure and function after reclamation of a tailing pond with aided phytostabilization. *Geoderma* **2015**, *245*, 1–10. [CrossRef]

74. Liu, J.; Yu, Z.; Yao, Q.; Sui, Y.; Shi, Y.; Chu, H.; Tang, C.; Franks, A.E.; Jin, J.; Liu, X.; et al. Ammonia–Oxidizing Archaea Show More Distinct Biogeographic Distribution Patterns than Ammonia–Oxidizing Bacteria across the Black Soil Zone of Northeast China. *Front. Microbiol.* **2018**, *9*, 171. [CrossRef]

75. Gao, S.; Guan, D.; Ma, M.; Zhang, W.; Li, J.; Shen, D. Effects of Fertilization on Bacterial Community Under the Condition of Continuous Soybean Monoculture in Black Soil in Northeast China. *Sci. Agric. Sin.* **2017**, *50*, 1271–1281.

76. Shen, J.; Zhang, L.; Guo, J.; Ray, J.L.; He, J. Impact of long–term fertilization practices on the abundance and composition of soil bacterial communities in Northeast China. *Appl. Soil Ecol.* **2010**, *46*, 119–124. [CrossRef]

77. Zhao, J.; Ni, T.; Li, J.; Lu, Q.; Fang, Z.; Huang, Q.; Zhang, R.; Li, R.; Shen, B.; Shen, Q. Effects of organic–inorganic compound fertilizer with reduced chemical fertilizer application on crop yields, soil biological activity and bacterial community structure in a rice–wheat cropping system. *Appl. Soil Ecol.* **2016**, *99*, 1–12. [CrossRef]

78. Aarti, C.; Khusro, A.; Agastian, P.; Darwish, N.M.; Al Farraj, D.A. Molecular diversity and hydrolytic enzymes production abilities of soil bacteria. *Saudi J. Biol. Sci.* **2020**, *27*, 3235–3248. [CrossRef] [PubMed]

79. Xu, Z.; Liu, J.; Lu, X.; Chen, X.; Zhang, B.; Zhang, X.; Zhou, W.; Yang, Y. The application of organic fertilizer improves the activity of the soil enzyme, increases the number and the species variety of bacteria in black soil. *Soil Fertil. Sci. China* **2020**, *4*, 50–55.

80. Zhou, B.; Zhao, L.; Wang, Y.; Sun, Y.; Li, X.; Xu, H.; Weng, L.; Pan, Z.; Yang, S.; Chang, X.; et al. Spatial distribution of phthalate esters and the associated response of enzyme activities and microbial community composition in typical plastic–shed vegetable soils in China. *Ecotoxicol. Environ. Saf.* **2020**, *195*, 110495. [CrossRef]

81. Bhuiyan, M.A.H.; Karmaker, S.C.; Bodrud–Doza, M.; Rakib, M.A.; Saha, B.B. Enrichment, sources and ecological risk mapping of heavy metals in agricultural soils of dhaka district employing SOM, PMF and GIS methods. *Chemosphere* **2021**, *263*, 128339. [CrossRef] [PubMed]

82. Hayat, R.; Ali, S.; Amara, U.; Khalid, R.; Ahmed, I. Soil beneficial bacteria and their role in plant growth promotion: A review. *Ann. Microbiol.* **2010**, *60*, 579–598. [CrossRef]

83. Ramos–Miras, J.J.; Roca–Perez, L.; Guzman–Palomino, M.; Boluda, R.; Gil, C. Background levels and baseline values of available heavy metals in Mediterranean greenhouse soils (Spain). *J. Geochem. Explor.* **2011**, *110*, 186–192. [CrossRef]

84. Qu, M.; Chen, J.; Huang, B.; Zhao, Y. Source apportionment of soil heavy metals using robust spatial receptor model with categorical land–use types and RGWR–corrected in–situ FPXRF data. *Environ. Pollut.* **2021**, *270*, 116220. [CrossRef]

85. Zhao, X.; Huang, J.; Lu, J.; Sun, Y. Study on the influence of soil microbial community on the long–term heavy metal pollution of different land use types and depth layers in mine. *Ecotoxicol. Environ. Saf.* **2019**, *170*, 218–226. [CrossRef] [PubMed]

86. Kuang, J.L.; Huang, L.N.; Chen, L.X.; Hua, Z.S.; Li, S.J.; Hu, M.; Li, J.T.; Shu, W.S. Contemporary environmental variation determines microbial diversity patterns in acid mine drainage. *ISME J.* **2013**, *7*, 1038–1050. [CrossRef]

87. Rahman, Z.; Singh, V.P. Assessment of heavy metal contamination and Hg–resistant bacteria in surface water from different regions of Delhi, India. *Saudi J. Biol. Sci.* **2018**, *25*, 1687–1695. [CrossRef] [PubMed]

88. Li, X.; Meng, D.; Li, J.; Yin, H.; Liu, H.; Liu, X.; Cheng, C.; Xiao, Y.; Liu, Z.; Yan, M. Response of soil microbial communities and microbial interactions to long–term heavy metal contamination. *Environ. Pollut.* **2017**, *231*, 908–917. [CrossRef]

89. Chodak, M.; Gołębiewski, M.; Morawska–Płoskonka, J.; Kuduk, K.; Niklińska, M. Diversity of microorganisms from forest soils differently polluted with heavy metals. *Appl. Soil Ecol.* **2013**, *64*, 7–14. [CrossRef]

International Journal of
*Environmental Research
and Public Health*

Article

Effects of Different Land Use Types and Soil Depths on Soil Mineral Elements, Soil Enzyme Activity, and Fungal Community in Karst Area of Southwest China

Jiyi Gong [1,†], Wenpeng Hou [2,3,†], Jie Liu [1], Kamran Malik [3], Xin Kong [1], Li Wang [1], Xianlei Chen [1], Ming Tang [1], Ruiqing Zhu [4], Chen Cheng [2,3], Yinglong Liu [2,3], Jianfeng Wang [2,3,5,6,*] and Yin Yi [1,*]

1. Key Laboratory of National Forestry and Grassland Administration on Biodiversity Conservation in Karst Mountainous Areas of Southwestern China, Guizhou Normal University, Guiyang 550025, China; 201307048@gznu.edu.cn (J.G.); lanzhoudaxue2022@163.com (J.L.); kongxin1232022@163.com (X.K.); gznu_wangli0521@163.com (L.W.); chenxianlei0321@163.com (X.C.); mingtang@gznu.edu.cn (M.T.)
2. State Key Laboratory of Grassland Agro-Ecosystems, Center for Grassland Microbiome, Lanzhou University, Lanzhou 730000, China; houwp19@lzu.edu.cn (W.H.); chengch20@lzu.edu.cn (C.C.); liuyl2020@lzu.edu.cn (Y.L.)
3. College of Pastoral Agriculture Science and Technology, Lanzhou University, Lanzhou 730000, China; malik@lzu.edu.cn
4. Qinghai Provincial Key Laboratory of Medicinal Plant and Animal Resources of Qinghai-Tibet Plateau, Academy of Plateau Science and Sustainability, School of Life Sciences, Qinghai Normal University, Xining 810008, China; zrq15002679150@163.com
5. Collaborative Innovation Center for Western Ecological Safety, Lanzhou University, Lanzhou 730000, China
6. State Key Laboratory of Plateau Ecology and Agriculture, Qinghai University, Xining 810016, China
* Correspondence: wangjf12@lzu.edu.cn (J.W.); yiyin@gznu.edu.cn (Y.Y.)
† These authors contributed equally to this work.

Citation: Gong, J.; Hou, W.; Liu, J.; Malik, K.; Kong, X.; Wang, L.; Chen, X.; Tang, M.; Zhu, R.; Cheng, C.; et al. Effects of Different Land Use Types and Soil Depths on Soil Mineral Elements, Soil Enzyme Activity, and Fungal Community in Karst Area of Southwest China. *Int. J. Environ. Res. Public Health* **2022**, *19*, 3120. https://doi.org/10.3390/ijerph19053120

Academic Editor: Giulia Maisto

Received: 2 February 2022
Accepted: 1 March 2022
Published: 7 March 2022

Publisher's Note: MDPI stays neutral with regard to jurisdictional claims in published maps and institutional affiliations.

Abstract: The current research was aimed to study the effects of different land use types (LUT) and soil depth (SD) on soil enzyme activity, metal content, and soil fungi in the karst area. Soil samples with depths of 0–20 cm and 20–40 cm were collected from different land types, including grassland, forest, *Zanthoxylum planispinum* land, *Hylocereus* spp. land and *Zea mays* land. The metal content and enzyme activity of the samples were determined, and the soil fungi were sequenced. The results showed that LUT had a significant effect on the contents of soil K, Mg, Fe, Cu and Cr; LUT and SD significantly affected the activities of invertase, urease, alkaline phosphatase and catalase. In addition, Shannon and Chao1 index of soil fungal community was affected by different land use types and soil depths. Ascomycota, Basidiomycota and Mortierellomycota were the dominant phyla at 0–20 cm and 20–40 cm soil depths in five different land types. Land use led to significant changes in soil fungal structure, while soil depth had no significant effect on soil fungal structure, probably because the small-scale environmental changes in karst areas were not the dominant factor in changing the structure of fungal communities. Additionally, metal element content and enzyme activity were related to different soil fungal communities. In conclusion, soil mineral elements content, enzyme activity, and soil fungal community in the karst area were strongly affected by land use types and soil depths. This study provides a theoretical basis for rational land use and ecological restoration in karst areas.

Keywords: karst areas; soil depth; land use types; soil metal elements; soil enzyme activity; soil fungal community and diversity

1. Introduction

Karst landform is a variety of peculiar landforms formed on the surface and underground under the continuous dissolution of a large number of soluble rocks by flowing water. Karst area accounts for about 12% of the world's land area [1]. Due to the unique geographical conditions, the southwest karst region centered on Guizhou Province in China has the largest continuous coverage area of about a 5.1 million km^2 in the world. It is also the most complete development type and typical karst ecosystem in the world [2]. However, the karst ecosystem is highly heterogeneous and fragile, and can cause rocky desertification [3]. Here, arable land is limited by rocks, and soil functions and ecosystem services are negatively affected by poorly managed land use patterns [4].

Land use, as a comprehensive reflection of human behaviors, is closely related to plant communities, soil nutrients and soil enzyme activities, resulting in differences in soil microbial characteristics [5]. Studies have confirmed that different land use patterns showed significant effects on soil nutrients in karst areas of Southwest China [6]. There were differences in community structure/diversity of soil fungi in different vegetation succession stages in the karst area and the soil bacterial community had complex responses to tillage patterns [7]. Both soil enzyme activity and microbial community structure depend on land use patterns. Therefore, in this case, land use conversion in the karst area has become a major issue. Additionally, different depths of the soil layer influence soil microorganisms, vegetation types, and litter quality, which leads to the difference in microbial community structure. Few studies claimed that microbial respiration activity, biomass and diversity were decreased with soil depth and the number of bacteria in the topsoil layers of three different land types was 2–20 times more than the lower layer in the karst area compared to the non-karst [8,9]. Hence, the change of soil nutrients from top to bottom in the karst area is very clear. In recent years, local farmers have been encouraged to grow different cash crops, to improve the ecological environment of the karst area. Therefore, exploring the effects of different land use patterns and soil layers is helpful for ecological restoration in this fragile ecological area.

Soil enzymes have a high catalytic capacity and take part in organic matter decomposition and cycling of nutrient in ecosystem [10]. Soil nutrients, microorganisms, vegetation types and management measures affect soil enzyme activities in different degrees [11–13]. More importantly, soil enzyme activity can represent the rate of nutrient uptake by microorganisms and plants. Enzymes are sensitive to reflect the early changes of soil quality caused by soil management [14]. Soil invertase plays a vital role in C decomposition, transformation, and soil bio-respiration [15]. Urease converts organic N to available N by hydrolyzing urea [16]. In addition, alkaline phosphatase plays an important role in organic phosphorus (P) mineralization and plant P nutrition, especially in calcareous soil with limited P [17]. Microbial activity is a key indicator to detect soil quality and to control land degradation which fully shows that soil microorganism is still an important research direction, especially in karst areas. However, the majority of the studies focused on the soil bacterial community in the karst area, such as the response of soil bacterial community structure to different disturbances [18] and the correlation between vegetation succession and bacterial metabolic diversity [19]. Fungi are the basic components of the soil microbial community, and it is necessary to investigate the relationship between soil fungal community and land use types.

Heavy metals are widespread on the earth's surface which are persistent, stable, and difficult to degrade [20]. In recent years, due to unreasonable exploitation of mineral resources, improper disposal of hazardous wastes, and the extreme vulnerability of groundwater systems, heavy metals in karst areas have been seriously diffused, posing a serious threat to the biological community [21,22]. Studies have shown that land use can directly or indirectly affect the content of heavy metals by changing soil properties [23,24]. Soil enzymes produced by microbial metabolism can be used as monitoring factors for heavy metals. Soil mineral elements are also important indicators to determine soil fertility and the healthy growth of plants. For instance, potassium and sodium are closely related to the soil microbial community in the karst ecosystem [25].

Land use conversion is a part of China's policy of "Grain for Green Project"; however, little is known about the effects of different land use patterns and soil depths on mineral elements content, enzyme activities and fungal communities in karst areas, Southwest

China. In the current study, the contents of soil mineral elements, soil enzyme activities and fungal communities in five land use types and two soil depths in Guizhou Province of China were investigated to provide a theoretical basis for soil management and ecological restoration in karst areas.

2. Materials and Methods

2.1. Study Area and Soil Sampling

Huajiang town is located in Guanling Buyi and Miao Autonomous county, Anshun city, Guizhou province, southwest China with an area of 294.9 km^2 (Altitude 1439 m, 105°34′ E, 25°43′ N), which is a typical karst landform. The study site is dominated by the humid subtropical monsoon climate, and the annual mean temperature and rainfall are 17 °C and 1200 mm per year, respectively, while the frost-free period is about 288 days. In November 2019, we selected five land use types, including: grassland (the main species are *Themeda japonica* and is located in 105°28′42″, 25°43′48″, altitude 851 m), secondary forest (the main species are *Liquidambar formosana* and is located in 105°28′46″, 25°43′48″, altitude 798 m); pepper field (the main cultured species are *Zanthoxylum planispinum* and is located in 105°39′58″, 25°40′07″, altitude 517 m), dragon fruit field (the main cultured species is *Hylocereus* spp. and locate in 105°39′41″, 25°40′33″, altitude 598 m) and maize field (the main cultured species are *Zea mays* and is located in 105°39′42″, 25°40′33″, altitude 798 m). There were 5 sampling points for each land use type, the area of each sampling point is 4 m × 4 m, and the distance between each sampling point was 5 m. After removing 1–2 cm of topsoil, soil samples were collected at depths of 0–20 cm and 20–40 cm. Each soil sample was mixed from four subplots (1 m × 1 m range) at 0–20 cm and 20–40 cm soil depths, respectively, and collected in dry, clean, sterile polyethylene bags. Next, all soil samples were passed through a 2 mm sieve to remove visible roots and stones. A portion of each soil sample was transferred in the 50 mL sterile centrifuge tube and placed in a liquid nitrogen tank. The samples were transported to the laboratory, and the sterile centrifuge tube was immediately stored in an ultra-low temperature refrigerator at −80 °C for subsequent soil enzyme activity measurement and soil DNA extraction. The other portion of the soil samples were kept for natural drying to determine the content of soil mineral elements.

2.2. Soil Mineral Elements Content Assay

The contents of soil mineral elements include potassium (K$^+$), calcium (Ca^{2+}), sodium (Na$^+$), magnesium (Mg^{2+}), iron (Fe^{3+}), copper (Cu^{2+}), zinc (Zn^{2+}), cadmium (Cd^{2+}), chromium (Cr^{2+}) and lead (Pb^{2+}) were determined by atomic absorption spectroscopy method after digestion according to the method described by Li et al. [26] The contents of soil mineral elements were measured by TRACE AI1200 atomic absorption spectrometer (Canada Aurora, Vancouver, BC, Canada).

2.3. DNA Extraction and Illumina Sequencing

Soil DNA (300 mg) was extracted using the power soil DNA extraction and separation kit according to the protocol of the manufacturer (MoBio, Carlsbad, CA, USA). The ITS1 region of the fungal rRNA gene was amplified by PCR as described by Zhong et al. [27]. The primers were ITS1 (5′-CTTGGTCATTTAGAGGAAGTAA-3′) and ITS2 (5′-GCTGCGTTCTTCATCGATGC-3′). The PCR procedure consisted of 27 cycles at 94 °C for 2 min, 94 °C for 30 s, 55 °C for 30 s, and 72 °C for 60 s, and 72 °C for 10 min. PCR products were sent to Genesky Biotechnologies Inc., Shanghai, 201,315 (China) for sequencing with an Illumina 2 × 250 bp platform.

2.4. Sequencing Data Processing and Analysis

The original sequence was filtered using the method described by Caporaso et al. [28] to eliminate the low-quality sequence, and then the ITS2 region was extracted by the same method used by Bengtsson-Palme et al. [29]. In the subsequent examination of potential chimeras, the uchime command in mothur version 1.31.2 [30] was used and compared with entries in the DNA based fungal species unified system related to the classification (unite) database [31]. Finally, after dereplicating and discarding all monomers, the non-chimeric sequences were aggregated into the operational taxonomic units (OTUs), and the OTUs were clustered based on the UPARSE pipeline using USEARCH version 8.0 (the similarity of OTU is 97% [32].

2.5. Soil Enzymes Assay

The activities of soil invertase (Inv), urease (Ure), and alkaline phosphatase (Alp)was measured by 3,5-Dinitrosalicylic acid colorimetry, sodium phenol-sodium hypochlorite and disodium diphenyl phosphate colorimetry, respectively according to description of Hou et al. [33]. The activity of soil catalase (Cat) was determined by Potassium permanganate titration according to Chao et al. [34].

2.6. Statistical Analyses

The data were analyzed using SPSS software 17.0 (IBM Inc., Armonk, NY, USA). The effects of land use type (LUT) and soil depth (SD) on the soil fungal alpha diversity (Chao1 and Shannon indices), soil mineral elements content and soil enzyme activity were analyzed by two-way ANOVA. Statistical significance was defined as $p = 0.05$ confidence level, and the mean was evaluated by the standard error. The Chao1, Shannon, the heatmap, PCoA, redundancy analysis (RDA), Variance Partitioning Analysis (VPA) and Spearman's rank correlation analysis were carried out in R (version 3.2.2). The Linear discriminant analysis (LDA) coupled with effect size measurement (LEfSe) analysis was performed using the OmicStudio tools at https://www.omicstudio.cn/tool (accessed on 27 December 2021).

3. Results

3.1. Differences in Content of Soil K, Na, Ca, Mg and Fe in Different Land Use Types and Soil Depths

Table S1 presented data on soil K, Na, Ca, Mg and Fe contents as influenced by land use types (LUT) and soil depth (SD). Our results showed that LUT have significant effects on soil K ($p < 0.001$), Na ($p = 0.029$), Ca ($p = 0.01$), Mg ($p < 0.001$) and Fe ($p < 0.001$) contents. However, SD had no remarkable role in the content of all soil mineral elements. Additionally, the LUT × SD interaction only had a clear effect on Ca ($p < 0.001$) and Mg contents ($p < 0.001$) (Table S1). Further, we found that there was no significant difference in the K, Na, and Mg content between 0–20 cm and 20–40 cm soil layers under the five different land use types (Figure 1b,c,e). The soil Ca content at 0–20 cm depths in grassland and *Zea mays* was 2.3-fold higher and 1.4-fold lower than that at 20–40 cm depth, respectively (Figure 1d). Interestingly, the Fe content was significantly 1.6 times higher only in the 0–20 cm *Zea mays* soil compared to the 20–40 cm soil layer (Figure 1f). Different land use types also have different effects on soil mineral elements content. Significant differences in soil K, Mg, and Fe contents were found between land use types at both soil depths; however, land use type played no clear role on Na content in different soil layers (Figure 1). The highest levels of K were found in *Zanthoxylum planispinum* soils, while Mg and Fe contents were highest in grassland soils at both soil depths. Meanwhile the contents of K in grassland soil and Fe in *Zanthoxylum planispinum* soil were the lowest, (Figure 1b,e,f). Whereas, at the deep soil layer (20–40 cm), the Ca content was highest in the grassland soil and lowest in the *Zea mays* soil (Figure 1d).

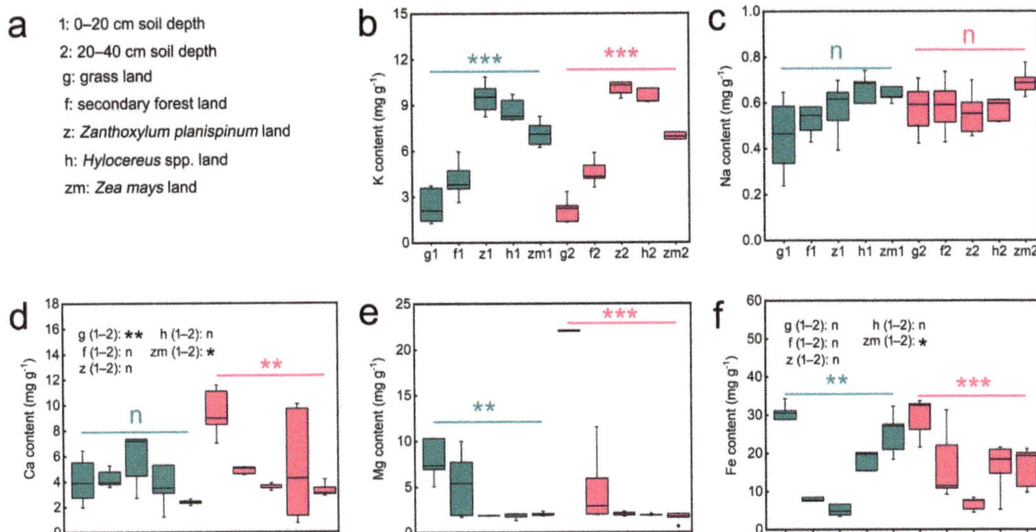

Figure 1. Effect of the different land use types on content of soil K, Na, Ca, Mg Fe in 0–20 cm and 20–40 cm soil depth. (**a**) the description of abbreviations, (**b**) K content, (**c**) Na content, (**d**) Ca content, (**e**) Mg content, (**f**) Fe content. ** and *** above green line and purple line indicate there are significant difference at $p < 0.05$, $p < 0.01$ and $p < 0.001$ level between the five different land use types in 0–20 cm and 20–40 cm soil depths, respectively; n indicates that there is no statistically significantly different among the five different land use types. * and ** after g/f/z/h/zm (1–2) indicate significant difference at $p < 0.05$ and $p < 0.01$ level between 0–20 cm and 20–40 cm depths in the five different land use types, respectively, n indicates not statistically significantly different between 0–20 cm and 20–40 cm soil depths in the five different land use types, respectively.

3.2. Differences in Content of soil Cr, Ni, Cu, Zn, Cd and Pb in Different Land Use Types and Soil Depths

As shown in Table S2, soil Cr ($p = 0.033$), Cu ($p < 0.001$), Zn ($p = 0.011$) and Cd ($p < 0.001$) contents were obviously influenced by LUT. However, soil heavy metal levels were not significant to SD. Interestingly, LUT × SD interaction only affected soil Cu content ($p = 0.026$) (Table S2). Subsequent studies have shown that in the five different land use types, there was no significant difference in heavy metal content between different layers (Figure 2). However, we found that all heavy metal contents in deep soil layers were significantly affected by land use types. The highest content of Cr, Ni, and Pb existed in secondary forest soil and the lowest existed in *Zea mays* soil (Figure 2a,b,f). The contents of Cu, Zn and Cd in grassland soil were significantly lower than the other land use types (Figure 2c–e). Moreover, soil Cu and Cd contents in *Zanthoxylum planispinum* and Zn content in the forest were the highest (Figure 2c–e). It should be noted that in the topsoil (0–20 cm), different land use types only affected the contents of Cu and Cd, and the contents of Cu and Cd in *Hylocereus* spp. soil were significantly higher than those in other land types (Figure 2c,e).

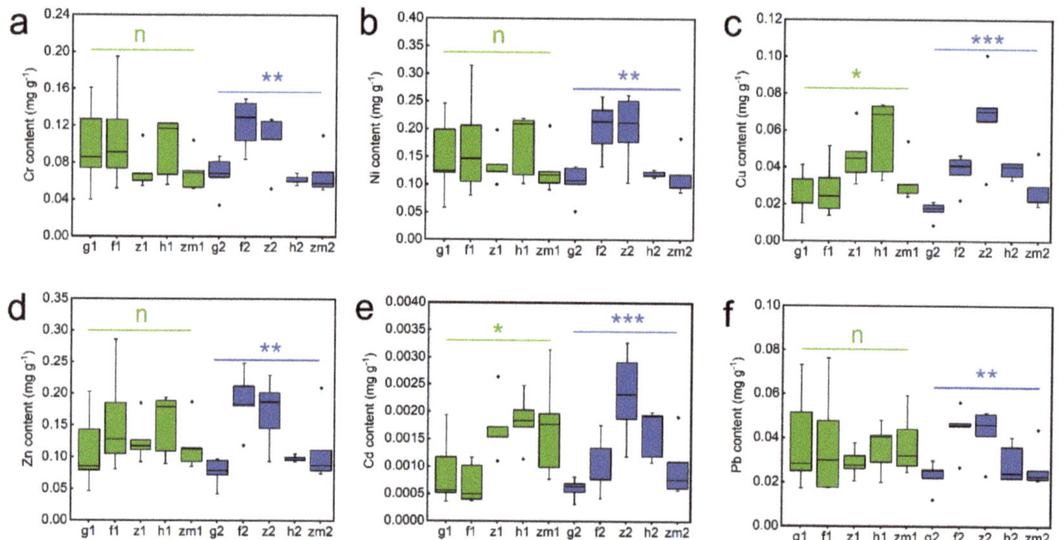

Figure 2. Effect of the different land use types on contents of Cr, Ni, Cu, Zn, Cd, Pb in 0–20 cm and 20–40 cm soil depth. (**a**) Cr content, (**b**) Ni content, (**c**) Cu content, (**d**) Zn content, (**e**) Cd content, (**f**) Pb content. *, ** and *** above green line and blue line indicate there are significant difference at $p < 0.05$, $p < 0.01$ and $p < 0.001$ level between the five different land use types in 0–20 cm and 20–40 cm soil depths, respectively; n indicates that there is no statistically significantly different among the five different land use types.

3.3. Differences in Soil Enzyme Activity in Different Land Use Types and Soil Depths

Statistical evaluation of the effects of land use types on enzyme activity were statistically significant ($p < 0.05$) as shown in Table S3, which the role of soil depth was considered. LUT significantly affected the activity of Inv ($p < 0.001$), Ure ($p < 0.001$), Alp ($p < 0.001$). SD and LUT × SD interaction all had extremely significant influence on the activity of the four enzymes ($p < 0.001$) (Table S3). With the increase in soil depth, the enzyme activities of different land use types all showed a downward trend. In grass land, *Zanthoxylum planispinum* and *Hylocereus* spp. soil, Inv activity in 0–20 cm soil decreased by 1.5 times, increased by 1.7 times and 3.6 times compared with that in 20–40 cm soil depths, respectively (Figure 3a). The activity of Ure at soil depths of 0–20 cm in grassland, forest, and *Zanthoxylum planispinum* was 1.7, 4.3 and 12.9 times higher than that in 20–40 cm soil, respectively, but it was 5.7 times lower in 0–20 cm soil compared to 20–40 cm depths in *Zea mays* soil (Figure 3b). The Alp activity in 0–20 cm depth was increased by 2.2, 1.7 and 1.6 times compared to 20–40 cm depth in forest land, *Hylocereus* spp. and *Zea* mays soil, respectively (Figure 3c). Similarly, Cat activity in topsoil of *Hylocereus* spp. and *Zea mays* soil was significantly higher than that in deep soil layers, which was 1.9 and 1.8 times, respectively (Figure 3d). Further, we studied the response of soil enzyme activities to land use types at different depths. At the two soil depths, different land use types had significant effects on soil enzyme activity. The activity of Inv and Ure was highest at 0–20 cm depths of *Zanthoxylum planispinum* lowest in grass land. In 20–40 cm soil layer, Inv activity was the highest in forest, and the lowest was found in *Hylocereus* spp. soil(Figure 3a). Urease activity was the lowest in 0–20 cm layer of *Zea* mays soil. On the contrary, in 20–40 cm layer, its activity was the highest in *Zea* mays soil and the lowest in grass land (Figure 3b). In shallow soil, the activity of Alp was the highest in forest, and the lowest in *Hylocereus* spp. Interestingly, the lowest activity of Alp and catalase in deep soil all existed in *Hylocereus* spp. soil (Figure 3c,d).

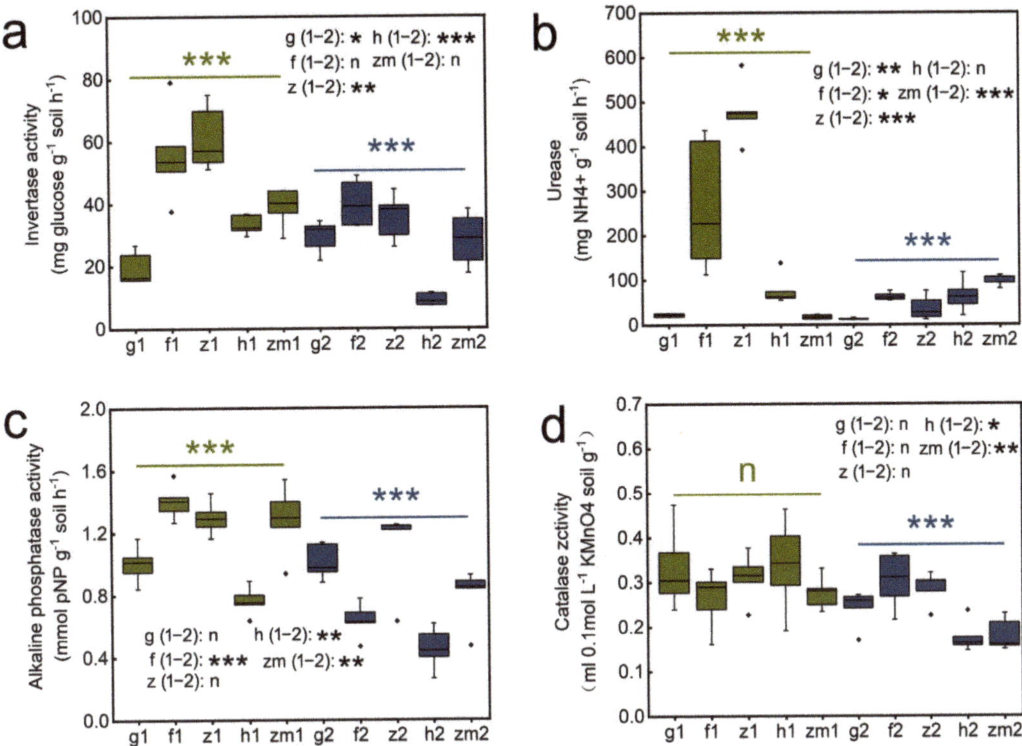

Figure 3. Effect of the different land use types on soil enzyme activity in 0–20 cm and 20–40 cm soil depth. (**a**) invertase (Inv) activity (**b**) urease (Ure) activity, (**c**) alkaline phosphatase (Alp) activity, (**d**) catalase (Cat) activity. *** above yellow line and blue line indicate there were significant difference at $p < 0.001$ level between the five different land use types in 0–20 cm and 20–40 cm soil depths, respectively; n indicates that there is no statistically significantly different among the five different land use types. *, ** and *** after g/f/z/h/zm (1–2) indicate significant difference at $p < 0.05$, $p < 0.01$ and $p < 0.001$ level between 0–20 cm and 20–40 cm depths at the five different land use types, respectively, n indicates not statistically significantly different between 0–20 cm and 20–40 cm soil depths at the five different land use types, respectively. The abbreviations are described in Figure 1a.

3.4. The Richness and Diversity of Soil Fungal Community

We found that LUT and SD had an obvious effect on the Shannon index ($p = 0.018$; $p = 0.004$, respectively, Table S4), but the Shannon index was not significantly affected by the interaction of LUT and SD. Similarly, LUT had a marked influence on the Chao 1 index ($p = 0.028$, $p = 0.03$, respectively, Table S4), and LUT × SD caused a significant influence on the Chao 1 ($p = 0.032$, Table S4). Our results also showed that Shannon and Chao 1 index were affected by different soil depth and land use types. In *Zea* mays, the Shannon and Chao 1 index in the topsoil layers are all significantly higher than that of the deep soil layers, and the Shannon and Chao 1 of topsoil layers in *Zea* mays land were 1.21 times and 1.42 times higher than those of deep soil layers, respectively (Figure 4a,b). However, the rest of the land use types did not change significantly with the change of soil depth. Additionally, Shannon at a soil depth of 0–20 cm in *Zanthoxylum planispinum* land was the highest (4.87 ± 0.15) and was the lowest in *Hylocereus* spp. land (3.86 ± 0.15) (Figure 4a). The same result was shown in the Chao 1. Chao 1 at 0–20 cm depth in *Zanthoxylum planispinum* land was the highest (579.20 ± 22.39) and *Hylocereus* spp. land was the lowest (385.00 ± 29.85) (Figure 4b).

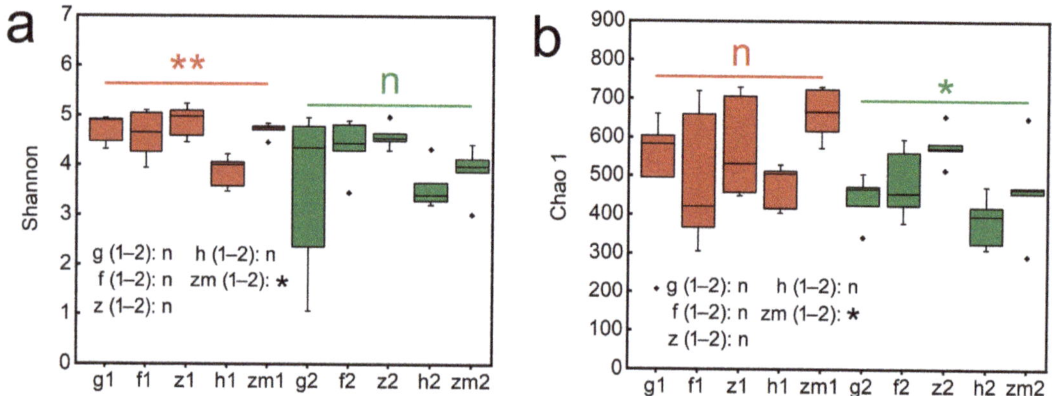

Figure 4. Effect of the different land use types on fungal alpha diversity, (**a**) Chao1, (**b**) Shanon in 0–20 cm and 20–40 cm soil depth. * and ** above red line and green line indicate there are significant difference at *p* < 0.05 and *p* < 0.01 level between the five different land use types in 0–20 cm and 20–40 cm soil depths, respectively; n indicates that there is no statistically significantly different among the five different land use types. * after g/f/z/h/zm (1–2) indicates significant difference at *p* < 0.05 level between 0–20 cm and 20–40 cm depths at the five different land use types, respectively, n indicates not statistically significantly different between 0–20 cm and 20–40 cm soil depths at the five different land use types, respectively. The abbreviations are described in Figure 1a.

3.5. Relative Abundance of Soil Fungal Community

Figure 5a,b showed the relative abundances of major fungal phyla and genera in 0–20 cm and 20–40 cm soil depth in five different land use types, respectively. The Ascomycota, Basidiomycota, and Mortierellomycota were all the dominant phyla at 0–20 cm and 20–40 cm soil depths in the five different land use types (Figure 5a). Interestingly, in 20–40 cm depth soil of grassland, the relative abundance of Basidiomycota was the highest (Figure 5a). From the level of genus, the *Fusarium*, *Mortierella* and *Tetracladium* were the dominant genus in 0–20 cm and 20–40 cm soil depths in forest and grassland (Figure 5b). *Fusarium* and *Mortierella* were the dominant genus in 0–20 cm and 20–40 cm soil depths in *Hylocereus* spp., *Zanthoxylum planispinum* and *Zea* mays (Figure 5b). Furthermore, *Preussia* was also the dominant genus at two soil depths of *Zea* mays (Figure 5b).

At the genus level, the heat map produced by R also showed the differences of soil fungal community aggregation patterns at 0–20 cm and 20–40 cm depths in different land use types (Figure 6). Compared with the other four land use types, the vast majority of the top 30 soil fungal genera in 0–20 cm and 20–40 cm depths soil in *Zea* mays land have higher absolute abundance (Figure 6). In contrast, the absolute abundance of most fungal genera in forest and grassland at 0–20 cm and 20–40 cm soil depths was significantly lower than that of the other three different land use types (Figure 6). Besides, we determined the beta diversity to study the effects of different land use types and soil depth on soil fungal communities (Figure 7). The first principal component explained 21.83% of the total variance, and the second principal component explained 10.79% of the variance. The points of five different land types were significantly dispersed, indicating that land use type caused the significant change of soil fungal community structure. However, except for *Zea* mays, the points of 0–20 cm depths soil and 20–40 cm depths soil were very concentrated, indicating the similarity of fungal community composition at the two soil depths, and the effect of soil depth on soil fungal structure was not significant.

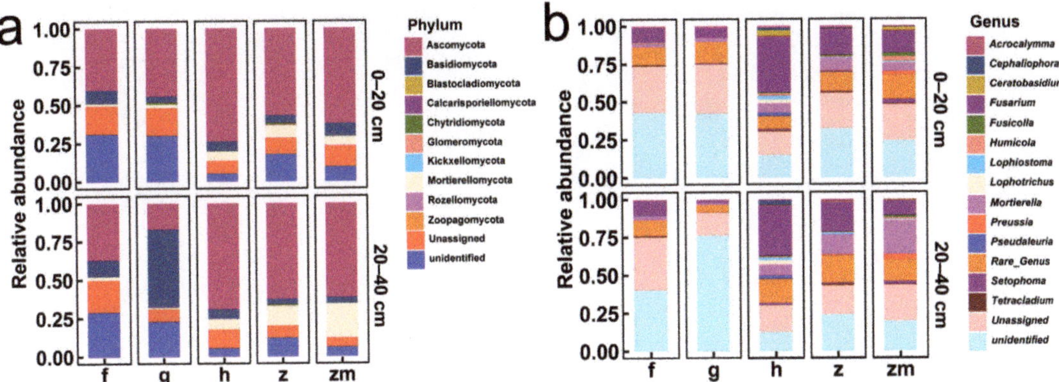

Figure 5. The relative abundance of major fungal phyla in all soil samples ((**a**,**b**) represent phylum level and genus level %, respectively). The abbreviations are described in Figure 1a.

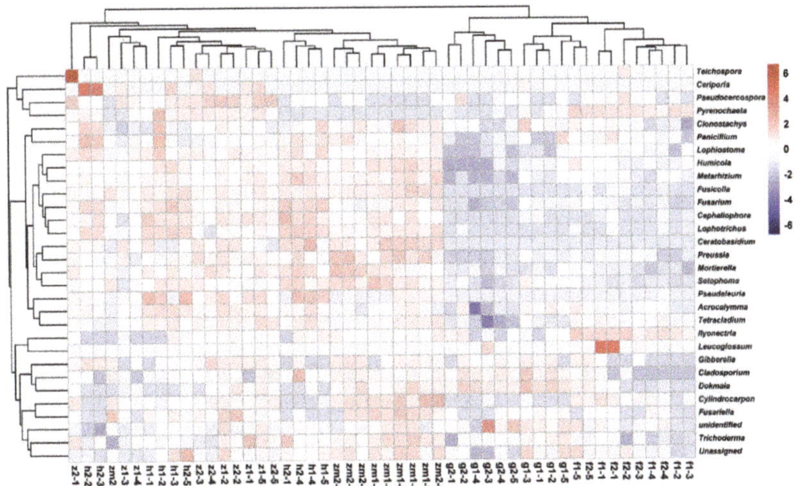

Figure 6. Heat maps of the 30 most abundant fungal genera in different land use types at 0–20 cm and 20–40 cm soil depths, the absolute abundance of fungi is expressed by color intensity. The abbreviations are described in Figure 1a.

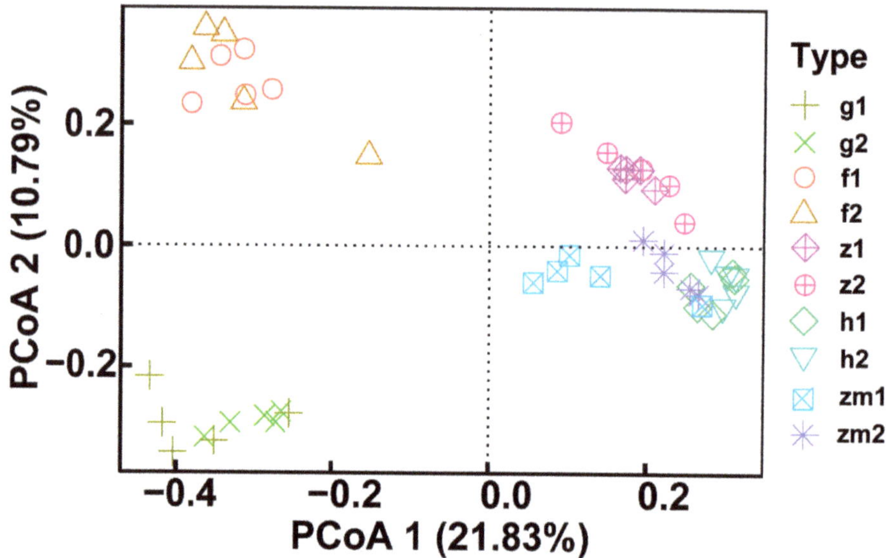

Figure 7. Effect of the different land use types on fungal beta diversity at 0–20 cm and 20–40 cm soil depths. Principal coordinate analysis (PCoA) based on Bray-Curtis of all soil fungal communities. The abbreviations are described in Figure 1a.

3.6. Comparison of Different Fungal Species at 0–20 cm and 20–40 cm Soil Depth in Different Land Use Types

Histogram of LDA value distribution of soil fungi at different depths in different land use types. Figure S1 showed the significantly different species with LDA score greater than the preset value (the default preset value was 3.0). In the 20–40 cm depth soil of grassland, the abundance of fungi of only one genus was significantly higher than that of 0–20 cm depth which was Ochroconis (Figure S1a). In the topsoil layers of *Zanthoxylum planispinum*, the relative abundance of Glomeromycota and Glomus was significantly higher than that at the soil depths of 20–40 cm (Figure S1b). In *Hylocereus* spp. land, the relative abundance of Tecladium in 0–20 cm soil was significantly higher than that in 20–40 cm soil (Figure S1c). In *Zea* mays soil, the relative abundance of Humicola in 0–20 cm soil layer was significantly higher than that in 20–40 cm soil layer, while the relative abundance of Gibberella in 20–40 cm soil layer was significantly higher than that in 0–20 cm soil layer (Figure S1d). We also found that there were no fungal groups with significant differences in relative abundance in different soil depths of forest.

3.7. Correlation of Soil Fungal Communities Composition and Diversity with Content of Soil Mineral Elements

At soil depths of 0–20 cm and 20–40 cm, there was significant relevance between soil fungal diversity and soil mineral contents of different land use types (Figure S2). We found that at depth of 0–20 cm soil, the contents of Cr, Ni, Cu ($p < 0.05$) and Zn, Cd and Pb ($p < 0.001$) in grassland were significantly correlated with Shannon index (Figure S2a); the content of Mg ($p < 0.05$) in forest was significantly related to Chao1 index (Figure S2c). Similarly, the contents of soil Ni, Cu, Cd and Pb ($p < 0.05$) were closely related to PCoA in *Zanthoxylum planispinum* (Figure S2e), and in *Hylocereus* spp., PCoA was closely related to Mg content ($p < 0.05$ Figure 8g). At soil depths of 20–40 cm, PCoA was closely related to K content in forest ($p < 0.05$ Figure S2d); in *Hylocereus* spp. soil, Chao1 was significantly correlated with K content, but PCoA was significantly correlated with contents of Cr ($p < 0.05$), Ni ($p < 0.01$), Zn ($p < 0.05$), Cd ($p < 0.05$) and Pb ($p < 0.01$ Figure S2f). We also found that Shannon and Chao 1 index was clearly related to K content ($p < 0.01$) and Ni content ($p < 0.05$) in *Hylocereus* spp. soil (Figure S2h). Morever, in *Zea* mayssoil, PCoA was significantly correlated with contents of Cd and Pb ($p < 0.05$ Figure S2j).

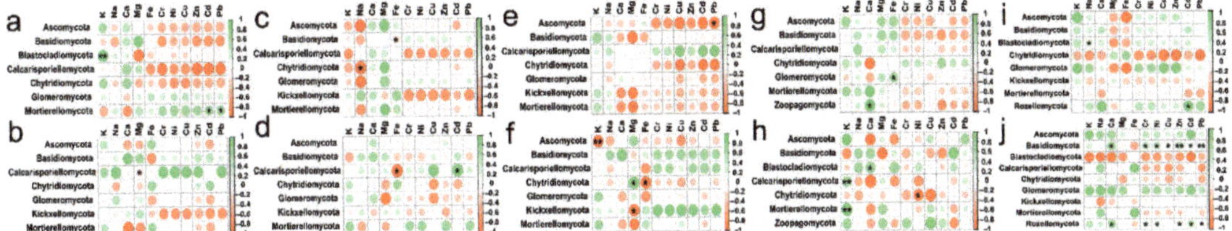

Figure 8. Spearman's Rank correlation coefficients between soil mineral elements content and soil fungal abundance at (**a**) 0–20 cm in grassland, (**b**) 20–40 cm in grassland, (**c**) 0–20 cm in forest, (**d**) 20–40 cm in forest, (**e**) 0–20 cm in *Zanthoxylum planispinum* soil, (**f**) 20–40 cm in *Zanthoxylum planispinum* soil, (**g**) 0–20 cm in *Hylocereus* spp. soil, (**h**) 20–40 cm in *Hylocereus* spp. soil, (**i**) 0–20 cm in *Zea mays* soil and (**j**) 20–40 cm in *Zea mays* soil. * and ** indicate obvious difference at $p < 0.05$ and $p < 0.01$ between soil mineral elements content and the abundance of soil fungal phyla.

At soil depths of 0–20 cm, there was a significant positive correlation between K content and Blastocladiomycota; Na content also had a positive effect on Blastocladiomycota and negative effect on Chytridiomycota; Ca content had a positive impact on Zoopagomycota; Fe content had a positive effect on Glomeromycota and negative effect on Basidiomycota; Content of Cd and Pb had a positive effect on Mortierellomycota and Rozellomycota; interestingly, the content of Cr, Ni, Cu and Zn had no significant effect on the types of fungi (Figure 8a,c,e,g,i). RDA also showed that in topsoil layers, contents of Ca, Fe and Cu were the main factors affecting the change of soil fungal community in *Hylocereus* spp. soil; K was an important factor to affect the changes of soil fungal community in *Zanthoxylum planispinum* soil; Cd and Na were primary factors to affect soil fungal community in *Zea mays* soil, and Mg and Ca were principal considerations affecting the changes of fungal community in grassland and *Hylocereus* spp. soil (Figure 9a). At 20–40 cm soil depths, K content had a strong negative effect on Ascomycota and a positive effect on Calcarisporiellomycota; Ca content had a negative effect on Blastocladiomycota and Basidiomycota; the content of other heavy metals had a strong negative effect on Basidiomycota and Rozellomycota (Figure 8b,d,f,h,j). In addition, in deeper soil, K was the important factor to affect the changes of soil fungal community in *Zea mays* soil; content of Ca, Fe, and Mg were the main factors affecting the change of soil fungal community in grassland and forest; and heavy metal content were important factors to affect soil fungal community in *Zanthoxylum planispinum* soil (Figure 9b).

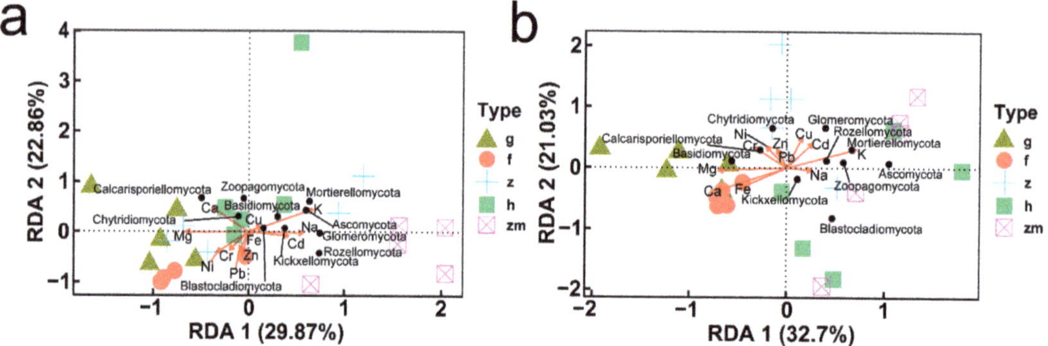

Figure 9. Redundancy analysis (RDA) of relationship between soil mineral elements content (**red arrows**) and the relative abundance of soil microbial phyla (**black points**) at (**a**) 0–20 cm soil depth and (**b**) 20–40 cm soil depth, respectively. The abbreviations are described in Figure 1a.

3.8. Correlation of Soil Fungal Communities Composition and Diversity with Soil Enzyme Activity

Similarly, our result showed that under the two soil depths, there was significant relevance between soil fungal diversity and soil enzyme activity of different land use types (Figure S3). At 0–20 cm soil depths, Shannon index was related to Inv activity in grassland ($p < 0.05$ Figure S3a) and Cat activity in *Zea* mays soil ($p < 0.05$ Figure S3i), and in *Zea* mays soil, Chao1 was also closely related to Cat activity ($p < 0.01$ Figure S3i). At soil depths of 20–40 cm, the activity of Cat was apparently correlated with Shannon index and PCoA ($p < 0.05$ Figure S3b); in *Zanthoxylum planispinum*, Shannon index was markedly correlated with Alp activity ($p < 0.01$ Figure S3d), and in *Hylocereus* spp. soil, Chao1 was significantly correlated with Inv activity ($p < 0.05$ Figure S3h).

In different soil depths, the response of fungal phylum level categories of different land use types to the change of soil enzyme activity was also different (Figure 10). In topsoil layers, we found that there was a clear positive correlation between Ure activity and Calcarisporiellomycota and Chytridiomycota; Cat activity had a strong positive effect on Mortierellomycota; interestingly, in forest Ure activity had a positive effect on Ascomycota and Chytridiomycota, but in *Hylocereus* spp. soil, Ure activity had a strong negative effect on Ascomycota; and Inv activity had no distinct effect on all the phyla levels (Figure 10a,c,e,g,i). In deep soil layers, Inv activity had a strong positive effect on Chytridiomycota; Ure activity had a positive effect on Calcarisporiellomycota and a negative effect on Ascomycota; Alp activity had a positive effect on Ascomycota and a passive effect on Chytridiomycota, Glomeromycota and Kickxellocymota; and Cat activity had a strong effect on Basidiomycota (Figure 10b,d,f,h,j). RDA analysis also showed that in 0–20 cm soil depth layer, the activities of Inv and Alp were the main factors affecting soil fungal communities in forest, *Zanthoxylum planispinum* and *Zea mays* soil; Cat activity was an important factor affecting soil fungal community in *Hylocereus* spp. soil, such as Zoopagomycota and Ascomycota; Ure mainly affected grassland soil Blastocladiomycota (Figure 11a). At 20–40 cm soil depths, the activities of Inv and Alp were the main factors affecting the fungal community in forest and *Zanthoxylum planispinum* soil; Cat was the main factor affecting soil fungal communities in grassland, forest, and *Zanthoxylum planispinum* land; Ure had a strong positive effect on *Zea mays* land soil, such as Rozelomycota, Mortierellomycota, and Kickxelomycota (Figure 11b).

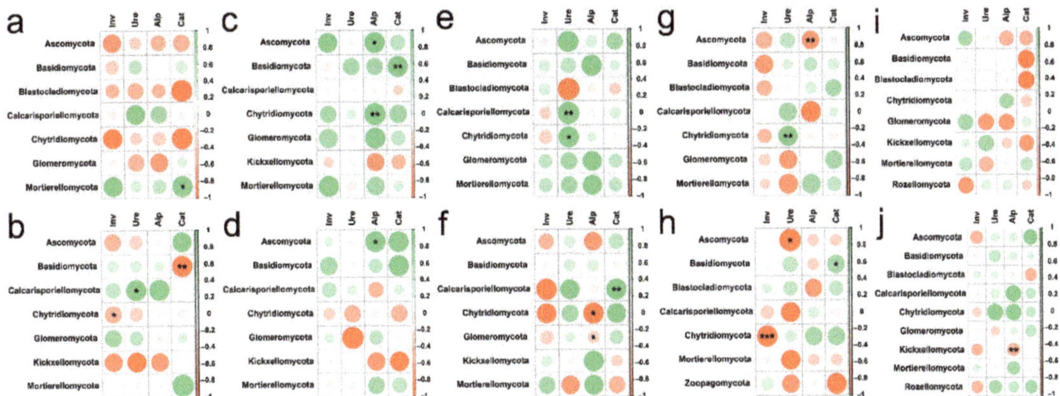

Figure 10. Spearman's Rank correlation coefficients between soil enzyme activity and soil fungal abundance at (**a**) 0–20 cm in grassland, (**b**) 20–40 cm in grassland, (**c**) 0–20 cm in forest, (**d**) 20–40 cm in forest, (**e**) 0–20 cm in *Zanthoxylum planispinum* soil, (**f**) 20–40 cm in *Zanthoxylum planispinum* soil, (**g**) 0–20 cm in *Hylocereus* spp. soil, (**h**) 20–40 cm in *Hylocereus* spp. soil, (**i**) 0–20 cm in *Zea mays* soil and (**j**) 20–40 cm in *Zea mays* soil. *,** and *** indicate obvious difference at $p < 0.05$, $p < 0.01$ and $p < 0.001$ between soil enzyme activity and the abundance of soil fungal phyla.

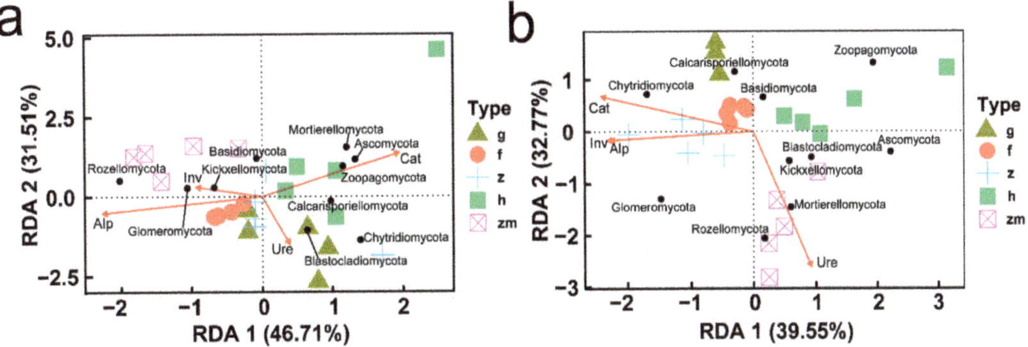

Figure 11. Redundancy analysis (RDA) of relationship between soil enzyme activity (**red arrows**) and the richness of soil microbial phyla (**black points**) at (**a**) 0–20 cm soil depth and (**b**) 20–40 cm soil depth, respectively. The abbreviations are described in Figure 1a.

The composition and diversity of soil fungal community in different types of land use and soil depth changed, and the change depended on soil mineral elements contents and soil enzyme activity. However, the contribution of these factors to the soil fungal community is still unclear. Therefore, the bonding contributions of soil mineral elements and soil enzyme activities in the fungal communities were also investigated by VPA analysis. The result revealed that land use types, soil depth, soil mineral elements content, and soil enzyme activity explained 12%, 0.9%, 1.2%, and 0.4% on the total variations of the fungal community, respectively (Figure 12).

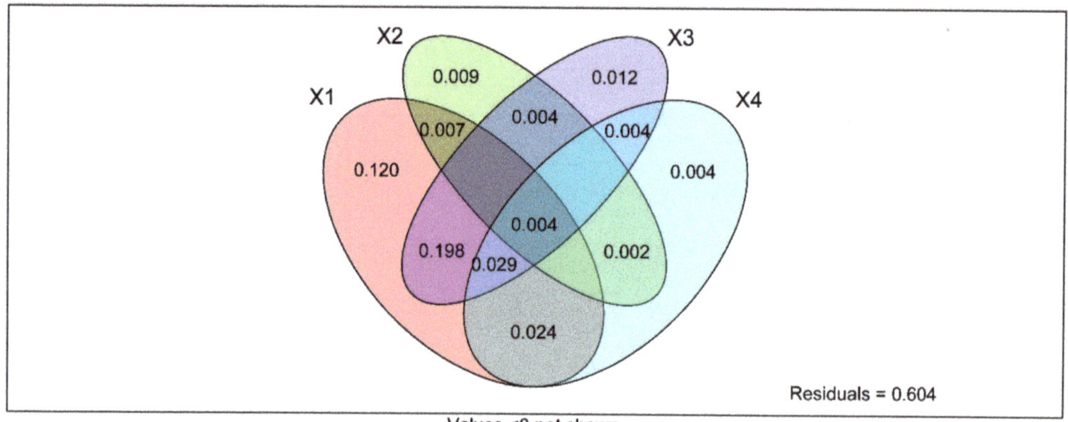

Figure 12. Variance distribution analysis (VPA) to determine the relative contributions of land use type, soil depth, soil metal element content and soil enzyme activity to soil fungal community structure and diversity. X1 = LUT, X2 = SD, X3 = soil metal element content, X4 = soil enzyme activity. Value represents the significance level when $p < 0.0$ not shown.

4. Discussion

Different land use types can lead to different ecosystem functions by influencing underground processes, and soil mineral elements can be affected by soil properties and land management and utilization [35,36]. It has been reported that grassland reclamation in the karst area reduced soil trace elements (Cu, Fe, Mo, B) and enriched in surface soil [37]. Li et al., found that the land use history changed the central trend and heterogeneity of soil properties, including C, N, C:N, Ca, and K [38]. Similarly, land use patterns affect plant litter, thus changing the microbial activities, leading to changes in soil nutrients, and informing a typical correlation among soil microorganisms, vegetation, soil nutrient, and mineral [39]. In the current study, in two different depths of soil K content of different land use types from high to low was followed *Zanthoxylum planispinum* land > *Hylocereus* spp. land > *Zea mays* land > forest land > grassland at two different depths of soil. Similar studies also observed that the K content in cultivated soil was higher than the other types of land [40]. *Zanthoxylum planispinum* and *Hylocereus* spp. are economic crops with a large amount of fertilizer, which may accelerate nutrient turnover, promote K accumulation, or increase availability [41]. Yu et al., also confirmed that chemical fertilizers can effectively increase the total K and P content in soil [42]. The natural karst environment is rich in Ca, which has a remarkable effect on the soil physical and chemical properties [43]. We found that the Ca content was the highest in grass land at soil depths of 20–40 cm, followed by secondary forest. Studies showed that the organic matter and humic acid of karst soil have strong adsorption and complexation to Ca [44,45]. A large amount of litter in grassland and secondary forest returned to the soil and provided a C source. The input of litter is related to the richness of the forest. The plant tissue falling from tulip poplar constituted a large amount of organic matter input into the soil, which contributes to the low variability of C and N concentration on a small scale, while the Ca accumulated in plant tissues enables the trees to absorb it from deep soil layers to maintain the effective calcium concentration in topsoil layers, which eventually cause the reduction in local soil Ca variability [46]. Similarly, the contents of Mg and Fe was highest in the grassland at two different soil depths, while they were reduced in different tillage systems and planting patterns. The possible reason for the decline may include: (1) the coarsening of soil particle composition resulted in the decrease in element adsorption, resulting in the leaching of mineral nutrients, (2) the decomposition of organic matter accelerated, and pH changed, which affected the deposition of elements, (3) Fe was mainly controlled by soil formation and weathering, and the degree of weathering was high in the farming area [41,47]. Our results showed that the five different land management types had no significant effect on Na content, similar to Yu et al. [41]. The differences in field management methods, such as fertilizer application quantity and mode can lead to variations in soil physical and chemical properties and nutrient availability under different tillage systems and planting modes [48]. Interestingly, we found that soil depth had no significant effect on mineral nutrient contents. However, the content of Ca and Fe in 0–20 cm soil layer was significantly higher than that in the 20–40 cm soil layer probably because the soil biological activities were frequent in the thick maize roots that contributed to deep absorption. Ca content in the deep soil layer was higher than the shallow layer of grassland, which may be related to the return of grass root biomass and the decrease in surface runoff [49].

Land use types correspond to the changes in soil management, vegetation types and microbial activity. These changes have an important impact on the migration, transformation, and enrichment of soil heavy metals [50,51]. Wilck et al. [52] investigated the soil contents of heavy metals in three land use patterns (cultivated land, forest, and grassland) in Slovakia, and found that the concentrations of heavy metals in forest soil was lowest due to complexity of the organic matter. Moreover, the concentrations of Cr, Cu, and nickel (Ni) were highest in cultivated soil, while Cd and Zn were highest in grassland soil [52]. In addition, it was reported that heavy metals under different land use types affect the soil microbial and enzyme mediated soil C/N cycle in karst areas [53]. Our results suggested that the interaction of LUT × SD only had a significant effect on Cu content, and Cd and Cu in different types of land and soil layers had significant statistical differences, indicating that Cd and Cu were the dominant factors controlling soil quality. The contents of Cu and Cd were highest in the top and lower soil layers of *Hylocereus* spp. and *Zanthoxylum planispinum* land while lowest in grassland, implying that agricultural activities including the use of chemical fertilizers, exacerbated the accumulation of heavy metals, which was similar to previous studies [54,55].

There was no significant difference in Cr, Ni, Zn, and Pb contents in 0–20 cm soil layer among the five different types of land use, but the content of these heavy metals was highest in forest in 20–40 cm soil layers, followed by *Zanthoxylum planispinum*, contrary to the reported studies. The studies showed that a large amount of humus in forest has functional groups and chelating quality, which reduces the bioavailability of heavy metals and increases their content. Since woody plants have more roots, the canopy intercepts and absorbs the deposited heavy metals [56]. Therefore, in the process of land use conversion, the environmental effects and the absorption differences of various plants at different spatial and temporal scales produce different results. In general, the contents of heavy metals in grassland, *Hylocereus* spp. and *Zea mays* soil decreased gradually with the increase in soil depth, showing obvious surface enrichment [42], but increased in forest and *Zanthoxylum planispinum* soil. It was reported that Cr content was the highest in deep soil of forest land and grassland, possibly due to the poor migration ability of Cr [57]. The sources and migration characteristics of different heavy metals depend on the type of land use. In a specific ecosystem like karst, the growth of soil microorganisms is affected by vegetation and human activities, and then eventually affects the expression and activity of enzymes [58]. Microorganisms are mostly distributed in the surface soil to decompose the surface litter and root exudates and get more organic matter input [59]. It was reported that soil enzyme activities in temperate grassland and tropical forest, decreased exponentially with depth [60]. The current study found that with the increase in soil depth, the enzyme activity in different types of land gradually decreased. These results were consistent with the findings of Stone et al. [60] and Gelsomino and Azzellino. [61], who believed that increment in the depth resulted in the decreased availability of active substrate and oxygen supply. However, the activities of invertase and urease in the topsoil of grassland and maize were lower than in the deep soil. Grassland as the initial stage of vegetation succession grows rapidly and develops roots. It absorbs a lot of nutrients and improves the decomposition of carbohydrates to stimulate the secretion of sucrase in deep soil [62]. Generally, land use can change soil enzyme activities through plants and microorganisms, or indirectly affect soil enzymes through soil characteristics [63]. In the shallow soil, the activities of Inv, Ure, and Alp were the highest in forest or *Zanthoxylum planispinum* soil, and similar phenomena were also observed in deep soil where activities of Inv, Alp, and Cat were the highest in forest or *Zanthoxylum planispinum* soiland the lowest in *Hylocereus* spp. soil. It was proved that nitrogen fixation of trees again, concurrently, the forest is rich in the litter and the more SOM content, the higher the water holding capacity and effective C, which is more conducive to the activities of microorganisms [64]. A study reported that activities of C and P-related acquisition enzymes increased with the improvement in N utilization, which might explain the increase in Alp and Inv activities in *Zanthoxylum planispinum* soil [64]. In addition, *Hylocereus* spp. was weak in promoting the conversion of invalid P, and effective P nutrition was appropriately supplemented during the cultivation process. However, Cat activity was not significantly different in the surface soil of each land type. Liu et al. [65] said that surface soil SOC and Ca form insoluble substances, resulting in the same inert c pool decomposition of the soil, which is not conducive to the production of oxidase by microorganisms.

In the present study, with the increase in soil depth, the fungal α diversity index (Shannon and Chao 1) showed a downward trend, and the difference was the most significant in the maize field which was consistent with previous studies [66,67]. We also found that except for *Zea mays* soil, the contribution rate of soil depth of the other four land use types to the change of soil fungal community was low. Although some studies showed that there was a little difference between fungal community composition/diversity and soil depth, in the *Zea mays* soil system in karst area, soil depth was the main driving factor of fungal community composition and diversity [68]. The factors affecting fungal diversity are soil management and nutrient level [69]. Our results showed that the Chao1 and Shannon indexes at five different land use types of soil fungal communities were significantly different at 0–20 cm and 20–40 cm soil depths, and the α diversity index was the highest in *Zanthoxylum planispinum* soil and lowest in *Hylocereus* spp. soil. Studies discussed that the ACE, Chao 1 and Shannon indexes of soil fungi in natural shrubs were greater than those in artificial forest and grassland which indicated the importance of vegetation restoration to microbial diversity [70]. The distance of PCoA also showed that land use type mainly had a significant impact on soil fungal community. Different land use types have obvious effects on the structure and diversity of soil fungal community in Karst and non-karst areas which were consistent with our results [71,72]. It was reported that under high-intensity management, soil permeability and nutrient content were increased, which provided a suitable growth environment for soil fungi and improved fungal diversity [73]. However, excessive C and N input caused by high-intensity fertilization could reduce soil microbial diversity, which might explain the lowest fungal diversity in *Hylocereus* spp. and *Zea mays* soil [74].

In different land use patterns, the characteristics of soil fungal communities changed in the karst area. The findings of Cheng et al., were consistent with our results that Ascomycota and Basidiomycota were the dominant phyla in different wetlands and cultivated land, and *Fusarium* was the most dominant genus in the corn field and paddy field in karst [71]. Most of the differences in soil fungal composition are related to soil properties. From grassland and forest to agricultural management land, the abundance of Basidiomycetes decreased, while the abundance of Ascomycota was increased. Ascomycota was mainly involved in the degradation of organic matter and the assimilation of root exudates [75]. Previous studies had investigated that the increase in litter was conducive to the transfer of Basidiomycota to Ascomycota [76]. As an important source of soil microbial nutrients, litter plays an important role in the composition and structure of the microbial community. In addition, the abundance of coccidiota in *Zanthoxylum planispinum* soiland other three types of soil increased significantly. Glomeromycota can form arbuscular mycorrhiza with plants and promote the host to absorb nutrients which indicated that crops need more nutrients [77]. Therefore, the relative abundance of different types of soil fungi was different, indicating the differences in root residues, secretions, and crop management of different plants, which will affect soil physical and chemical properties and then change the species composition and structure of microorganisms.

In the current study, heavy metal content and α and β Diversity of soil fungi were significantly negatively correlated at the soil depths of 0–20 cm in grassland and pepper field, and the results were the same at depths of 20–40 cm soil of *Zanthoxylum planispinum*. Some studies have shown that soil heavy metals can reduce the soil fungal diversity [78]. Mg and K contents had a clear positive effect on soil fungal diversity in forest and *Hylocereus* spp. soil, which may be related to the lack of related elements in these two land types in karst areas. Plenty of studies had confirmed that soil carbon, nitrogen, phosphorus, and pH were the main driving factors to change soil fungal community, it was not clear whether soil mineral nutrients would affect the soil fungal community structure [79]. The results of RDA analysis demonstrated that the relationship between fungal phylum communities and soil mineral content in different land use types was different. In different soil depths of *Zea mays* and *Hylocereus* spp. soil, the dominant fungi were significantly correlated with K and Na. Pan et al., found that total K was positively correlated with Ascomycota and Basidiomycota, and soil K could provide nutrients for soil microorganisms [80]. There was a significant positive correlation between Basidiomycetes and content of Mg, Ca and Fe at the depths of 20–40 cm soil in grassland, which was consistent with previous studies on the relationship between soil fungal community composition and soil Fe content [81]. The correlation between soil mineral nutrients and fungal community structure in karst areas was different in five land types, which may be caused by different soil environments in the karst region.

Fungi are considered to be the main producers of soil enzymes. It was reported that the relationship between soil fungal community diversity and soil enzyme activity [82,83]. In our study, soil fungi with different use types and depths were correlated with different kinds of enzymes. Inv and Alp mainly affect the fungal community in forest and *Zanthoxylum planispinum* soil. Studies have shown that alkaline phosphatase was mainly related to the fungal composition of grassland topsoil and mature forest subsoil [83]. Rotation and fertilization will increase the enzymes involved in organic carbon, nitrogen mineralization and decomposition. Our results showed that Ure and Inv activity mainly affected the composition and diversity of fungi in *Zanthoxylum planispinum* and *Zea mays* soil [84]. Therefore, farmland management can regulate the effect of soil fungal communities on soil enzyme activity. Similarly, we also found that Inv was positively correlated with Glomeromycota and Kickxellomycota. Therefore, these two fungi might be the main fungal species for the decomposition of soil organic matter in forest and grassland. However, some studies have shown that the relativity between soil enzyme activity and abiotic factors is greater than that with fungal community, so it is necessary to further study the relationship between soil enzyme and abiotic factors such as soil physical and chemical properties [85].

5. Conclusions

Rational land use is conducive to improving soil nutrients and soil fungal communities in karst areas. In this study, land use type and soil depth significantly affected soil mineral elements contents, soil enzyme activity, and fungal community. Specifically, land use types had significant effects on the contents of soil K, Mg, Fe, Cu and Cr; however, soil depth had no significant effect on soil mineral elements contents. Both land use type and depth significantly affected the invertase, urease, alkaline phosphatase, and catalase activity. In addition, Shannon and Chao1 index of soil fungal community was affected by different land use types and depths. Ascomycota, Basidiomycota, and Mortierellomycota were the dominant phyla at 0–20 cm and 20–40 cm soil depths on five different land types. However, soil depth had no significant effect on the soil fungal structure. This might be because small-scale environmental variation in the karst areas was not the dominant factor in changing the fungal community structure. Soil mineral elements content, enzyme activity, and soil fungal community in the karst area were strongly affected by land use type and soil depth.

Our findings provided a theoretical basis for the rational use of limited land in Karst's fragile ecological environment. More importantly, in recent years, the local government has encouraged farmers to grow different cash crops, which has led to changes in land types and management. The current research can help to guide the selection of appropriate land use types for crop planting and nutrient management in karst areas based on soil fungal community structure.

Supplementary Materials: The following are available online at https://www.mdpi.com/article/10.3390/ijerph19053120/s1, Figure S1: LDA of soil fungi with statistical difference between 0–20 cm and 20–40 cm soil depth in different land use types; Figure S2: Spearman's Rank correlation coefficients between soil mineral elements content and soil fungal community variety; Figure S3: Spearman's Rank correlation coefficients between soil enzyme activity and soil fungal community variety; Table S1: Results of two-way ANOVA for the effects of land use types (LUT) and soil depths (SD) on soil contents of K, Na, Ca, Mg Fe ; Table S2: Results of two-way ANOVA for the effects of land use types (LUT) and soil depths (SD) on soil contents of Cr, Ni, Cu, Zn, Cd, Pb; Table S3: Results of two-way ANOVA for the effects of land use types (LUT) and soil depths (SD) on enzyme activities; Table S4: Results of two-way ANOVA for the effects of land use types (LUT) and soil depths (SD) on Shannon and Chao1.

Author Contributions: Conceptualization, Y.Y.; experimental design and methodology, J.W. and Y.Y.; resource, J.G., W.H., J.L. and X.K.; investigation, J.G., W.H., X.K., L.W., X.C., C.C. and Y.L.; statistical analyses, J.G., J.W., M.T., J.L., K.M. and R.Z.; writing-original draft preparation, J.G. and W.H., writing-review and editing, K.M. and J.W.; supervision, Y.Y.; project administration, Y.Y. and J.W.; funding acquisition, Y.Y. and J.W. All authors have read and agreed to the published version of the manuscript.

Funding: This research was financially supported by Program for the Joint Fund of the National Natural Science Foundation of China and the Karst Science Research Center of Guizhou province (Grant No. U1812401), the Fundamental Research Funds for the Central Universities (lzujbky-2021-ey01, lzujbky-2021-kb12) in Lanzhou University, Changjiang Scholars and innovative Research Team in University (IRT_17R50), the Open Project of State Key Laboratory of Plateau Ecology and Agriculture, Qinghai University (2021-KF-02), Lanzhou University "Double First-Class" guiding special project-team construction fund-scientific research start-up fee standard (561119206), Technical service agreement on research and development of beneficial microbial agents for Alpine Rhododendron (071200001).

Institutional Review Board Statement: Not applicable.

Informed Consent Statement: Not applicable.

Data Availability Statement: Not applicable.

Conflicts of Interest: The authors state no conflict of interest.

References

1. Sweeting, M.M. Karst in China. In *Karst in China Series: Springer Series in Physical Environment*; Springer: Berlin/Heidelberg, Germany, 1995; Volume 15.
2. Green, S.M.; Dungait, J.; Tu, C.; Buss, H.L.; Sanderson, N.; Hawkes, S.J.; Xing, K.; Yue, F.; Hussey, V.L.; Peng, J. Soil functions and ecosystem services research in the Chinese karst Critical Zone. *Chem. Geol.* **2019**, *527*, 119107. [CrossRef]
3. Yang, X.Q.; Hu, B.Q. Quality Characteristics of Soils in Karst Rocky-Desertified Areas With Ecosystem Under Restoration Succession-A Case Study of Chengjiang Subwatershed,Du'an County, Guangxi. *J. Ecol. Rural Environ.* **2009**, *25*, 1–5.
4. Richardson, M.; Kumar, P. Critical Zone services as environmental assessment criteria in intensively managed landscapes. *Earth's Future* **2017**, *5*, 617–632. [CrossRef]
5. Vitousek, P.M.; Aber, J.D.; Howarth, R.W.; Likens, G.E.; Matson, P.A.; Schindler, D.W.; Schlesinger, W.H.; Tilman, D.G. Human alteration of the global nitrogen cycle: Sources and consequences. *Ecol. Appl.* **1997**, *7*, 737–750. [CrossRef]
6. Xiao, S.; Wei, Z.; Ye, Y.; Jie, Z.; Wang, K. Soil aggregate mediates the impacts of land uses on organic carbon, total nitrogen, and microbial activity in a Karst ecosystem. *Sci. Rep.* **2017**, *7*, 41402. [CrossRef] [PubMed]
7. Chen, X.; Su, Y.; He, X.; Wei, Y.; Wei, W.; Wu, J. Soil bacterial community composition and diversity respond to cultivation in Karst ecosystems. *World J. Microbiol. Biotechnol.* **2012**, *28*, 205–213. [CrossRef]

8. He, S.; Guo, L.; Niu, M.; Miao, F.; Jiao, S.; Hu, T.; Long, M. Ecological diversity and co-occurrence patterns of bacterial community through soil profile in response to long-term switchgrass cultivation. *Sci. Rep.* **2017**, *7*, 3608. [CrossRef] [PubMed]

9. Wang, J.F.; Xie, S.Y.; Feng, H.F.; Yuan, W.H.; Wang, C.X. Characteristic Study of Soil Microbe under Different Land-use Types in Chongqing Karst Region. *J. Environ. Sci. Manag.* **2010**, *35*, 150–154.

10. Makoi, J.; Ndakidemi, P.A. Selected soil enzymes: Examples of their potential roles in the ecosystem. *Afr. J. Biotechnol.* **2018**, *7*, 181–191.

11. Aon, M.A.; Colaneri, A.C. Temporal and spatial evolution of enzymatic activities and physico-chemical properties in an agricultural soil. *Appl. Soil Ecol.* **2011**, *18*, 255–270. [CrossRef]

12. Wang, X.; Fan, J.; Xing, Y.; Xu, G.; Wang, H.; Jian, D.; Wang, Y.; Zhang, F.; Li, P.; Li, Z. The Effects of Mulch and Nitrogen Fertilizer on the Soil Environment of Crop Plants. *Adv. Agron.* **2019**, *153*, 121–173.

13. Wu, J.; Wang, H.; Li, G.; Ma, W.; Xu, G. Vegetation degradation impacts soil nutrients and enzyme activities in wet meadow on the Qinghai-Tibet Plateau. *Sci. Rep.* **2020**, *10*, 21271. [CrossRef] [PubMed]

14. Dick, R.P.; Kandeler, E. Enzymes in soils. *Encycl. Soils Environ.* **2005**, 448–456.

15. Jílková, V.; Jandová, K.; Kukla, J. Responses of microbial activity to carbon, nitrogen, and phosphorus additions in forest mineral soils differing in organic carbon content. *Biol. Fertil. Soils* **2021**, *57*, 513–521. [CrossRef]

16. Maithani, S.; Pal, M.; Maity, A.; Pradhan, M. Isotope selective activation: A new insight into the catalytic activity of urease. *RSC Adv.* **2017**, *7*, 31372–31376. [CrossRef]

17. Khadem, A.; Raiesi, F. Response of soil alkaline phosphatase to biochar amendments: Changes in kinetic and thermodynamic characteristics. *Geoderma* **2019**, *337*, 44–54. [CrossRef]

18. Zhang, X.; Liu, S.; Li, X.; Wang, J.; Ding, Q.; Wang, H.; Tian, C.; Yao, M.; An, J.; Huang, Y. Changes of soil prokaryotic communities after clear-cutting in a karst forest: Evidences for cutting-based disturbance promoting deterministic processes. *FEMS Microbiol. Ecol.* **2016**, *92*, fiw026. [CrossRef]

19. Cardinale, B.J.; Wrigh, J.P.; Cadotte, M.W.; Carroll, I.T.; Hector, S. Impacts of plant diversity on biomass production increase through time because of species complementarity. *Proc. Nat. Acad. Sci. USA* **2007**, *104*, 18123–18128. [CrossRef]

20. Narottam, S.; Mollah, M.Z.I.; Alam, M.F.; Rahman, S.M. Seasonal investigation of heavy metals in marine fishes captured from the Bay of Bengal and the implications for human health risk assessment. *Food Control* **2016**, *70*, 110–118.

21. Liu, R.; Zhang, Z.; Shen, J.; Wang, Z. Community characteristics of bryophyte in Karst caves and its effect on heavy metal pollution: A case study of Zhijin Cave, Guizhou Province. *Biodivers. Sci.* **2018**, *26*, 1277–1288. [CrossRef]

22. Ping, C.; Ruan, Y.; Wang, S.; Liu, X.; Lian, B. Effects of organic mineral fertiliser on heavy metal migration and potential carbon sink in soils in a karst region. *Chin. J. Geochem.* **2017**, *36*, 539–543.

23. Jiao, W.; Ouyang, W.; Hao, F.; Liu, B.; Wang, F. Geochemical variability of heavy metals in soil after land use conversions in Northeast China and its environmental applications. *Environ. Sci. Processes Impacts* **2014**, *16*, 924–931. [CrossRef]

24. Chrastný, V.; Komárek, M.; Procházka, J.; Pechar, L.; Vaněk, A.; Penížek, V.; Farka, J. 50 years of different landscape management influencing retention of metals in soils. *J. Geochem. Explor.* **2012**, *115*, 59–68. [CrossRef]

25. Qi, D.; Wieneke, X.; Zhou, X.; Jiang, X.; Xue, P. Succession of plant community composition and leaf functional traits in responding to karst rocky desertification in the Wushan County in Chongqing, China. *Community Ecol.* **2017**, *18*, 157–168. [CrossRef]

26. Li, Y.; Han, C.; Sun, S.; Zhao, C. Effects of Tree Species and Soil Enzyme Activities on Soil Nutrients in Dryland Plantations. *Forests* **2021**, *12*, 1153. [CrossRef]

27. Zhong, Y.; Yan, W.; Wang, R.; Wang, W.; Shangguan, Z.P. Decreased occurrence of carbon cycle functions in microbial communities along with long-term secondary succession. *Soil Biol. Biochem.* **2018**, *123*, 207–217. [CrossRef]

28. Caporaso, J.G.; Kuczynski, J.; Stombaugh, J.; Bittinger, K.; Bushman, F.D.; Costello, E.K.; Fierer, N.; Peña, A.G.; Goodrich, J.K. QIIME allows analysis of high-throughput community sequencing data. *Nat. Methods* **2010**, *7*, 335–336. [CrossRef]

29. Bengtsson-Palme, J.; Ryberg, M.; Hartmann, M.; Branco, S.; Wang, Z.; Godhe, A. Improved software detection and extraction of ITS1 and ITS2 from ribosomal ITS sequences of fungi and other eukaryotes for analysis of environmental sequencing data. *Methods Ecol. Evol.* **2013**, *4*, 914–919. [CrossRef]

30. Schloss, P.D.; Westcott, S.L.; Ryabin, T.; Hall, J.R.; Hartmann, M.; Hollister, E.B.; Lesniewski, R.A.; Oakley, B.B.; Parks, D.H. Introducing mothur: Open-source, platform-independent, community-supported software for describing and comparing microbial communities. *Appl. Environ. Microb.* **2009**, *75*, 7537–7541. [CrossRef]

31. Kõljalg, U.; Nilsson, R.H.; Abarenkov, K.; Tedersoo, L.; Taylor, A.; Bahram, M.; Bates, S.; Bruns, T.; Bengtsson-Palme, J.; Callaghan, T. Towards a unified paradigm for sequence-based identification of Fungi. *Mol. Ecol.* **2013**, *22*, 5271–5277. [CrossRef]

32. Edgar, R.C. UPARSE: Highly accurate OTU sequences from microbial amplicon reads. *Nat. Methods* **2013**, *10*, 996–998. [CrossRef] [PubMed]

33. Hou, W.P.; Wang, J.F.; Nan, Z.B. *Epichloë gansuensis* endophyte-infection alters soil enzymes activity and soil nutrients at different growth stages of *Achnatherum inebrians*. *Plant Soil* **2020**, *455*, 227–240. [CrossRef]

34. Zhang, C.; Liu, G.B.; Xue, S.; Lin, Z. Rhizosphere soil microbial activity under different vegetation types on the Loess Plateau, China. *Geoderma* **2011**, *161*, 115–125. [CrossRef]

35. Macdonald, C.A.; Thomas, N.; Robinson, L.; Tate, K.R.; Ross, D.J.; Dando, J.; Singh, B.K. Biochemical and molecular responses of the soil microbial community after afforestation of pastures with *Pinus radiata*. *Soil Biol. Biochem.* **2009**, *41*, 1642–1651. [CrossRef]

36. Liao, Q.; Nan, Z.; Wang, S.; Huang, H.; Ding, H. Spatial variability and abundance evaluation of available microelements in the middle reaches of Heihe River. *J. Arid. Land Resour. Environ.* **2012**, *13*, 187.

37. Chen, C.; Yang, F.; Liu, H.; Yao, H.; Song, G. Effects and evaluation of soil trace elements after grassland converted into cropland in Guizhou karst area. *Trans. Chin. Soc. Agric. Eng.* **2013**, *29*, 230–237.

38. Li, J.; Richter, D.D.; Mendoza, A.; Heine, P. Effects of land-use history on soil spatial heterogeneity of macro- and trace elements in the Southern Piedmont USA. *Geoderma* **2010**, *156*, 60–73. [CrossRef]

39. Song, M.; Zou, D.; Hu, D.U.; Peng, W.; Zeng, F.; Tan, Q.; Fan, F. Characteristics of soil microbial populations in depressions between karst hills under different land use patterns. *Chin. J. Appl. Ecol.* **2013**, *24*, 2471–2478.

40. Feyisa, D.; Kissi, E.; Kebebew, Z. Rethinking Eucalyptus globulus Labill. Based Land Use Systems in Smallholder Farmers Livelihoods: A Case of Kolobo Watershed, West Shewa, Ethiopia. *Nephron Clin. Pract.* **2018**, *37*, 57–68. [CrossRef]

41. Yu, Y.H.; Yang, D.; Zhong, X. Characteristics of Soil Affinity Elements of Typical Land Use Types in the Rocky Desertification Area of Central Guizhou. *Earth Environ.* **2019**, *47*, 429–435.

42. Yang, Y.; Hu, D.; Song, T.; Peng, W.; Zeng, F.; Wang, K.; Lu, S.Y.; Fan, F.; Lu, C.Y. Characteristics of soil fertility in different ecosystems in depressions between karst hills. *Acta Ecol. Sin.* **2013**, *33*, 7455–7466. [CrossRef]

43. Chen, T.; Wei, X.; Guan, G.; Li, Z.Y. Impact of Different Land Use Types on Soil Calcium in Northern Guangdong. *Trop. Geogr.* **2014**, *15*, 61–68.

44. Chen, J.R.; Cao, J.H.; Liang, Y.; Yang, H. Relationship of the humus components and the calcium form with the development of limestone soil. *Carsol. Sin.* **2012**, *31*, 7–11.

45. Xie, L.P.; Wang, S.J.; Xiao, D.A. Ca Covariant Relation in Plant-soil System in a Small Karst Catshment. *Earth Environ.* **2007**, *1*, 26–32.

46. Fraterrigo, J.M.; Turner, M.G.; Pearson, S.M.; Dixon, P. Effects of Past Land Use on Spatial Heterogeneity of Soil Nutrients in Southern Appalachian Forests. *Ecol. Monogr.* **2005**, *75*, 215–230. [CrossRef]

47. Han, M.R.; Song, T.Q.; Peng, W.X.; Huang, G.Q.; Shi, W.W. Compositional characteristics and roles of soil mineral substances in depressions between hills in karst region China. *J. Appl. Ecol.* **2012**, *23*, 685–693.

48. Liu, X.L.; He, Y.Q.; Zhang, H.L.; Schroder, J.K.; Li, C.L.; Zhou, J.; Zhang, Z.Y. Impact of Land Use and Soil Fertility on Distributions of Soil Aggregate Fractions and Some Nutrients. *Pedosphere* **2010**, *20*, 666–673. [CrossRef]

49. Negasa, D.J. Effects of Land Use Types on Selected Soil Properties in Central Highlands of Ethiopia. *Appl. Environ. Soil Sci.* **2020**, *2020*, 7026929. [CrossRef]

50. Liu, S.; Fu, J.P.; Cai, X.; Zhou, J.; Dang, Z.; Zhu, R. Effect of Heavy Metals Pollution on Ecological Characteristics of Soil Microbes: A Review. *Ecol. Environ. Sci.* **2018**, *27*, 1173–1178.

51. Du, J.Y.; Tu, C.L.; Sheng, M.Y.; Cui, L.F.; Chen, Z.Y.; Zhang, L.K. Effect of Land Use Change on Microbial Community Structure in Central Guizhou Province. *J. Sichuan Agric. Univ.* **2018**, *36*, 350–356.

52. Wilcke, W.; Krauss, M.; Kobza, J. Concentrations and forms of heavy metals in Slovak soils. *J. Plant Nutr. Soil Sci.* **2010**, *168*, 676–686. [CrossRef]

53. Li, Q.; Hu, Q.; Zhang, C.; Jin, Z. Effects of Pb, Cd, Zn, and Cu on Soil Enzyme Activity and Soil Properties Related to Agricultural Land-Use Practices in Karst Area Contaminated by Pb-Zn Tailings. *Pol. J. Environ. Stud.* **2018**, *27*, 2623–2632. [CrossRef]

54. Jia, Y.N.; Yuan, D.X. Effects of Land Use Changes on Trace Elements of Karst Soil in Shuicheng Basin. *J. Soil Sci.* **2007**, *6*, 1174–1177.

55. Thinh, N.V.; Akinori, O.; Hoang, N.T.; Anh, N.D.; Yen, T.T.; Kiyoshi, K. Arsenic and Heavy Metal Contamination in Soils under Different Land Use in an Estuary in Northern Vietnam. *Int. J. Environ. Res. Public Health* **2016**, *13*, 1091.

56. Ouyang, W.; Shan, Y.; Hao, F.; Lin, C. Differences in soil organic carbon dynamics in paddy fields and drylands in northeast China using the CENTURY model. *Agric. Ecosyst. Environ.* **2014**, *194*, 38–47. [CrossRef]

57. Zheng, R.; Zhao, J.; Zhou, X.; Chao, M.A.; Wang, L.; Gao, X. Land Use Effects on the Distribution and Speciation of Heavy Metals and Arsenic in Coastal Soils on Chongming Island in the Yangtze River Estuary, China. *Psedosphere* **2016**, *56*, 74–84. [CrossRef]

58. Sun, C.L.; Wang, Y.W.; Wang, C.J.; Li, Q.J.; Wu, Z.H.; Yuan, D.S.; Zhang, J.L. Effects of land use conversion on soil extracellular enzyme activity and its stoichiometric characteristics in karst mountainous. *Acta Ecol. Sin.* **2021**, *41*, 4140–4149.

59. Gocke, M.I.; Huguet, A.; Derenne, S.; Kolb, S.; Dippold, M.A.; Wiesenberg, G. Disentangling interactions between microbial communities and roots in deep subsoil. *Sci. Total Environ.* **2017**, *575*, 135–145. [CrossRef]

60. Stone, M.M.; Deforest, J.L.; Plante, A.F. Changes in extracellular enzyme activity and microbial community structure with soil depth at the Luquillo Critical Zone Observatory. *Soil Biol. Biochem.* **2014**, *75*, 237–247. [CrossRef]

61. Gelsomino, A.; Azzellino, A. Multivariate analysis of soils: Microbial biomass, metabolic activity, and bacterial-community structure and their relationships with soil depth and type. *J. Plant Nutr. Soil Sci.* **2011**, *174*, 381–394. [CrossRef]

62. Hao, C.; Pan, L.; Li, W.; Yang, L.; Li, D. Determinants of soil extracellular enzyme activity in a karst region, southwest China. *Eur. J. Soil Biol.* **2017**, *80*, 69–76.

63. Wallenius, K.; Rita, H.; Mikkonen, A.; Lappi, K.; Lindstr, M.K.; Hartikainen, H.; Raateland, A.; Niemi, R.M. Effects of land use on the level, variation and spatial structure of soil enzyme activities and bacterial communities. *Soil Biol. Biochem.* **2011**, *43*, 1464–1473. [CrossRef]

64. Keeler, B.L.; Hobbie, S.E.; Kellogg, L.E. Effects of Long-Term Nitrogen Addition on Microbial Enzyme Activity in Eight Forested and Grassland Sites: Implications for Litter and Soil Organic Matter Decomposition. *Ecosystems* **2009**, *12*, 1–15. [CrossRef]

65. Liu, L.; Chen, H.; Li, D.; Liang, S. Changes of soil hydrolytic and oxidized enzyme activities under the process of vegetation restoration in a karst area, southwest china. *Acta Sci. Circumstantiae* **2017**, *37*, 3528–3534.

66. Schlatter, D.C.; Kendall, K.; Bryan, C.; Huggins, D.R.; Timothy, P. Fungal community composition and diversity vary with soil depth and landscape position in a no-till wheat-based cropping system. *FEMS Microbiol. Ecol.* **2018**, *7*, fiy098. [CrossRef] [PubMed]
67. Zhao, H.; Zheng, W.; Zhang, S.; Gao, W.; Fan, Y. Soil Microbial Community Variation with Time and Soil Depth in Eurasian Steppe (Inner Mongolia, China). *Ann. Microbiol.* **2021**, *71*, 21. [CrossRef]
68. Ko, D.; Yoo, G.; Yun, S.T.; Jun, S.C.; Chung, H. Bacterial and fungal community composition across the soil depth profiles in a fallow field. *J. Ecol. Environ.* **2017**, *41*, 34. [CrossRef]
69. Qin, H.; Li, C.X.; Ren, Q. Effects of different land use patterns on soil bacterial and fungal biodiversity in the hydro-fluctuation zone of the Three Gorges Reservoir region. *Acta Ecol. Sin.* **2017**, *37*, 3494–3504.
70. Yang, Y.; Dou, Y.; Huang, Y.; An, S. Links between Soil Fungal Diversity and Plant and Soil Properties on the Loess Plateau. *Front. Micro.* **2017**, *8*, 2198. [CrossRef]
71. Cheng, Y.Y.; Jin, Z.J.; Wang, X.T.; Jia, Y.H.; Zhou, J.B. Effect of Land-use on Soil Fungal Community Structure and Associated Functional Group in Huixian Karst Wetland. *China Environ. Sci.* **2020**, *41*, 4294–4304.
72. Wang, G.; Liu, Y.; Cui, M.; Zhou, Z.; Zhou, J. Effects of secondary succession on soil fungal and bacterial compositions and diversities in a karst area. *Plant Soil* **2021**. [CrossRef]
73. Sui, X.; Zhang, R.T.; Xu, N. Fungal community structure of different degeneration Deyeuxia angustifoliawetlands in Sanjiang Plain. *Environ. Sci.* **2016**, *37*, 3598–3605.
74. Geisseler, D.; Scow, K.M. Long-term effects of mineral fertilizers on soil microorganisms-a review. *Soil Biol. Biochem.* **2014**, *75*, 54–63. [CrossRef]
75. Mylène, H.; Patricia, L.; Julien, G.; Zahar, H. Plant host habitat and root exudates shape fungal diversity. *Mycorrhiza* **2018**, *28*, 451–463.
76. Peng, D.; Xuan, Y.; Le, H.; Liu, J.; Zhong, Z. Effects of stand age and soil properties on soil bacterial and fungal community composition in Chinese pine plantations on the Loess Plateau. *PLoS ONE* **2017**, *12*, e0186501.
77. Haug, I.; Lempe, J.; Homeier, J.; Wei, M.; Setaro, S.; Oberwinkler, F.; Kottke, I. *Graffenrieda emarginata* (Melastomataceae) forms mycorrhizas with Glomeromycota and with a member of the *Hymenoscyphus ericae* aggregate in the organic soil of a neotropical mountain rain forest. *Can. J. Bot.* **2004**, *82*, 340–356. [CrossRef]
78. Chodak, M.; Gołębiewski, M.; Morawska-Płoskonka, J.; Kuduk, K.; Niklińska, M. Diversity of microorganisms from forest soils differently polluted with heavy metals. *Appl. Soil Ecol.* **2013**, *64*, 7–14. [CrossRef]
79. Xue, C.; Penton, C.R.; Zhu, C.; Chen, H.; Duan, Y.; Peng, C.; Guo, S.; Ling, N.; Shen, Q. Alterations in soil fungal community composition and network assemblage structure by different long-term fertilization regimes are correlated to the soil ionome. *Biol. Fertil. Soils* **2018**, *54*, 95–106. [CrossRef]
80. Pan, X.; Zhang, S.; Zhong, Q.; Gong, G.; Wang, G.; Guo, X.; Xu, X. Effects of soil chemical properties and fractions of Pb, Cd, and Zn on bacterial and fungal communities. *Sci. Total Environ.* **2020**, *715*, 136901–136904. [CrossRef]
81. Benedicte, B.; María, U.J.; Zimmerman, J.K.; Thompson, J.; Jonathan, W.L. Long-lasting effects of land use history on soil fungal communities in second-growth tropical rain forests. *Ecol. Appl.* **2016**, *26*, 1881–1895.
82. Huang, M.; Fu, H.; Kong, X.; Ma, L.; Liu, C.; Fang, Y.; Zhang, Z.; Song, F.; Yang, F. Effects of Fertilization Methods on Chemical Properties, Enzyme Activity, and Fungal Community Structure of Black Soil in Northeast China. *Diversity* **2020**, *12*, 476. [CrossRef]
83. Zhang, Y.; Cao, H.; Zhao, P.; Wei, X.; Shi, M. Vegetation Restoration Alters Fungal Community Composition and Functional Groups in a Desert Ecosystem. *Front. Environ. Sci.* **2021**, *9*, 589068. [CrossRef]
84. Ai, C.; Zhang, S.; Zhang, X.; Guo, D.; Zhou, W.; Huang, S. Distinct responses of soil bacterial and fungal communities to changes in fertilization regime and crop rotation. *Geodera* **2018**, *319*, 156–166. [CrossRef]
85. Wang, J.; Yuan, Y.; Zhang, M.; Dai, X.; He, H.; Li, H.; Li, Y. Impact of degradation and restoration on soil fungi and extracellular enzyme activity in alpine rangelands on the Tibetan Plateau. *Arch. Agron. Soil Sci.* **2021**, *14*, 1917–1929. [CrossRef]

International Journal of
Environmental Research
and Public Health

Article

Effects of Seven-Year Fertilization Reclamation on Bacterial Community in a Coal Mining Subsidence Area in Shanxi, China

Li Li [1,2], Tingliang Li [1,2,*], Huisheng Meng [1,2], Yinghe Xie [1,2], Jie Zhang [1,2] and Jianping Hong [1,2,*]

[1] College of Resources and Environment, Shanxi Agricultural University, Jinzhong 030801, China; lili_306@163.com (L.L.); huishengmeng@126.com (H.M.); xieyinghe@163.com (Y.X.); zhangjie880124@foxmail.com (J.Z.)

[2] National Experimental Teaching Demonstration Center for Agricultural Resources and Environment, Shanxi Agricultural University, Jinzhong 030801, China

* Correspondence: litingliang021@126.com (T.L.); hongjpsx@163.com (J.H.)

Abstract: The restoration of soil fertility and microbial communities is the key to the soil reclamation and ecological reconstruction in coal mine subsidence areas. However, the response of soil bacterial communities to reclamation is still not well understood. Here, we studied the bacterial communities in fertilizer-reclaimed soil (CK, without fertilizer; CF, chemical fertilizer; M, manure) in the Lu'an reclamation mining region and compared them with those in adjacent subsidence soil (SU) and farmland soil (FA). We found that the compositions of dominant phyla in the reclaimed soil differed greatly from those in the subsidence soil and farmland soil ($p < 0.05$). The related sequences of *Acidobacteria*, *Chloroflexi*, and *Nitrospirae* were mainly from the subsided soil, whereas those of *Alphaproteobacteria*, *Planctomycetes*, and *Deltaproteobacteria* were mainly derived from the farmland soil. Fertilization affected the bacterial community composition in the reclaimed soil, and bacteria richness and diversity increased significantly with the accumulation of soil nutrients after 7 years of reclamation ($p < 0.05$). Moreover, soil properties, especially SOM and pH, were found to play a key role in the restoration of the bacterial community in the reclaimed soil. The results are helpful to the study of soil fertility improvement and ecological restoration in mining areas.

Keywords: coal mining; soil reclamation; bacterial community; bacterial diversity; high-throughput sequencing

Citation: Li, L.; Li, T.; Meng, H.; Xie, Y.; Zhang, J.; Hong, J. Effects of Seven-Year Fertilization Reclamation on Bacterial Community in a Coal Mining Subsidence Area in Shanxi, China. *Int. J. Environ. Res. Public Health* **2021**, *18*, 12504. https://doi.org/10.3390/ijerph182312504

Academic Editor: Paul B. Tchounwou

Received: 21 October 2021
Accepted: 24 November 2021
Published: 27 November 2021

Publisher's Note: MDPI stays neutral with regard to jurisdictional claims in published maps and institutional affiliations.

1. Introduction

China is one of the largest coal-producing countries in the world [1]. However, the large-scale and high-intensity exploitation of coal resources has caused a series of ecological and environmental problems, such as soil erosion, declines in soil quality, aggravation of land degradation, and imbalance of the soil ecosystem [2–4]. In particular, land subsidence caused by underground coal mining can lead to drastic disturbances of soil structure and remarkable variation in soil microbial communities [5,6], which greatly reduce soil fertility, crop productivity, and the stability of the soil ecosystem [4,7,8]. These have serious impacts on the sustainable development of agriculture in mining areas. Land reclamation is an effective method to solve the conflict between coal mining and land resource protection and to alleviate the contradiction between humans and land in the coal mining area.

Restoring soil fertility is the emphasis for land reclamation and ecological restoration in coal mining subsidence areas. Soil microorganisms, one of the most important parts of the soil ecosystem, are essential in soil formation, nutrient cycling (such as carbon, nitrogen, and phosphorus), and ecological balance [9–11]. The abundance, diversity, and activity of soil bacteria can be used as effective indicators of soil quality due to their high sensitivity to environmental changes and soil nutrient status [12–14]. It has been reported that bacterial community stability in subsidence soil was dramatically disrupted by coal mining activities, resulting in reductions in total bacterial biomass and diversity [15–18]. To some extent, it is not only necessary to increase soil nutrients but also more important to restore soil microbial activities and communities for the sustainability of reclaimed soil ecosystems [19]. For a long time, research on reclaimed soil has mainly focused on the improvement of soil physicochemical status and vegetation characteristics [20,21]. However, there are few studies on microbial population, diversity, and function in reclaimed soil.

The sustainability of terrestrial agroecosystems depends to a great extent on soil bacterial diversity for sustaining soil biological activity and crop productivity [22,23]. Fertilization can effectively improve soil nutrient conditions and affect soil microbial communities [24,25]. Generally, the application of organic fertilizer is beneficial to soil microbial communities [26–28], while the long-term application of chemical fertilizer can decrease soil microbial diversity in farmland [24,29,30]. At present, fertilization is the most effective way to restore cultivated land and improve soil in mining subsidence areas. Different fertilization methods have different effects on soil physical and chemical properties and microbial community. Therefore, an in-depth study on the changes in microbial community composition in reclaimed soil caused by fertilization will help to further understand how soil fertility affects the changes in microbial communities during the restoration of disturbed land to farmland.

The present study was carried out to investigate the response of bacterial communities to land reclamation with different fertilizers. Illumina high-throughput sequencing technology was used to compare the bacterial community structure and diversity in reclaimed soil with those in adjacent unreclaimed soil (from adjacent farmland and subsided land) from a mining area (in the same edaphic-climatic area) in Shanxi Province, China. We hypothesized that fertilization activity in the process of reclamation could improve soil nutrients and increase bacterial community diversity, and the variation of bacterial community structure may be related to changes in physicochemical properties, such as pH, soil organic matter, available nitrogen, phosphorus, and potassium. This study could identify an effective and appropriate method for the rapid restoration of soil fertility and provide a valuable reference for the soil restoration of coal mining subsidence areas with similar climatic and soil conditions.

2. Materials and Methods

2.1. Experimental Site Description

The field experimental area is located in the Lu'an coal mine (36°28′12″ N, 113°00′53″ E), Xiangyuan county in Shanxi, China. This region has a warm and semi-humid continental monsoon climate with a frost-free period of 160 d. The average annual temperature is about 9.5 °C, with monthly mean minimum temperatures occurring in January (−8.1 °C) and monthly mean maximum temperatures in July (23.4 °C). The mean annual precipitation is approximately 532.8 mm and mainly occurs from July to September. The soil type of the research area is calcareous cinnamon soil with silty loam, which is classified as Luvisols according to the World Reference Base (FAO) system [31]. In this region, coal mining has triggered the goaf in underground mines and formed land subsidence since the 1970s. Land consolidation, including topsoil stripping, land leveling, and backfilling, was carried out using large loaders before the reclamation in the autumn of 2008. The leveled land was divided into separate plots for fertilizer reclamation.

2.2. Experimental Treatments and Soil Sampling

The field experiment in this reclaimed site included three fertilization amended treatments: without fertilizer (CK), chemical fertilizer (CF), and manure (M) treatment. The chemical fertilizer was applied as urea, calcium superphosphate, and potassium chloride in CF. Decomposed chicken manure (27.8% organic matter, 1.68% N, 1.54% P_2O_5, and 0.82% K_2O) was provided as an organic amendment at a rate of 12,000 kg·ha^{-1} in the M treatment. An equal amount of 201 kg N·ha^{-1}, 185 kg P_2O_5·ha^{-1}, and 98.5 kg K_2O·ha^{-1} was applied before corn sowing in the fertilizer treatments. The treatments were arranged in a randomized complete block design with three replicates, and the size of each plot was 100 m^2. Maize (*Zea mays* L.) was continuously planted in each plot from 2009 to 2015. According to local farming practices, crops are sown on or about May 1 and harvested on October 1. In addition, two unreclaimed treatments (SU and FA) were selected as controls for the reclaimed site. SU is a neighboring unclaimed subsidence site, and its surface vegetation is sparse and naturally growing weed. FA is another adjacent farmland that has been disturbed by coal mining and has been planting maize using local traditional fertilizing practices for many years.

Soil samples were collected after the maize harvest in October 2015. A total of fifteen individual soil samples (5 treatments × 3 replicates) represent three reclaimed treatments (CK, CF, and M) and two controls (SU and FA). All the soil samples were taken using a hand auger (5 cm diameter) at a depth of 0–20 cm after the superficial vegetation was removed, and each sample was a composite of six subsamples randomly collected from the five treatments. After mixing thoroughly, the homogeneous composite soil samples were enclosed in sterile plastic bags and transferred to the laboratory on ice. The samples were sieved through a 2.0 mm mesh and immediately divided into two parts: one part was stored at −80 °C for further molecular analysis, and the other was air-dried for chemical determination.

2.3. Selected Soil Properties Analysis

Soil pH was measured with a soil–water mixture (1:1) using a glass combination electrode [32]. Soil organic matter (SOM) was measured according to the method described by Strickland and Sollins [33]. The Mason-jar diffusion method by Bremner [34] was used to determine the soil alkali-hydrolyzable nitrogen (AN). Available phosphorus (AP) was analyzed by resin extraction following a protocol modified from Hedley and Stewart [35]. Available potassium (AK) was extracted with ammonium acetate and determined by flame photometry [36].

2.4. DNA Extraction, PCR Amplification, Illumina MiSeq Sequencing, and Sequencing Data Processing

Soil microbial DNA was extracted from approximately 1 g of soil samples using the TIANamp Genomic DNA Kit (TIANGEN Biotech, Beijing, China, Cat. No.: DP304) according to the manufacturer's instructions. The integrity of the extracted DNA was assessed by agarose gel electrophoresis (1%), and the concentration and purification of the DNA (2 µL) were determined using NanoDrop ND-1000 microspectrophotometry (NanoDrop Technologies, Wilmington, DE, USA). The bacterial primer set of forward primer 341F (5′-CCTACGGGNBGCASCAG-3′) and reverse primer 806R (5′-GACTACNVGGGTATCTAATCC-3′) was used to amplify the 16S rDNA gene sequence in the V3–V4 hypervariable region (465 bp). PCRs were carried out in triplicate, 25 µL reactions with 2.5 µL of Ex Taq buffer (Takara Bio Inc., Kusatsu, Japan, Takara code: RR001B), 1.5 µL of 2.5 mM Mg^{2+}, 2 µL of 2.5 mM dNTPs, 0.25 µL Ex Taq DNA Polymerase (Takara Bio Inc., Kusatsu, Japan, Takara Code: RR001B), 16.75 µL of double-distilled water, 10 µM of each primer, and approximately 20 ng of DNA template. The amplification program consisted of an initial denaturation step of 94 for 2 min, followed by 30 cycles of denaturation at 94 for 30 s, annealing at 50 for 30 s, and elongation at 72 for 30 s, with a final extension at 72 for 5 min. Replicate reaction mixtures of the same sample were assembled within a PCR tube. After visualization on agarose gels (1% in TBE buffer) containing ethidium bromide, PCR products were purified using the QIAquick PCR Purification Kit (QIAGEN, Hilden, Germany, Cat. No.: 28106) and quantified with a NanoDrop ND-1000 spectrophotometer (NanoDrop Technologies, Wilmington, DE, USA). Purified amplicons were pooled in equimolar concentrations and paired-end sequenced on the Illumina MiSeq TM System platform according to the manufacturer's protocols.

Sequence analysis was conducted using quantitative insights into the microbial ecology (QIIME) pipeline (version 1.7.0), as previously described by Fadrosh et al. [37]. After the barcodes and primers were trimmed, and low-quality sequences were removed (<Q20), the remaining high-quality reads were clustered into operational taxonomic units (OTUs)

based on their sequence similarity at 97%. Community richness and diversity indices based on the number of OTUs and rarefaction curves were obtained using the Mothur software (version 1.34.0, Pat Schloss, Michigan, USA). Prior to the data analysis, the alpha-diversity indices of the bacterial community, including Good's coverage, Chao1, ACE, and the Shannon index, were calculated based on an appropriate subsample depth.

2.5. Statistical Analysis

Statistical analyses were performed by one-way analysis of variance (ANOVA) using the PASW Statistics program (version 18.0 for windows). The means were segregated using Duncan's multiple comparison test with a significance level of $p < 0.05$. Pearson's correlation analysis was conducted to evaluate the correlations between soil physicochemical and microbiological characteristics. To compare bacterial community structures across all soil samples, principal coordinate analysis (PCoA) and cluster analysis were performed based on the unweighted UniFrac distance matrix [38]. Redundancy analysis (RDA) was carried out to examine the relationship between abundant phyla (proteobacterial classes) and soil physicochemical characteristics [39].

3. Results

3.1. Soil Properties

The result showed that fertilizer reclamation clearly affected the soil nutrient amounts. As shown in Table 1, the subsidence soil showed the lowest nutrient amounts. The amounts of SOM in the SU treatment were 36.56% of that in the FA treatment, and AN, AP, and AK in the SU treatment accounted for 33.97%, 16.37%, and 61.14%, respectively. After the 7-year fertilizer reclamation, the amounts of SOM, AN, AP, and AK were consistently increased in the reclamation-treated soil compared with the subsidence soil. The SOM and AN amounts in the CF and M treatments were significantly higher than those in the CK treatment ($p < 0.05$). Significantly higher SOM and AN amounts ($p < 0.05$) were obtained in the M treatment compared with the CF treatment. The highest pH was observed in the subsidence soil, and fertilizer reclamation decreased the soil pH (Table 1). However, there was no significant difference in soil pH between CF treatment and M treatment.

Table 1. Soil chemical characteristics in different treatments.

Treatments	pH	SOM (g·kg⁻¹)	AN (mg·kg⁻¹)	AP (mg·kg⁻¹)	AK (mg·kg⁻¹)
SU	8.06 ± 0.06 [a]	9.74 ± 0.28 [e]	16.51 ± 0.58 [e]	3.18 ± 0.12 [c]	123.30 ± 4.61 [b]
CK	7.91 ± 0.03 [b]	11.37 ± 0.59 [d]	24.68 ± 1.01 [d]	4.65 ± 0.31 [c]	135.30 ± 2.31 [b]
CF	7.84 ± 0.02 [bc]	14.06 ± 0.46 [c]	27.13 ± 1.46 [c]	17.75 ± 1.54 [b]	233.57 ± 15.32 [a]
M	7.76 ± 0.02 [c]	18.15 ± 0.51 [b]	35.60 ± 0.68 [b]	19.46 ± 1.47 [b]	236.36 ± 19.58 [a]
FA	7.85 ± 0.03 [bc]	26.64 ± 0.40 [a]	48.59 ± 2.33 [a]	35.60 ± 0.68 [a]	201.7 ± 15.78 [a]

Values followed by different lowercase letters (a–e) are significantly different ($p < 0.05$) according to Duncan's multiple comparison test; SOM: organic matter; AN: alkali-hydrolyzable nitrogen; AP: available phosphorus; AK: available potassium; CK: reclaimed soil sampled in no-fertilizer treatment; CF: reclaimed soil sampled in chemical fertilizer treatment; M: reclaimed soil sampled in manure treatment. SU: subsided soil sampled in an adjacent site; FA: soil sampled in another adjacent farmland.

3.2. Soil Bacterial Abundance and Diversity

A total of 138,894 high-quality sequences (average read length of 440 bp) were obtained from all the soil samples. These optimized sequences were clustered into OTUs by Mothur software. As shown in Table 2, the Sobs values in all treatments were in the range of 3672–11,030, and the lowest Sobs was found in the SU treatment. There was no significant difference in Sobs between the reclaimed soil (CK, CF, and M) and farmland soil (FA), but they were significantly higher than those in the subsided soil (SU). Venn analysis (Figure 1) showed that 805 Sobs were found in all five treatments, accounting for 7.2–21.92% of each treatment, respectively. A total of 3478 Sobs were detected in the CK, CF, and M treatments, which were 31.53–35.59% of their total number. In addition, 5131 Sobs were detected in the CF and M treatments, accounting for 46.52% and 46.56% of their total Sobs, respectively.

The unique Sobs out of all the treatments accounted for 21.52–88.82%, indicating that there were differences in the composition of the soil bacterial community among the treatments.

Table 2. Estimated number of observed Sobs, coverage, richness, and diversity in different treatments.

Treatments	Sobs	Coverage	Richness and Diversity Indices		
			Chao1	ACE	Shannon
SU	3672 ± 195 [b]	0.97 ± 0.01 [a]	4455 ± 323 [c]	4358 ± 319 [c]	10.02 ± 0.24 [d]
CK	10803 ± 106 [a]	0.96 ± 0.02 [a]	13864 ± 174 [b]	14078 ± 242 [b]	11.37 ± 0.05 [c]
CF	11021 ± 731 [a]	0.94 ± 0.01 [ab]	15190 ± 463 [a]	15621 ± 413 [a]	11.60 ± 0.05 [b]
M	11030 ± 337 [a]	0.95 ± 0.00 [a]	15330 ± 216 [a]	15845 ± 170 [a]	11.39 ± 0.12 [c]
FA	9772 ± 288 [a]	0.91 ± 0.02 [b]	14430 ± 313 [a]	14988 ± 302 [a]	11.77 ± 0.04 [a]

Values followed by different lowercase letters (a–d) are significantly different ($p < 0.05$) according to Duncan's multiple comparison test; Sobs: the species of OTU that can be detected; Coverage: Good's nonparametric coverage estimator; ACE: abundance-based coverage estimator; Shannon: nonparametric Shannon diversity index. CK: reclaimed soil sampled in no-fertilizer treatment; CF: reclaimed soil sampled in chemical fertilizer treatment; M: reclaimed soil sampled in manure treatment. SU: subsided soil sampled in an adjacent site; FA: soil sampled in another adjacent farmland.

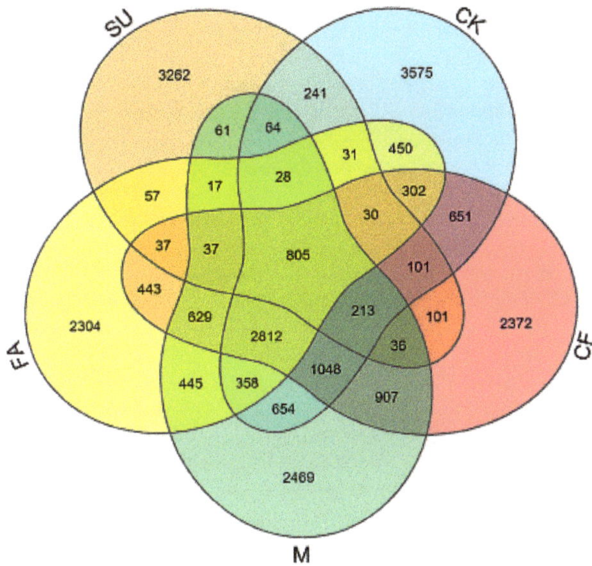

Figure 1. OTU Venn analysis of different treatments. OTU: Operational Taxonomic Units; CK: reclaimed soil sampled in no-fertilizer treatment; CF: reclaimed soil sampled in chemical fertilizer treatment; M: reclaimed soil sampled in manure treatment. SU: subsided soil sampled in an adjacent site; FA: soil sampled in another adjacent farmland.

The result of bacterial community diversity is presented in Table 2. The SU treatment showed the lowest Chao1, ACE, and Shannon index (alpha-diversity indices), which were significantly different from those in the reclaimed soil. After reclamation, soil bacterial diversity and abundance in the reclaimed soil were significantly higher than those in the subsidence soil ($p < 0.05$). However, Chao1 and ACE showed no significant difference among the CF, M, and FA treatments, which were significantly higher ($p < 0.05$) than those in the CK treatment. The Shannon index of the FA treatment was the highest (11.77) and significantly higher than those of the reclamation treatments (CK, CF, and M). In addition, the Shannon–Weiner curve showed similar trends (FA > CF > M > CK > SU) in terms of high species richness at 97% similarity (Figure S1). Good's coverage values in all samples ranged from 91% to 97% at a similarity cutoff of 97%, indicating that the current numbers of sequence reads were sufficient to capture the bacterial diversity in these soils.

3.3. Soil Bacterial Taxa Community Composition

Based on the Illumina platform analysis, sequences from all soil samples were classified into 54 different phyla, 138 classes, 213 orders, 244 families, 423 genera, and 218 species (Figure S2). Proteobacteria were the most abundant phyla, accounting for 20.15%–31.66% in these five treatments (Figure 2a). Furthermore, classes of *Alphaproteobacteria*, *Betaproteobacteria*, *Deltaproteobacteria*, and *Gammaproteobacteria* were detected in this study, and *Alphaproteobacteria* was the most abundant phylum, accounting for 41.48% of total *Proteobacteria* sequences. Other predominant phyla were *Actinobacteria* (27.80%), *Acidobacteria* (7.96%), *Bacteroidetes* (7.42%), *Chloroflexi* (7.21%), and *Gemmatimonadetes* (6.62%), accounting for 81.30% of the bacterial sequences. Additionally, *Firmicutes* (2.81%), *Planctomycetes* (2.01%), *Cyanobacteria* (1.50%), *Nitrospirae* (1.25%), *TM7* (1.07%), and *Verrucomicrobia* (1.01%) were present in soil samples with lower relative abundances, which occupied 9.65% of bacterial sequences (Figure S2 and Figure 2a).

The distribution of predominant bacterial phyla (classes) between the different treatments is illustrated in Figure 2a. Although similar main phyla (classes) existed in the selected treatments, the relative abundance of taxa in these treatments was different. No significant differences in the abundance of *Gemmatimonadetes* and *Verrucomicrobia* were observed in all treatments (Table S1). The highest relative abundance of *Acidobacteria* (9.37%), *Chloroflexi* (9.85%), and *Nitrospirae* (5.88%), as well as the lowest abundance of *Alphaproteobacteria* (4.16%), *Gammaproteobacteria* (3.85%), and *Actinobacteria* (16.12%), was detected in the SU treatment. *Alphaproteobacteria* (15.32%), *Planctomycetes* (2.44%), and *Deltaproteobacteria* (6.34%) showed the highest relative abundance in the FA treatment, while *Firmicutes* showed the lowest abundance. In the reclaimed soil, *Bacteroidetes* and *Firmicutes* were remarkably distinct ($p < 0.05$) in the CK treatment as compared with those in the CF and M treatments. Moreover, the respective abundances of the top 10 genera were examined to compare the distribution of bacterial genera in the different treatments (Figure 2b). It was found that the most bacterial taxa at the phylogenetic of genera differed greatly among the selected treatments, except for *Streptomyces* and *Bacillus* (Table S1). The lowest ($p < 0.05$) relative abundance of *Sphingomonas* (0.12%), *Rhodoplanes* (0.24%), *Skermanella* (0.09%), *Steroidobacter* (0.24%), and *Lentzea* (0.14%) was obtained in the SU treatment. The abundance of *Kaistobacter* in the M treatment (2.42%) was significantly ($p < 0.05$) higher than that in the FA treatment (1.62%). The highest ($p < 0.05$) abundance of *Lentzea* (1.16%) and *Balneimonas* (0.94%) was observed in the CF and FA treatments, respectively. Furthermore, the abundance of *Balneimonas* and *Lentzea* were significantly distinct ($p < 0.05$) between the CF and M treatments.

PCoA analysis based on the unweighted UniFrac distance metric was conducted to estimate β-diversity, which clearly revealed the variation of bacterial community among the treatments (Figure 3a). The first and second principal components explained 47.96% (PC1) and 11.25% (PC2) of the variance, respectively. As illustrated in the results of PCoA, the soil bacterial community in the SU treatment (with lower soil nutrient content) was separated from that in the CK, CF, M, and FA treatments (with higher soil nutrient content) along the PC1 axis. The soil bacterial community in the farmland soil (with higher soil OM, AN, and AP) was separated from that in the reclamation soil (CK, CF, and M treatments) along the PC2 axis. In addition, the CF and M treatments were clustered together and located in the first quadrant. It meant that the soil from the CF and M treatments had some of the same Sobs but at different levels. Moreover, Bray–Curtis analysis of dissimilarity (ANOSIM, R = 0.778, $p = 0.001$) showed that the differences between groups were larger than within groups, thus there was dissimilarity between the groups.

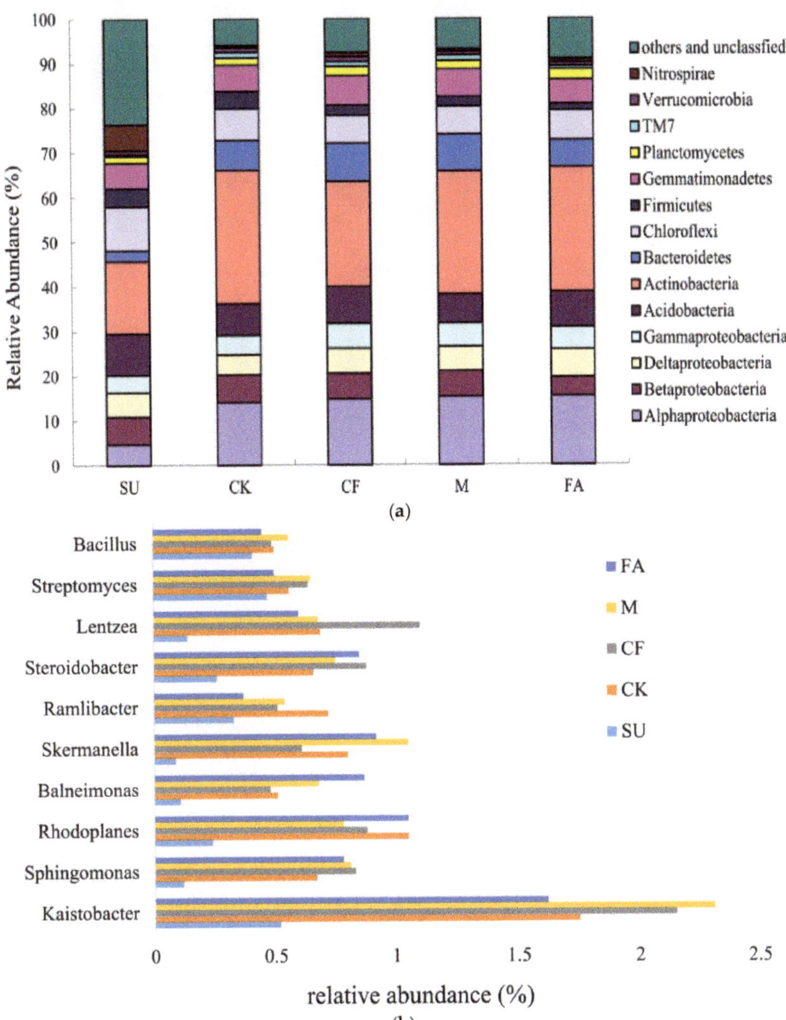

Figure 2. (**a**) Relative abundance of the dominant bacteria phyla (top 10) in all different treatments; (**b**) relative abundance of the dominant bacteria genera (top 10) in all different treatments. Relative abundances (>1%) are based on the proportional frequencies of those DNA sequences that could be classified at the phylum (proteobacterial class) level. Sequences not classified to any known phylum and phylogenetic groups accounting for ≤1% of all classified sequences are summarized in the artificial group "others and unclassified". CK: reclaimed soil sampled in no-fertilizer treatment; CF: reclaimed soil sampled in chemical fertilizer treatment; M: reclaimed soil sampled in manure treatment. SU: subsided soil sampled in an adjacent site; FA: soil sampled in another adjacent farmland. TM7: the phylum candidatus *Saccharibacteria*.

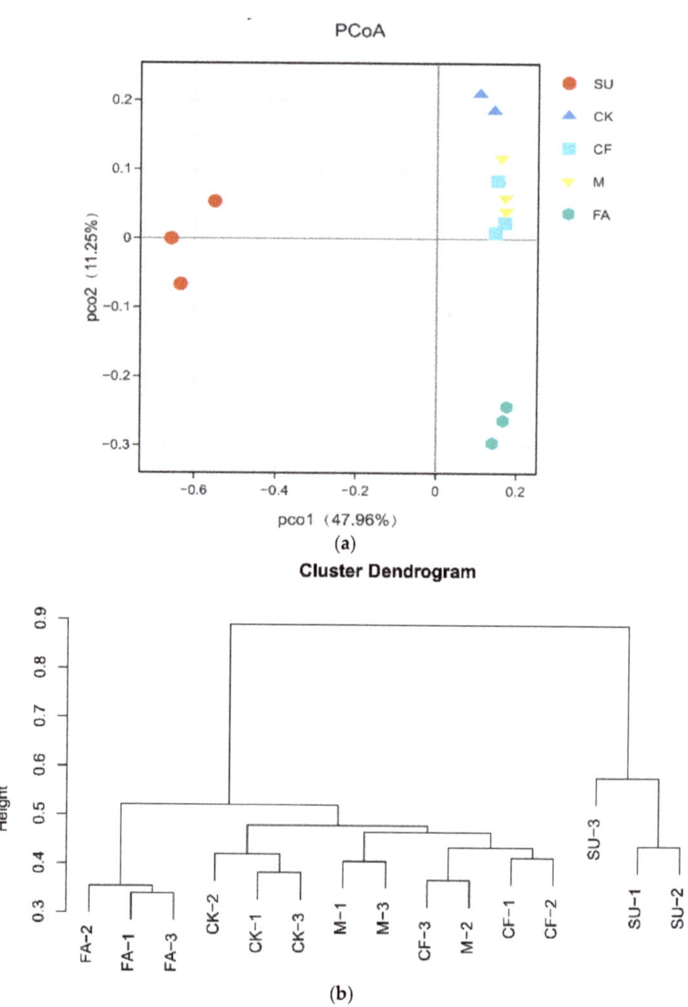

Figure 3. (**a**) Principal coordinate analysis (PCoA) based on unweighted UniFrac distances of soil bacterial communities sampled from different treatments; (**b**) similarity trees based on Bray–Curtis distance indices were calculated by OTUs at a distance of 3% using the hierarchical clustering analysis of bacterial communities for soil samples. CK: reclaimed soil sampled in no-fertilizer treatment; CF: reclaimed soil sampled in chemical fertilizer treatment; M: reclaimed soil sampled in manure treatment. SU: subsided soil sampled in an adjacent site; FA: soil sampled in another adjacent farmland.

A hierarchical cluster analysis based on the beta distance matrix was conducted to compare the similarity of the soil bacterial community in the different treatments. The results showed that soil samples (three replicates) from the subsidence site were grouped together and differed from other samples (Figure 3b). The other samples were clustered into two main groups: one group consisting of reclaimed soil samples from the CK, CF, and M treatments and the other consisting of samples from the FA treatment. In addition, the cluster tree revealed that the bacterial communities in the soil of the CF treatment were similar to that of the M treatment, which was different from that of the CK treatment. Overall, the results of cluster analysis were in line with PCoA.

3.4. Relationship between Soil Properties and Bacterial Community Composition

The results of Pearson's correlation analysis showed that soil properties were highly related to bacterial abundance and diversity. As shown in Table 3, soil pH showed significant negative correlations with Chao1 ($r = -0.826$, $p < 0.01$), ACE ($r = -0.824$, $p < 0.01$), and Shannon index ($r = -0.745$, $p < 0.01$). In contrast, SOM, AN, and AP showed significant positive correlations with Chao1 ($r = 0.545$–0.606, $p < 0.05$), ACE ($r = 0.564$–0.659, $p < 0.05$, $p < 0.01$), and Shannon index ($r = 0.559$–0.623, $p < 0.05$). AK was positively correlated with Chao1 ($r = 0.644$, $p < 0.01$) and ACE ($r = 0.656$, $p < 0.01$).

Table 3. Pearson's correlation coefficients between soil physicochemical characteristics and Sobs, and diversity indices.

Pearson	Sobs	Chao1	ACE	Shannon
pH	−0.768 **	−0.826 **	−0.824 **	−0.745 **
SOM	0.399	0.545 *	0.564 *	0.616 *
AN	0.463	0.572 *	0.647 **	0.559 *
AP	0.377	0.606 *	0.659 **	0. 623 *
AK	0.508	0.644 **	0.656 **	0.409

** Correlation is significant at the 0.01 level; * Correlation is significant at the 0.05 level. pH: potential of hydrogen; SOM: organic matter; AN: alkali-hydrolyzable nitrogen; AP: available phosphorus; AK: available potassium.

Results of the RDA analysis are shown in Figure 4. Selected soil property factors could explain 78.96% of the variation. The first and second axes of RDA explained 64.4% and 9.6% of total variation in our data, respectively. The bacterial community of the FA treatment was related to higher SOM, AN, and AP contents and lower pH, as shown by the vectors, while that of the SU treatment was associated with higher pH and lower soil nutrients (SOM, AN, and AP). As shown in Figure 4, the main group of phyla in the CK, M, and CF treatments was closer to that in the FA treatment along the first axis with decreasing pH and increasing soil nutrient contents. Furthermore, abundant phyla of the M and CF treatments were more alike their closer distance than the other treatments. The contributions of the selected physicochemical factors followed this trend: pH > SOM > AP > AN > AK, and their contributions were 39.34%, 27.59%, 18.27%, 13.80% and 11.27%. It is indicated that the structure of the bacterial community was closely correlated with soil properties and mainly shaped by soil pH and SOM. Pearson's correlation was calculated between the most abundant bacterial phyla (*Proteobacteria* classes) and soil environmental factors (Table S2). We found that the relative abundance of dominant phyla (*Proteobacteria* classes), such as *Alphaproteobacteria*, *Gammaproteobacteria*, *Bacteroidetes*, and *Planctomycetes*, showed significantly negative correlations with soil pH and positive correlations with AK ($p < 0.05$; $p < 0.01$). *Chloroflexi*, *Firmicutes*, and *Nitrospirae* showed significantly positive correlations with pH and negative correlations with SOM, AN, and AK ($p < 0.05$; $p < 0.01$). *Alphaproteobacteria* and *Planctomycetes* were positively correlated with SOM and AN ($p < 0.05$; $p < 0.01$), and *Actinobacteria* was positively correlated with AN ($p < 0.05$). Additionally, *Acidobacteria*, *Gemmatimonadetes*, *Verrucomicrobia*, and *TM7* had no significant correlation with the selected soil factors.

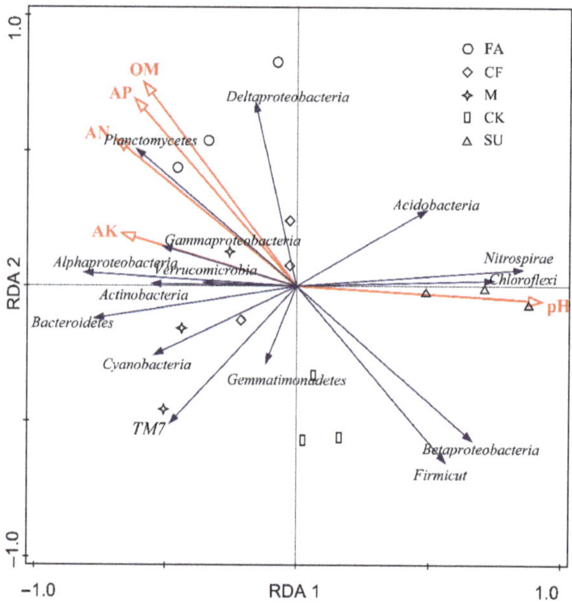

Figure 4. Redundancy analysis (RDA) of abundant phyla (*proteobacterial* classes) and selected soil edaphic properties such as pH, SOM, AN, AP, and AK for individual samples from three sites. SOM: organic matter; AN: alkali-hydrolyzable nitrogen; AP: available phosphorus; AK: available potassium; CK: reclaimed soil sampled in no-fertilizer treatment; CF: reclaimed soil sampled in chemical fertilizer treatment; M: reclaimed soil sampled in manure treatment. SU: subsided soil sampled in an adjacent site; FA: soil sampled in another adjacent farmland. TM7: the phylum candidatus *Saccharibacteria*. The red arrow: soil property factors; The blue arrow: the main group of bacteria phyla.

4. Discussion

4.1. Effects of Reclamation on Soil Properties and Bacterial Community

Coal mining disturbance and reclamation can change original soil physical and chemical characteristics, consequently affecting soil microbial community structure and diversity [16,40,41]. In general, coal mining activity and preliminary engineering reclamation, including stripping, reconstruction, and tillage, lead to the disturbance of vegetation and original surface soil structure, resulting in the degradation of soil bacterial diversity. In our study, the lower soil nutrients and bacterial community diversity shown in the subsided soil were mainly attributable to the deficient management and high erosion rates in the post-mined soil. After 7-year consecutive fertilizer reclamation, *Proteobacteria*, *Actinobacteria*, *Acidobacteria*, *Bacteroidetes*, *Chloroflexi*, and *Gemmatimonadetes* were the main bacterial phyla (Figure S2). In particular, *Proteobacteria* and *Actinobacteria*, as the copiotrophic groups living in nutrient-rich conditions, were the most dominant phyla in the reclaimed soil (Figure 2a), which is in agreement with previous studies [42–44]. This consistency of dominant phyla in different mining areas indicates that these bacteria play an important role in soil improvement in mining areas and have a wide range of adaptability to the soil environment of the mining area.

In this study, we found *Alphaproteobacteria* was the largest subgroup of *Proteobacteria* in the reclaimed soil. Moreover, the populations of *Rhizobiales*, *Rhodospirillales*, and *Sphingomonadales* affiliated with the alpha-subclass were also observed in significant proportions in the reclaimed soil (Table S1). The bacterial community of those functional species played an important role in C, N cycles and in maintaining the integrity of the coal mine ecosystem [45]. This is consistent with our findings of higher SOM and AN in the reclaimed soil.

In contrast, we found *Chloroflexi* (9.85%) was the advantageous population in the subsidence soil (Table S1) and showed lower abundance in the reclaimed soil with improvement in soil nutrients. This is mainly because this bacterium is a kind of autotrophic bacterium that does not depend on the nutrient supply in the environment and has a survival advantage in barren soil [32]. We also found a higher abundance of *Acidobacteria* (9.37%) in the subsidence soil (Table S1). As a type of slow-growing oligotroph [46] with a rich diversity of metabolism and function, they prefer oligotrophic living environments with poor available carbon sources [47]. Our results demonstrated that the accumulation of soil available nutrients promoted copiotrophic bacteria but negatively affected the oligotrophic groups [8].

Reclamation significantly increased the nutrient contents (SOM, AN, AP, and AK) of reclaimed soil, improved the abundance and diversity of bacteria, and promoted the restoration of the soil microbial community in the mining area [2,16,18,48]. In the present study, bacterial communities were restored after 7 years of reclamation with the increase in soil fertility, but it still differed from the adjacent farmland soil. In the reclaimed soil, the bacterial community composition of the organic fertilizer and chemical fertilizer treatments showed higher similarity (PCoA and cluster analysis). We found that *Actinomycetes* in manure treatment were slightly more abundant than those in the chemical fertilizer treatment (Table S1). Furthermore, *Lentzea* (affiliated to *Actinomycetales*) in the manure treatment were also significantly higher than those in the chemical fertilizer treatment (Table S1). This may be because manure contains more carbon and nitrogen sources for the growth of *Actinomycetes* than chemical fertilizers. In addition, the organic matter in manure can improve the aggregation and water-holding capacity of the soil, which is beneficial to the growth of *Actinomycetes*. We also observed that soil treated with manure had a higher abundance of *Balneimonas* (affiliated with *Rhizobiales*) than soil treated with chemical fertilizer (Table S1), probably because the dramatic increase in organic matter in manure could promote the intensive reproduction of this microbial taxon [49].

The results of PCoA and cluster analysis confirmed the distinct difference in bacterial community composition existed between the reclaimed and subsided soil (Figure 3a,b). It should be noted that the CK treatment (planting maize without fertilization for 7 years) significantly increased SOM and AN compared with the SU treatment, which resulted in elevated bacterial richness and diversity indices. This is mainly because the input of previous maize residues and the exudates released by maize roots through consolidation directly provided the energy source for microorganisms [50]. Additionally, aboveground maize can also alter the soil bacterial community by affecting the quality and quantity of microbial metabolic substrates [51]. These results indicated that surface vegetation restoration was also an important factor affecting bacterial communities in the reclaimed soil. However, the influences of various plants on soil microorganisms and soil physicochemical properties are different [52,53]. Different results could be obtained if other plants were selected in the soil reclamation of the mining area in this study.

Fertilization can improve soil quality and accelerate soil maturation, which further affects the diversity and richness of bacteria [25]. The application of organic fertilizer can improve soil structure and function and promote microbial richness and abundance [26–28,45], which is consistent with our results. In our study, nutrient contents and bacterial richness (Chao1 and ACE) in soil treated with manure were higher than those treated with chemical fertilizer (Tables 1 and 2), which may be related to the input of organic matter. In addition, it is probably because the original bacteria from manure contributed to the increase in bacterial species. Conversely, the CF treatment had remarkably higher bacterial diversity (Shannon index) than the M treatment (Table 2), mainly due to the mineral nutrient supplied by chemical fertilizer being sufficient to induce the corresponding bacterial population in the root system and increase the soil bacterial diversity. In addition, a dramatic increase in organic matter contained in mature crops could promote the intensive reproduction of some microbial groups (*Proteobacteria*, *Actinobacteria*, and *Bacteroidetes*), resulting in the decrease in soil bacterial diversity [49]. It has been reported that the long-term application of chemical fertilizer can destroy soil structure, cause soil acidification, reduce soil enzyme activity, and decrease microbial biomass and diversity [24,29,30]. In this study, fertilizers also significantly increased the abundance and diversity of soil bacteria, which may be related to the lower nutrient levels in the soil before fertilization treatment. It should be pointed out that, under equal nutrient conditions, much more manure will be applied than chemical fertilizer because of its fewer available nutrients. Therefore, the detrimental effects of heavy metal accumulation and antibiotic residues on soil bacteria caused by the extensive application of manure [54–56] cannot be ignored.

4.2. Effects of Soil Properties on Bacterial Community Composition

Soil microorganisms are closely related to soil properties [13,16,25,57]. Many studies have reported that improved soil fertility resulted in higher bacterial abundance and diversity [45,58,59]. In this experiment, we found that Chao1, ACE, and Shannon index were positively correlated with SOM, AN, AP, and AK ($p < 0.01$ or $p < 0.05$). It is indicated that the accumulation of soil organic matter and nutrients could be a good explanation for the higher bacterial richness and diversity indices in the reclaimed soil (Table 3). We also found that most of the abundant phyla or classes were significantly correlated with one or more selected soil properties, emphasizing the critical role of soil organic matter and nutrients in shaping the abundance and diversity of the soil bacterial community.

Previous studies have indicated that pH, SOM, and AN are the main factors affecting the composition of bacterial communities in soil [60–63], which is consistent with our study. In this study, pH was the major factor affecting the bacterial community structure in the reclaimed soil. After 7 years of reclamation, soil pH was significantly reduced. Moreover, the application of chemical fertilizer and manure decreased the soil pH, but there was no significant difference with no fertilizer (CK). This may be due to the short reclamation time (≤ 7 years) resulting in the insufficient effect of fertilization on soil pH (Table 1). Furthermore, it may relate to our experimental calcareous soil [8]. In this experiment, bacterial diversity increased with the decrease in soil pH (Table 2). It was found that there was a positive correlation between soil pH and *Nitrospirae* because *Nitrospirae* are sensitive to soil acidity and usually have a high abundance in alkaline soil [8,64]. Moreover, the abundance of *Actinobacteria* and *Bacteroidetes* had a negative correlation with the soil pH (Table S2), indicating that the bacteria were most abundant in agricultural soil near neutral pH [63,64]. It was also found that pH had a closer relationship with Chao1, ACE, and Shannon index (Table 3) than soil nutrients (SOM, AN, and AP) did. It is indicated that the soil bacterial diversity and community structure are shaped more by changes in soil pH than by direct nutrient addition (Zhang et al. 2017). In addition, soil pH could affect the structure of soil bacterial communities by altering many other environmental factors [42,65].

SOM is another major factor that was found to influence the soil bacterial community. It was observed that there were significant positive correlations with SOM, Chao1, ACE, and Shannon index ($p < 0.05$, Table 3). These results suggested that the restoration of bacterial diversity occurred gradually with the accumulation of soil organic carbon. Soil organic matter is considered to be one of the most common indicators of soil quality. The increase in soil organic matter can promote soil aggregation and improve soil physical properties (e.g., soil structure, bulk density, and water storage) and nutrients, which contributed to the growth and restoration of bacteria in the reclaimed soil. In addition, as a reservoir of carbon and nitrogen sources, the successive decomposition of SOM can produce diverse substrates for microbiota, thus contributing to the improvement of bacterial community diversity [45]. It was also pointed out that SOM had a significant positive correlation with the abundance of *Proteobacteria*, *Actinobacteria*, and *Bacteroidetes* ($p < 0.05$, RDA). These copiotrophic bacteria have a higher abundance in reclaimed soil for their fast growth rates in nutrient-rich conditions [46].

The aim of reclamation in the mining land is to re-establish a productive, healthy, and sustainable ecosystem suitable for post-mining land use. The restoration of soil fertility by increasing soil nutrients and enriching microbial populations is an effective method for ecological restoration in mining areas [66]. However, the reestablishment of the soil microbial community in mining subsidence areas not only depends on the reclamation practices but also the reclamation time. In general, soil nutrients are gradually accumulated with the increase in reclamation time, and soil fertility can approach a relatively stable or predisturbance level after almost 20 years of reclamation [2,48,59]. Studies on mining reclamation suggest that the most important recovery phase of microbial community occurs between 5 and 20 years after reclamation, and the difference is mainly associated with several critical factors, including reclamation practice, soil properties, climatic conditions, and vegetation [16,18,51]. In our study, soil nutrient and bacterial community diversity were significantly improved after 7 years of reclamation but did not reach the level of adjacent farmland soil. Continuous monitoring of soil nutrients and microbial communities in selected sites will be still needed in the future.

5. Conclusions

In summary, coal mining subsidence greatly reduced soil fertility and changed the bacterial community. Reclamation could promote the soil bacterial diversity and community composition in coal mining areas by improving soil physical and chemical characteristics. According to our hypothesis, the soil bacterial richness and diversity were elevated with the improvement of soil nutrients after 7 years of reclamation, but it still did not reach the level of the adjacent farmland soil. Moreover, fertilization (organic fertilizer and chemical fertilizer) could significantly increase the soil bacterial abundance, while inorganic fertilizer had a more obvious impact on bacterial diversity. In addition, soil physicochemical factors affected by the soil fertilizer remediation, especially SOM and pH, are critical in shaping the main bacterial populations.

Supplementary Materials: The following are available online at https://www.mdpi.com/article/10.3390/ijerph182312504/s1, Figure S1: Shannon wiener curves of OUT cluster at 97% sequence identity across different soil samples, Figure S2: The composition of the bacterial community taxa in all soil samples. Table S1: Relative abundance of dominant bacterial community taxa in different treatment, Table S2: Pearson's correlation coefficients between soil properties and dominant bacterial community taxa.

Author Contributions: L.L., T.L., H.M. and J.H. participated in this study's design, data collection and analyses. L.L. and J.Z. played a special role in writing the manuscript. T.L., Y.X. and J.H. provided scientific insights and detailed comments on the paper. All authors have read and agreed to the published version of the manuscript.

Funding: This research was funded by the National Natural Science Foundation of China (No. u1710255 and No. 41907215) and the Youth Science Foundation of Shanxi Province (No. 2013021032-2).

Institutional Review Board Statement: Not applicable.

Informed Consent Statement: Not applicable.

Conflicts of Interest: The authors declare no conflict of interest.

References

1. Xu, Z.J.; Zhang, Y.; Yang, J.; Liu, F.W.; Bi, R.T.; Zhu, H.F.; Lv, C.J.; Yu, J. Effect of Underground Coal Mining on the Regional Soil Organic Carbon Pool in Farmland in a Mining Subsidence Area. *Sustainability* **2019**, *11*, 4961. [CrossRef]
2. De Quadros, P.D.; Zhalnina, K.; Davis-Richardson, A.G.; Drew, J.C.; Menezes, F.B.; de O. Camargo, F.A.; Triplett, E.W. Coal mining practices reduce the microbial biomass, richness and diversity of soil. *Appl. Soil Ecol.* **2016**, *98*, 195–203. [CrossRef]
3. Chen, Y.; Zhang, J.; Zhou, A.; Yin, B. Modeling and analysis of mining subsidence disaster chains based on stochastic Petri nets. *Nat. Hazards* **2018**, *92*, 19–41. [CrossRef]
4. Ma, K.; Zhang, Y.X.; Ruan, M.Y.; Guo, J.; Chai, T.Y. Land Subsidence in a Coal Mining Area reduced Soil Fertility and Led to Soil Degradation in Arid and Semi-Arid Regions. *Int. J. Environ. Res. Public Health* **2019**, *16*, 3929. [CrossRef] [PubMed]
5. Shi, P.L.; Zhang, Y.X.; Hu, Z.Q.; Ma, K.; Wang, H.; Chai, T.Y. The response of soil bacterial communities to mining subsidence in the west China Aeolian sand area. *Appl. Soil Ecol.* **2017**, *121*, 1–10. [CrossRef]

6.   Guo, Y.; Liu, X.; Tsolmon, B.; Chen, J.; Bao, Y. The influence of transplanted trees on soil microbial diversity in coal mine subsidence areas in the Loess Plateau of China. *Glob. Ecol. Conserv.* **2019**, *21*, e00877. [CrossRef]
7.   Guebert, M.D.; Gardner, T.W. Macropore flow on a reclaimed surface mine: Infiltration and hillslope hydrology. *Geomorphology* **2001**, *39*, 151–169. [CrossRef]
8.   Wang, X.Y.; Li, Y.; Wei, Y.; Meng, H.S.; Cao, Y.Z.; Lead, J.R.; Hong, J.P. Effects of fertilization and reclamation time on soil bacterial communities in coal mining subsidence areas. *Sci. Total Environ.* **2020**, *739*, 139882. [CrossRef]
9.   Brussaard, L.; de Ruiter, P.C.; Brown, G.G. Soil biodiversity for agricultural sustainability. *Agr. Ecosyst. Environ.* **2007**, *121*, 233–244. [CrossRef]
10.   Wagg, C.; Bender, S.F.; Widmer, F.; Heijden, M. Soil biodiversity and soil community composition determine ecosystem multifunctionality. *Proc. Natl. Acad. Sci. USA* **2014**, *111*, 5266–5270. [CrossRef]
11.   Trap, J.; Bonkowski, M.; Plassard, C.; Villenave, C.; Blanchart, E. Ecological importance of soil bacterivores for ecosystem functions. *Plant Soil* **2016**, *398*, 1–24. [CrossRef]
12.   Romaniuk, R.; Giuffré, L.; Costantini, A.; Nannipieri, P. Assessment of soil microbial diversity measurements as indicators of soil functioning in organic and conventional horticulture systems. *Ecol. Indic.* **2011**, *11*, 1345–1353. [CrossRef]
13.   Singh, J.S.; Pandey, V.C.; Singh, D.P. Efficient soil microorganisms: A new dimension for sustainable agriculture and environmental development. *Agric. Ecosyst. Environ.* **2011**, *140*, 339–353. [CrossRef]
14.   Tu, J.; Qiao, J.; Zhu, Z.; Li, P.; Wu, L. Soil bacterial community responses to long-term fertilizer treatments in Paulownia plantations in subtropical China. *Appl. Soil Ecol.* **2018**, *124*, 317–326. [CrossRef]
15.   Anderson, J.D.; Ingram, L.J.; Stahl, P.D. Influence of reclamation management practices on microbial biomass carbon and soil organic carbon accumulation in semiarid mined lands of Wyoming. *Appl. Soil Ecol.* **2008**, *40*, 387–397. [CrossRef]
16.   Dangi, S.R.; Stahl, P.D.; Wick, A.F.; Ingram, L.J.; Buyer, J.S. Soil Microbial Community Recovery in Reclaimed Soils on a Surface Coal Mine Site. *Soil Sci. Soc. Am. J.* **2012**, *76*, 915–924. [CrossRef]
17.   Dimitriu, P.A.; Prescott, C.E.; Quideau, S.A.; Grayston, S.J. Impact of reclamation of surface-mined boreal forest soils on microbial community composition and function. *Soil Biol. Biochem.* **2010**, *42*, 2289–2297. [CrossRef]
18.   Cheng, Z.; Zhang, F.; Gale, W.J.; Wang, W.; Sang, W.; Yang, H. Effects of reclamation years on composition and diversity of soil bacterial communities in Northwest China. *Can. J. Microbiol.* **2018**, *64*, 28–40. [CrossRef]
19.   Kohler, J.; Caravaca, F.; Azcon, R.; Diaz, G.; Roldan, A. Suitability of the microbial community composition and function in a semiarid mine soil for assessing phytomanagement practices based on mycorrhizal inoculation and amendment addition. *J. Environ. Manag.* **2016**, *169*, 236–246. [CrossRef]
20.   Hahn, A.S.; Quideau, S.A. Long-term effects of organic amendments on the recovery of plant and soil microbial communities following disturbance in the Canadian boreal forest. *Plant Soil* **2012**, *363*, 331–344. [CrossRef]
21.   Mummey, D.L.; Stahl, P.D.; Buyer, J.S. Soil microbiological properties 20 years after surface mine reclamation: Spatial analysis of reclaimed and undisturbed sites. *Soil Biol. Biochem.* **2002**, *34*, 1717–1725. [CrossRef]
22.   Edmeades, D.C. The long-term effects of manures and fertilisers on soil productivity and quality: A review. *Nutr. Cycl. Agroecosyst.* **2003**, *66*, 165–180. [CrossRef]
23.   Torsvik, V.; Øvreås, L. Microbial diversity and function in soil: From genes to ecosystems. *Curr. Opin. Microbiol.* **2002**, *5*, 240–245. [CrossRef]
24.   Ding, J.; Jiang, X.; Ma, M.; Zhou, B.; Guan, D.; Zhao, B.; Zhou, J.; Cao, F.; Li, L.; Li, J. Effect of 35 years inorganic fertilizer and manure amendment on structure of bacterial and archaeal communities in black soil of northeast China. *Appl. Soil Ecol.* **2016**, *105*, 187–195. [CrossRef]
25.   Sun, L.; Xun, W.; Huang, T.; Zhang, G.; Gao, J.; Ran, W.; Li, D.; Shen, Q.; Zhang, R. Alteration of the soil bacterial community during parent material maturation driven by different fertilization treatments. *Soil Biol. Biochem.* **2016**, *96*, 207–215. [CrossRef]
26.   Gomez, E.; Ferreras, L.; Toresani, S. Soil bacterial functional diversity as influenced by organic amendment application. *Bioresour. Technol.* **2006**, *97*, 1484–1489. [CrossRef] [PubMed]
27.   Chang, E.H.; Chung, R.S.; Tsai, Y.H. Effect of different application rates of organic fertilizer on soil enzyme activity and microbial population. *Soil Sci. Plant Nutr.* **2007**, *53*, 132–140. [CrossRef]
28.   Zhuang, X.; Han, Z.; Bai, Z.; Zhuang, G.; Shim, H. Progress in decontamination by halophilic microorganisms in saline wastewater and soil. *Environ. Pollut.* **2010**, *158*, 1119–1126. [CrossRef]
29.   Zhong, W.; Gu, T.; Wang, W.; Zhang, B.; Lin, X.; Huang, Q.; Shen, W. The effects of mineral fertilizer and organic manure on soil microbial community and diversity. *Plant Soil* **2010**, *326*, 511–522. [CrossRef]
30.   Zhou, J.; Guan, D.W.; Zhou, B.K.; Zhao, B.S.; Ma, M.C.; Qin, J.; Jiang, X.; Chen, S.F.; Cao, F.M.; Shen, D.L.; et al. Influence of 34-years of fertilization on bacterial communities in an intensively cultivated black soil in northeast China. *Soil Biol. Biochem.* **2015**, *90*, 42–51. [CrossRef]
31.   Driessen, P.; Deckers, J.; Spaargaren, O.; Nachtergaele, F. *Lecture Notes on the Major Soils of the World*; FAO: Rome, Italy, 2001; pp. 265–270.
32.   Li, X.Y.; Deng, Y.; Li, Q.; Lu, C.; Wang, J.; Zhang, H.; Zhu, J.; Zhou, J.; He, Z. Shifts of functional gene representation in wheat rhizosphere microbial communities under elevated ozone. *ISME J.* **2013**, *7*, 660–671. [CrossRef] [PubMed]
33.   Strickland, T.C.; Sollins, P. Improved Method for Separating Light and Heavy-Fraction Organic Material from Soil. *Soil Sci. Soc. Am. J.* **1987**, *51*, 1390–1393. [CrossRef]

34. Bremner, J.M. Nitrogen Total. In *Methods of Soil Analysis Part 3: Chemical Methods*; SSSA Book Series, 5; Sparks, D.L., Ed.; Soil Science Society of America: Madison, WI, USA, 1996; pp. 1085–1122.

35. Hedley, M.J.; Stewart, J.W.B. Method to measure microbial phosphate in soils. *Soil Biol. Biochem.* **1982**, *14*, 377–385. [CrossRef]

36. Mehlich, A. Mehlich 3 soil test extractant: A modification of Mehlich 2 extractant. Commun. *Soil Sci. Plant. Anal.* **1984**, *15*, 1409–1416. [CrossRef]

37. Fadrosh, D.W.; Ma, B.; Gajer, P.; Sengamalay, N.; Ott, S.; Brotman, R.M.; Ravel, J. An improved dual-indexing approach for multiplexed 16S rRNA gene sequencing on the Illumina MiSeq platform. *Microbiome* **2014**, *2*, 6. [CrossRef]

38. Lozupone, C.; Lladser, M.; Knights, D.; Stombaugh, J.; Knight, R. UniFrac: An effective distance metric for microbial community comparison. *ISME J.* **2011**, *5*, 169–172. [CrossRef] [PubMed]

39. Sheik, C.S.; Mitchell, T.W.; Rizvi, F.Z.; Rehman, Y.; Faisal, M.; Hasnain, S.; McInerney, M.J.; Krumholz, L.R. Exposure of soil microbial communities to chromium and arsenic alters their diversity and structure. *PLoS ONE* **2012**, *7*, e40059. [CrossRef]

40. Gasch, C.; Huzurbazar, S.; Stahl, P. Measuring soil disturbance effects and assessing soil restoration success by examining distributions of soil properties. *Appl. Soil Ecol.* **2014**, *76*, 102–111. [CrossRef]

41. Poncelet, D.; Cavender, N.; Cutright, T.; Senko, J. An assessment of microbial communities associated with surface mining-disturbed overburden. *Environ. Monit. Assess.* **2013**, *186*, 1917–1929. [CrossRef]

42. Rastogi, G.; Osman, S.; Vaishampayan, P.A.; Andersen, G.L.; Stetler, L.D.; Sani, R.K. Microbial diversity in uranium mining-impacted soils as revealed by high-density 16S microarray and clone library. *Microb. Ecol.* **2010**, *59*, 94–108. [CrossRef] [PubMed]

43. Lauber, C.L.; Hamady, M.; Knight, R.; Fierer, N. Pyrosequencing-based assessment of soil pH as a predictor of soil bacterial community structure at the continental scale. *Appl. Environ. Microbiol.* **2009**, *75*, 5111–5120. [CrossRef]

44. Zhan, J.; Sun, Q.Y. Development of microbial properties and enzyme activities in copper mine wasteland during natural restoration. *Catena* **2014**, *116*, 86–94. [CrossRef]

45. Li, Y.Y.; Chen, L.; Wen, H.; Zhou, T.; Zhang, T.; Gao, X. 454 Pyrosequencing Analysis of Bacterial Diversity Revealed by a Comparative Study of Soils from Mining Subsidence and Reclamation Areas. *J. Microbiol. Biotechnol.* **2014**, *24*, 313–323. [CrossRef]

46. Kielak, A.; Pijl, A.S.; van Veen, J.A.; Kowalchuk, G.A. Phylogenetic diversity of Acidobacteria in a former agricultural soil. *ISME J.* **2009**, *3*, 378. [CrossRef] [PubMed]

47. Ward, N.L.; Challacombe, J.F.; Janssen, P.H.; Henrissat, B.; Coutinho, P.M.; Wu, M.; Xie, G.; Haft, D.H.; Sait, M.; Badger, J.; et al. Three genomes from the phylum Acidobacteria provide insight into the lifestyles of these microorganisms in soils. *Appl. Environ. Microbiol.* **2009**, *75*, 2046–2056. [CrossRef]

48. Li, Y.Y.; Wen, H.; Chen, L.; Yin, T. Succession of bacterial community structure and diversity in soil along a chronosequence of reclamation and re-vegetation on coal mine spoils in China. *PLoS ONE* **2014**, *9*, e115024. [CrossRef]

49. Tian, W.; Wang, L.; Li, Y.; Zhuang, K.; Li, G.; Zhang, J.; Xiao, X.; Xi, Y. Responses of microbial activity, abundance, and community in wheat soil after three years of heavy fertilization with manure-based compost and inorganic nitrogen. *Agric. Ecosyst. Environ.* **2015**, *213*, 219–227. [CrossRef]

50. Grayston, S.J.; Campbell, C.D.; Bardgett, R.D.; Mawdsley, J.L.; Clegg, C.D.; Ritz, K.; Griffiths, B.S.; Rodwell, J.S.; Edwards, S.J.; Davies, W.J.; et al. Assessing shifts in microbial community structure across a range of grasslands of differing management intensity using CLPP, PLFA and community DNA techniques. *Appl. Soil Ecol.* **2004**, *25*, 63–84. [CrossRef]

51. Li, Y.Y.; Chen, L.; Wen, H. Changes in the composition and diversity of bacterial communities 13 years after soil reclamation of abandoned mine land in eastern China. *Ecol. Res.* **2015**, *30*, 357–366. [CrossRef]

52. Fang, X.; Yu, D.; Zhou, W.; Zhou, L.; Dai, L. The effects of forest type on soil microbial activity in Changbai Mountain, Northeast China. *Ann. Forest Sci.* **2016**, *73*, 473–482. [CrossRef]

53. Wei, Y.; Yu, L.F.; Zhang, J.C.; Yu, Y.C.; Deangelis, D.L. Relationship Between Vegetation Restoration and Soil Microbial Characteristics in Degraded Karst Regions: A Case Study. *Pedosphere* **2011**, *21*, 132–138. [CrossRef]

54. Reichel, R.; Radl, V.; Rosendahl, I.; Albert, A.; Amelung, W.; Schloter, M.; Sören, T.B. Soil microbial community responses to antibiotic-contaminated manure under different soil moisture regimes. *Appl. Microbiol. Biotechnol.* **2014**, *98*, 6487–6495. [CrossRef]

55. Sun, J.; Zhang, Q.; Zhou, J.; Wei, Q. Pyrosequencing technology reveals the impact of different manure doses on the bacterial community in apple rhizosphere soil. *Appl. Soil Ecol.* **2014**, *78*, 28–36. [CrossRef]

56. Unger, I.M.; Goyne, K.W.; Kennedy, A.C.; Kremer, R.J.; McLain, J.; Williams, C.F. Antibiotic Effects on Microbial Community Characteristics in Soils under Conservation Management Practices. *Soil Sci. Soc. Am. J.* **2013**, *77*, 100. [CrossRef]

57. Francioli, D.; Schulz, E.; Lentendu, G.; Wubet, T.; Buscot, F.; Reitz, T. Mineral vs. Organic Amendments: Microbial Community Structure, Activity and Abundance of Agriculturally Relevant Microbes Are Driven by Long-Term Fertilization Strategies. *Front. Microbiol.* **2016**, *7*, 1446. [CrossRef]

58. Cui, J.; Liu, C.; Li, Z.; Wang, L.; Chen, X.; Ye, Z.; Fang, C. Long-term changes in topsoil chemical properties under centuries of cultivation after reclamation of coastal wetlands in the Yangtze Estuary, China. *Soil Tillage Res.* **2012**, *123*, 50–60. [CrossRef]

59. Sun, Y.G.; Li, X.; Mander, Ü.; He, Y.L.; Jia, Y.; Ma, Z.G.; Guo, W.Y.; Xin, Z.J. Effect of reclamation time and land use on soil properties in Changjiang River Estuary, China. *Chin. Geogr. Sci.* **2011**, *21*, 403. [CrossRef]

60. Hartmann, M.; Frey, B.; Mayer, J.; Mäder, P.; Widmer, F. Distinct soil microbial diversity under long-term organic and conventional farming. *ISME J.* **2015**, *9*, 1177–1194. [CrossRef] [PubMed]

61. Sun, R.; Zhang, X.X.; Guo, X.; Wang, D.; Chu, H. Bacterial diversity in soils subjected to long-term chemical fertilization can be more stably maintained with the addition of livestock manure than wheat straw. *Soil Biol. Biochem.* **2015**, *88*, 9–18. [CrossRef]

62. Zeng, J.; Liu, X.; Ling, S.; Lin, X.; Chu, H. Nitrogen fertilization directly affects soil bacterial diversity and indirectly affects bacterial community composition. *Soil Biol. Biochem.* **2016**, *92*, 41–49. [CrossRef]
63. Zhang, Y.; Shen, H.; He, X.; Thomas, B.Z.; Lupwayi, N.; Hao, X.; Thomas, M.; Shi, X. Fertilization Shapes Bacterial Community Structure by Alteration of Soil pH. *Front. Microbiol.* **2017**, *8*, 1325. [CrossRef] [PubMed]
64. Wu, Y.; Zeng, J.; Zhu, Q.; Zhang, Z.; Lin, X. pH is the primary determinant of the bacterial community structure in agricultural soils impacted by polycyclic aromatic hydrocarbon pollution. *Sci. Rep.* **2017**, *7*, 40093. [CrossRef] [PubMed]
65. Rousk, J.; Baath, E.; Brookes, P.C.; Lauber, C.L.; Lozupone, C.; Caporaso, J.G.; Knight, R.; Fierer, N. Soil bacterial and fungal communities across a pH gradient in an arable soil. *ISME J.* **2010**, *4*, 1340–1351. [CrossRef]
66. Sheoran, V.; Sheoran, A.; Poonia, P. Soil Reclamation of Abandoned Mine Land by Revegetation: A Review. *Int. J. Soil Sediment Water* **2010**, *3*, 13.

MDPI

St. Alban-Anlage 66

4052 Basel

Switzerland

www.mdpi.com

International Journal of Environmental Research and Public Health Editorial Office

E-mail: ijerph@mdpi.com

www.mdpi.com/journal/ijerph